NORTH CAROLINA
STATE BOARD OF COMMUNITY COLLEGES
LIBRARIES
ASHEVILLE-BUNCOMBE TECHNICAL COLLEGE

SO-BYS-431

Discarded
Date   SEP 1 7 2024

Manual of Soil Laboratory Testing

Manual of Soil Laboratory Testing

Volume 1: Soil Classification and Compaction Tests

K. H. Head, *MA(Cantab), C.Eng, FICE*

Engineering Laboratory Equipment Limited

A HALSTED PRESS BOOK

JOHN WILEY & SONS
New York- -Toronto

Published in the U.S.A. and Canada
by Halsted Press, a Division of
John Wiley & Sons, Inc., New York

© Engineering Laboratory Equipment Limited, 1980

Library of Congress Cataloging in Publication Data

Head, K   H
    Manual of soil laboratory testing.

    "A Halsted Press book."
    CONTENTS: v. 1. Soil classification and
testing.
    1. Soils - - Testing - - Laboratory manuals.
I. Title.
TA710.5.H4   620.1'91'0287   80-12258
ISBN 0-470-26973-1

Filmset by Mid-County Press, London SW15
Printed in Great Britain

Foreword

This book makes an outstanding contribution to the technology of soil testing, a vital factor in a well-planned soil mechanics site investigation.

Mr K. H. Head has, with painstaking skill, shown to his readers, by word and illustration, that accurate laboratory techniques and reporting on materials testing are essential for the quality control needed by the engineer in the planning and construction of any civil engineering project being carried out.

The author, quite rightly, highlights the need for good technical training and well-designed equipment, and the guidance which this book will give to the engineer and technician will, I believe, be invaluable.

T. G. Clark
Managing Director
Engineering Laboratory Equipment Ltd.
Eastman Way,
Hemel Hempstead,
Hertfordshire, England

ACKNOWLEDGEMENTS

The author would like to express his appreciation to the following organisations and individuals for their assistance in the preparation of this book.

British Standards Institution, for permission to make use of diagrams in BS 1377:1975 in the preparation of Figs. 2.19; 3.12, 3.18; 4.32, 4.36, 4.39, 4.40; 5.10; 6.7, 6.8, 6.16, 6.17, 6.18, 6.19 and Tables 2.5; 4.6, 4.13. Also for permission to quote from Document No. 76/11937, Code of Practice for Site Investigation (revision of CP2001), from which Fig. 2.6 and Table 7.1 have been prepared.

Transport and Road Research Laboratory, for permission to reproduce Figs. 6.30 and 6.31, and to quote from their Laboratory Reports.

Building Research Establishment for permission to quote from their Current Papers CP3/68, CP23/77 and Digest No. 174, from which Table 5.6 has been prepared.

Shell Chemicals U.K. Ltd., for supplying data on industrial waxes.

Photus Ltd., Bracknell, and Ray Buksh, Kings Langley, for photography.

Mr R. Glossop for his review and valuable criticism of the text and Mr R. C. Rooney for comments on Chapter 5.

Miss M. A. Sawyer for typing the calculation sheets and Mr D. Roberts for preparing the drawings.

The author also wishes to thank his colleagues at Soil Mechanics Ltd., Bracknell, and at Engineering Laboratory Equipment Ltd., Hemel Hempstead, for their assistance and cooperation.

Preface

This book is intended primarily as a working manual for laboratory technicians and others engaged on the testing of soils in a laboratory for building and civil engineering purposes. It is based on my own experience over many years both in managing large soil testing laboratories and in the instruction of technicians and engineers in test procedures. I have made a special effort to explain those points of detail which are often the cause of difficulty or misunderstanding. The step-by-step presentation of procedures, the use of flow diagrams, and the setting out of test data and calculations are provided for this purpose, especially for the newcomer to soil testing.

Volume 1 presents details of the methods and equipment used in soil classification and compaction tests, the former including relevant chemical tests. Most of these procedures are covered by British Standards, the most important being BS 1377:1975, 'Methods of test for Soils for civil engineering purposes', which this manual augments by its essentially practical approach. Reference is also made to certain US (ASTM) Standards.

A knowledge of mathematics, physics and chemistry up to GCE 'O' level standard is assumed, but some of the basic principles which are significant in soil testing are explained where appropriate. I hope that the sections giving background information, general applications and basic theory will enable technicians to obtain a better appreciation of the purpose and significance of the tests they perform. The inclusion of a chapter on soil description is intended as an introduction to that important aspect of soil mechanics, and may perhaps stimulate an interest in the broader topic of geology.

I should like to express my sincere thanks to Dr A. C. Meigh, who initiated this series of publications, and to Mr N. B. Hobbs for his most helpful and detailed comments on the original manuscript. In addition, I greatly appreciate the continued support and encouragement provided by Mr T. G. Clark, Mr I. K. Nixon and their colleagues at Engineering Laboratory Equipment Limited during the final stages of preparation of this work. Many others, too numerous to mention individually, have willingly given their support in various ways, and to them all I am very grateful.

I hope that this book will be well used in the laboratory, and I should welcome any comments and criticisms from those who use it.

K. H. Head

Contents

Summary of test procedures described in Volume 1

Test designation	Section	Abbreviated reference (see below)
Chapter 2		
Moisture content:		
Oven drying	2.5.2	BS Test 1(A)
Sand-bath	2.5.3	BS Test 1(B)
Liquid limit:		
Cone	2.6.4	BS Test 2(A)
Casagrande	2.6.6	BS Test 2(B)
One-point Casagrande	2.6.7	BS Test 2(C)
One-point cone	2.6.8	Clayton and Jukes (1978)
Plastic limit	2.6.5	BS Test 3
Shrinkage limit:		
TRRL method	2.7.2	*SMRE* (1952)
ASTM method	2.7.3	ASTM D427–39
Linear shrinkage	2.7.4	BS Test 5
Puddle clay	2.8.2	Nixon (1956)
Free swell	2.8.3	Gibbs and Holtz (1956)
Sticky Limit	2.8.4	Terzaghi and Peck (1948)
Chapter 3		
Density:		
Measurement	3.5.2	Soil Mechanics Ltd.
Tube	3.5.3	Soil Mechanics Ltd.
Water displacement	3.5.4	BS Test 15(F)
Weighing in water	3.5.5	BS Test 15(E)
Specific gravity:		
Density bottle	3.6.2	BS Test 6(B)
Gas jar	3.6.4	BS Test 6(A)
Pycnometer	3.6.5	BS 1377: 1967 Test 6(B)
Maximum density:		
Clean sands	3.7.2	Kolbuszewski (1948)
Silty soils	3.7.3	Soil Mechanics Ltd
Gravelly soils	3.7.6	Soil Mechanics Ltd
Maximum density:		
Dry	3.7.4	Kolbuszewski (1948)
In water	3.7.5	Kolbuszewski (1948)
Gravelly soils	3.7.6	Soil Mechanics Ltd
Chapter 4		
Sieving:		
Dry:simple	4.6.1	BS Test 7(B)
composite	4.6.2	BS Test 7(B)
very coarse soils	4.6.3	Soil Mechanics Ltd
Wet: fine soils	4.6.4	BS Test 7(A)
gravelly soils	4.6.5	BS Test 7(A)
cohesive soils	4.6.6	BS Test 7(A)
'Boulder clay'	4.6.7	Soil Mechanics Ltd

Test designation	Section	Abbreviated reference (see below)
Sedimentation:		
Pipette	4.8.1 and 4.8.2	BS Test 7(C)
Hydrometer	4.8.1 and 4.8.3	BS Test 7(D)
Chapter 5		
pH value:		
Indicator papers	5.5.1	(Supplier)
Colorimetric	5.5.2	BS Test 11(B)
Lovibond	5.5.3	(Supplier)
Electrometric	5.5.4	BS Test 11(A)
Sulphate content:		
Gravimetric: total	5.6.2 and 5.6.5	BS Test 9
water soluble	5.6.3 and 5.6.5	BS 1377:1967 Test 9(A)
groundwater	5.6.4 and 5.6.5	BS Test 1377:1967 Test 9(A)
Ion-exchange soil	5.6.6	BS Test 10
groundwater	5.6.7	BS Test 10
Organic content:		
Loss on ignition	5.7.2	*SMRE* (1952)
Peroxide oxidation	5.7.3	*SMRE* (1952)
Dichromate oxidation	5.7.4	BS Test 8
Carbonate content:		
Calcimeter: standard	5.8.1	Collins (1906); *SMRE* (1952)
simplified	5.8.2	Collins (1906)
Chloride content:		
Qualitative	5.9.1	Bowden (1968)
Volhard's method	5.9.2	BS 812:1976, Part 4
Mohr's method	5.9.3	Bowden (1968)
Total dissolved solids	5.10.2	BS 2690:1970, Part 9, Section 5
Use of indicator papers	5.10.3	(Supplier)
Chapter 6		
'Ordinary' compaction	6.5.2	BS Test 12
'Heavy' compaction	6.5.3	BS Test 13
Compaction by vibration	6.5.4	BS Test 14
Compaction of stony soils	6.5.6	(from BS Test 16)
Dietert compaction	6.6.2 and 6.6.3	Little (1948)
Harvard compaction	6.6.4	*ASTM STP* 479
Moisture condition	6.6.5	Parsons (1976)
Compactability	6.6.6	Pike and Acott (1975)

BS: British Standard 1377: 1975 (see Section 1.1.4).
ASTM: American Society for Testing and Materials.
SMRE: Soil Mechanics for Road Engineers (TRRL).
(Supplier): Supplier's or manufacturer's instruction leaflets.
References are listed in full at the end of each chapter.

Chapter 1

Equipment, techniques and safety

1.1 INTRODUCTION

1.1.1 Soils as Engineering Materials

Soils, in the geotechnical sense, can be regarded as engineering materials. Their physical characteristics can be determined by experiment, and the application of methods of analysis enables these properties to be used to predict their likely behaviour under defined working conditions. But unlike other engineering materials such as metals and concrete, over which control can be exercised during manufacture, soils are naturally occurring materials, which more often than not have to be used in their natural condition. Even when some kind of processing is possible, either *in situ* or by using excavated material, the soil can be modified only to a limited extent by relatively simple procedures on site. Perhaps the most important exception is the use of certain types of clay for the manufacture of bricks, but this is outside the scope of geotechnology.

The variety of soils is very wide indeed, and it is probably true to say that no two sites have identical soil conditions. Consequently many of the procedures used for determining soil characteristics consist of empirical methods derived from practical experience.

1.1.2 Purpose of Soil Testing

The physical properties of soils are usually determined by carrying out tests on samples of soil in a laboratory. These tests can be divided into two main categories:

(1) Classification tests, which indicate the general type of soil and the engineering category to which it belongs.
(2) Tests for the assessment of engineering properties, such as shear strength, compressibility and permeability.

The parameters determined from laboratory tests, taken together with descriptive data relating to the soil, are required by soil engineers for many purposes. The more usual applications are as follows:

(1) Data acquired from classification tests are applied to the identification of soil strata when the subsurface conditions of a site are being investigated — the process known as site investigation.
(2) Other test data enable the engineering properties of soils to be quantified in numerical terms, which can then be used as the basis of analysis on which the recommendations of a site investigation report are based.
(3) Test data may be used for the confirmation of assumptions which have been based on previous experience and engineering judgement.
(4) Criteria for the acceptance of a soil used in construction (possibly after a processing operation) can be drawn up in the light of available test results.
(5) Laboratory tests are needed as part of the control measures which are applied during

construction of earthworks or excavations, especially for ensuring that the design criteria are met.

(6)   The findings of a site investigation can be supplemented by further testing as construction proceeds, as, for instance, when new ground is being opened up.

The laboratory test procedures which have been evolved for the classification of soils, especially in Britain, are the subject of Volume 1. Tests for the determination of engineering properties will be dealt with in Volumes 2 and 3.

1.1.3   Advantages of Laboratory Testing

In a site investigation for a construction project, the field operations, which include studies of the geology and history of the site, subsurface exploration and *in situ* testing, are of prime importance. The determination of the ground characteristics by *in situ* testing can take into account large-scale effects, such as soil fabric, structure and discontinuities of strata, which cannot be represented in small laboratory specimens. Nevertheless the measurement of soil properties by means of laboratory tests offers a number of advantages, as follows:

(1)   Full control of the test conditions, including boundary conditions, can be exercised.
(2)   Laboratory testing generally permits a greater degree of accuracy of measurements than does field testing.
(3)   Control can be exercised over the choice of material which is to be tested.
(4)   A test can be run under conditions which are similar to, or which differ from, those prevailing *in situ*, as may be appropriate.
(5)   Changes in conditions can be simulated, as can the conditions which are likely to occur during or after completion of construction.
(6)   Tests can be carried out on soils which have been broken down and reconstituted, or processed in other ways.

1.1.4   General Applications

During the last half-century the evaluation of soil properties from reliable test procedures has led to a closer understanding of the nature and probable behaviour of soils as engineering materials. Some of the resulting benefits in the realm of civil engineering construction have been:

(1)   Reduction of uncertainties in the analysis of foundations and earthworks.
(2)   Economies in design due to the use of lower factors of safety.
(3)   Exploitation of difficult sites.
(4)   Erection of structures, and below-ground construction, which would not have been feasible without this knowledge.
(5)   Increased economy in the use of soils as construction materials (for instance, in earth dams and embankments).

1.1.5   Definitions

The following definitions are based on those found in the *Shorter Oxford English Dictionary* and the *Concise Oxford Dictionary*, expanded to cover usage in this book.

TEST(from Latin *testum*, tile or earthen pot (cupel) originally used for treating or trying metals in, especially gold or silver alloys). 1. Critical examination or trial by which the quality of anything may be determined. 2. The action or process of examining a substance under known conditions in order to determine its identity or that of one of its constituents. 3. The action by which the physical properties of materials are tested in order to determine their ability to satisfy particular requirements.

Definition (3) is the one which is applicable to the testing of soils, but (1) is also relevant, and (2) applies to classification and chemical testing.

LABORATORY (from Latin *laborare*, labour). Room or building set apart for experiments in natural science (originally in chemistry).

The 'experiments' with which we are concerned consist of standard tests for the classification of soils and the determination of their engineering properties.

SOIL (from Latin *solum*, ground). The earth or ground; the face or surface of the earth.

Soil can be defined in different ways for different purposes, and the above definition is too broad for engineering applications. An accepted definition of soil in the geotechnical sense is as follows: Any naturally occurring deposit forming the outer part of the earth's crust, consisting of an assemblage of discrete particles (usually mineral, sometimes with organic matter) that can be separated by gentle mechanical means, together with variable amounts of water and gas (usually air).

This definition is discussed in greater detail in Chapter 7.

SAMPLE (from Middle English *essample*, example). 1. A small separated part which illustrates the properties of the mass from which it is taken. 2. A relatively small quantity of material from which the quality of the mass which it represents may be inferred.

Material taken from the ground as being representative of a particular deposit or stratum is referred to in this book as a 'sample'.

SPECIMEN (from Latin *specere*, to look at). 1. An example of something from which the character of the whole may be inferred. 2. A part of something taken as representative of the whole. 3. A part of portion of some substance serving as an example of the thing in question for purposes of investigation or scientific study.

A portion of the original sample which is actually used for testing purposes is usually referred to as a 'specimen' when the material remains virtually undisturbed. However, the words 'sample' and 'specimen' are often used synonymously.

1.1.6 Scope of Book

GENERAL

This Manual is concerned with the *testing* of *soils* in a *laboratory*, these words being used in the sense defined above. The laboratory may range from a large fully equipped establishment to a small rudimentary testing centre set up on the site of an investigation or a construction project. Tests which are carried out on soil *in situ* are not included in this book. Since it is intended as a working manual, it is addressed primarily to those who are responsible for carrying out the tests.

PROCEDURES COVERED

This volume deals with standard laboratory classification, chemical and compaction tests for soils. The tests are described in Chapters 2–6, and reference is made wherever possible to British Standard 1377:1975 'Methods of test for soils for civil engineering purposes'. This important document is referred to in this volume as the British Standard, or BS. Other British Standards, or earlier editions of BS 1377, are referred to by their full title.

The test procedures described here are based on standard practice specified in the British Standard, where relevant. Where a test is not covered by the British Standard, reference is made to the appropriate source.

LIMITATIONS

The British Standard lays down standards of good practice which should be observed in soil testing. However, these procedures, and those described in this Manual, are based on normal British practice with sedimentary soils — that is, soils which were laid down under water and which form the majority of soils found in temperate zones. When other types of soil, such as residual soils as frequently encountered in tropical regions, are dealt with, special procedures may be necessary to obtain reliable and consistent test results. This applies particularly to the treatment of the soil before testing, and to the selection and preparation of a test specimen.

SUGGESTED APPROACH

It is essential that the laboratory technician be able to carry out tests with care and accuracy and to recognised standard procedures. This requires a knowledge of good testing techniques and an understanding of the correct procedures for the preparation of soil samples for testing. These topics are presented in Sections 1.3–1.5, and should be studied before the tests described in Chapters 2–6 are proceeded with. Important matters relating to safety in the laboratory are outlined in Section 1.6.

The summary of laboratory equipment given in Section 1.2 provides a list for reference purposes, together with comments on the characteristics and use of some of the items which are common to many different tests.

Chapter 7 provides an introduction to the description of soils in the laboratory. The engineering description of soils is an art which is acquired gradually over a period of years, but reasonable competence in giving reliable laboratory descriptions of soils can soon be gained by observing the correct basic principles and by applying common sense.

The Appendix provides a summary and a brief explanation of the metric (SI) units adopted by British Standards and used in this book. It also includes a summary of the symbols used, and a quick reference to other miscellaneous useful data.

GENERAL ARRANGEMENT

Each of the main chapters starts with a general introduction to the topic, followed by a list of definitions as applicable in this book. A section on theory presents sufficient theoretical background to enable the tests to be understood. This is followed by an outline, in general terms, of some of the more important applications of the results of the tests to engineering practice. The main emphasis of the book, however, is on the detailed procedures to be followed in preparing samples for and carrying out tests in the laboratory. Comments on general practical matters appertaining to the tests, details of the apparatus required, a list of the procedural stages, and step-by-step detailed procedures are included. Finally, the calculation and plotting of graphs and presentation of results are described, together with typical examples.

UNITS AND TERMINOLOGY

In this volume metric (SI) units of measurement are used throughout. In the few instances where obsolescent test equipment is referred to, the original Imperial measurements are also given.

In tests which are covered by a British Standard the terminology and symbols used are compatible with the BS. Otherwise the notation is as listed in the Appendix, except in those instances where it is separately defined.

REFERENCES

References are listed at the end of each chapter, under the names of authors arranged alphabetically. The British Standards listed at the end of Chapter 1 are quoted frequently throughout the book, but are not listed elsewhere.

1.2 LABORATORY EQUIPMENT

1.2.1 Scope

Equipment required for the laboratory tests covered by this volume is summarised in this section. Instruments used for making measurements of various kinds are listed first. Balances, ovens and ancillary items, which are required for almost all tests, are each discussed separately. Major items of mechanical and electrical equipment common to various tests are described, but reference is made to the appropriate chapter for apparatus needed only for a particular test.

   Other items required in a soil testing laboratory are listed under glassware and ceramic ware; hardware (i.e. metalware, plastics ware, etc.); small tools; chemicals and miscellaneous consumables; and cleaning materials.

1.2.2 Measuring Instruments

In soil testing, as in all laboratory work, it is necessary to take measurements of different kinds, and to record them. The devices used for making measurements in the performance of tests described in this volume are discussed in the following pages. However, measurements made with the use of instruments are not the only kind of observations necessary when testing soils. Visual observations requiring description in words can be equally important. Examples are the appearance and 'feel' of a soil (discussed in Chapter 7) and the behaviour of a sample during test. These aspects should not be overlooked, even though physical measurements are the main function of the tests.

   The tests described in this book for the determination of soil properties involve measurements of the fundamental quantities of length, volume, mass, fluid density, time and temperature. The types of devices required for these measurements are summarised in Tables 1.1–1.5, which include the 'capacity' or 'range', and 'resolution', of most of them. Capacity or range indicates the maximum capacity for which the instrument is designed, or the extent of the available scale readings. Resolution is the size of the smallest interval marked on the scale of the instrument. Readings can often be estimated to within one-half or one-fifth of a marked division, but while this is sometimes necessary, the apparent gain in accuracy is not always justifiable if the instrument is not sensitive enough to respond to so small an increment.

   Most of these instruments are shown in Figs. 1.1–1.9.

1.2.3 Balances

MEASUREMENT OF MASS

Mass can be measured to a greater degree of accuracy than can any other physical quantity in normal laboratory work. The accuracy of the balances referred to in Table 1.3, expressing resolution as a percentage of the capacity, ranges from 0.05% (1 part in 2000) for platform scales to 0.00006% (1 part in 1 600 000) for a sensitive analytical balance. Several different types of balance are necessary in a soil laboratory in order to cater for accurate weighing over the range from a fraction of a gram up to perhaps 100 kg.

SELECTION OF BALANCES

One selection of balances suitable for soil testing is represented by the first five items listed in Table 1.3, and illustrated in Figs. 1.3 to 1.7. Other types are available, the choice depending on circumstances, but, in general, at least four different types are desirable to cover the range of weighings normally required.

Table 1.1. INSTRUMENTS FOR MEASURING LENGTH

Instrument	Refer Fig. No.	Range	Resolution (mm)
Pocket tape	1.1(a)	10 m	1
Metre stick	1.1(b)	1 m	1
Steel rule	1.1(c)	300 mm	0.05
Pocket steel rule	1.1(d)	150 mm	0.5
Slide (Vernier) calipers	1.1(e)	150 mm	0.1
Depth gauge		150 mm	0.1
External calipers	1.1(f)		Measured by steel rule
Internal calipers	1.1(g)		
Micrometer	1.1(h)	25 mm	0.01
Micrometer		75–100 mm	0.01
Dial gauge	1.1(j)	15 mm	0.005
Dial gauge	1.1(k)	50 mm	0.01
Penetration gauge on cone LL apparatus	2.11	20 mm	0.1

Table 1.2. APPARATUS FOR MEASURING VOLUME AND FLUID DENSITY

Item	Refer to Fig. No.	Capacity	Resolution (ml)
Glass measuring cylinder	1.2(a)	2 litres	20
	1.2(b)	1 litre	10
	1.2(c)	500 ml	5
	1.2(d)	250	2
	1.2(e)	100	1
		50	0.5
	1.2(f)	25	0.5
Plastics measuring cylinder	1.2(g)	2 litres	
	1.2(h)	1 litre	
Glass beaker	1.2(j)	800 ml	limited
	1.2(k)	400	intermediate
	1.2(m)	200	markings
	1.2(n)	75	
Plastics beaker	1.2(p)	800 ml	
	1.2(q)	400	
Burette	1.2(r)	100 ml	0.2
Pipette	1.2(s)	50 ml	
	1.2(t)	25	no interme-
	1.2(u)	10	diate markings
	1.2(v)	5	
		2	
Gas jar	1.2(w); 3.17	1 litre	
Density bottle	1.2(x); 3.15	50 ml	
		Range	
Hydrometer (soil type)	4.36; 4.37	0.995–1.030 g/ml	0.000 5 g/ml

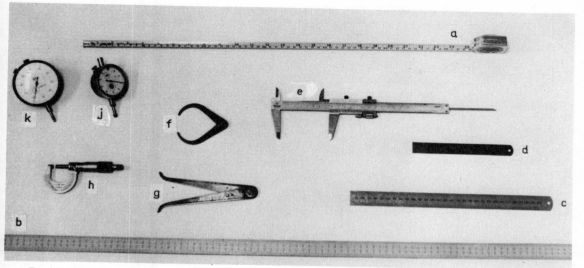

Fig. 1.1   Instruments for measuring length (see Table 1.1)

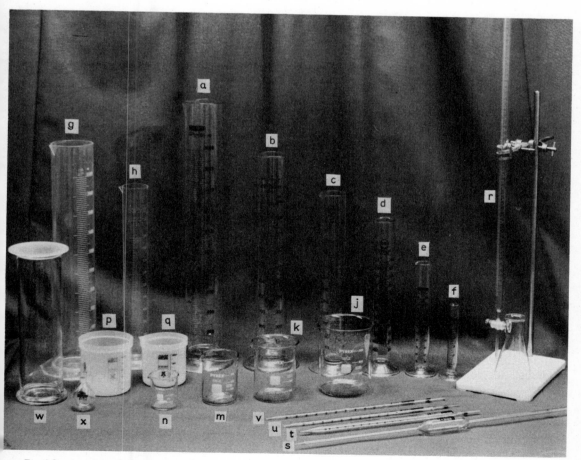

Fig. 1.2   Apparatus for measuring volume

Table 1.3. BALANCES AND SCALES FOR MEASURING MASS (see Section 1.2.3)

Category	Type	Refer to Fig. No.	Capacity	Resolution	Accuracy (%)
Heavy	Platform scales	1.3	100 kg	50 g	0.05
Coarse	Semi-self-indicating scales	1.4	25 kg	1 g	0.004
Medium-coarse	Semi-self-indicating scales	1.5	7 kg	0.5 g	0.007
Automatic	Top pan balance	1.6	1200 g	0.01 g	0.000 8
Fine	Analytical balance	1.7	160 g	0.1 mg	0.000 06
Fine	Chain dial balance	1.8	250 g	1 mg	0.000 4

Table 1.4. MEANS OF MEASURING TIME

Item	Scale range	Resolution
Stop-watch	30 min	0.2 s
Timer clock	1 h	1 s
Time switch (sieve shaker)	1 h	1 min
Wall clock	12 h	1 min
Calendar	1 year	1 day

Table 1.5. INSTRUMENTS FOR ENVIRONMENT MEASUREMENTS AND CONTROL

Item	Refer to Fig. No.	Range (°C)	Resolution or accuracy (°C)
Mercury thermometer	1.9(a)	0–250	1
	1.9(b)	0–110	1
	1.9(c)	0–50	0.5
Thermocouple		1000	20
Maximum/minimum thermometer	1.9(d)	−20–50	1
Wet and dry bulb thermometer	1.9(3)	−5–50	0.5
Water-bath control		15–50	0.1
Fortin barometer		670–820 torr (approx)	0.05 torr

The automatic balance and the fine analytical balance both require connection to an electricity supply, either to the mains via a transformer or to a 12 V battery. A chain dial balance (Fig. 1.8) is independent of electricity and is a relatively simple and inexpensive alternative to the analytical type of balance. While it is sufficiently accurage enough for many soil testing purposes, and is suitable for use in a small site laboratory, it does not have the high sensitivity needed for chemical tests and for the pipette sedimentation test.

SETTING UP A BALANCE

The principles to be followed in setting up a balance are outlined below. Details will vary with each type of balance, and the manufacturer's instructions or recommendations should be followed.

Fig. 1.3   *Heavy balance: platform scales,* 100 *kg* × 50 *g*

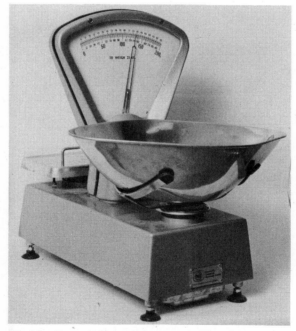

Fig. 1.4   *Coarse balance: semi-automatic,* 25 *g* × 1 *g*

(1)   Balances should be sited in areas away from walkways, doors and other sources of vibration such as machinery and lifts; and preferably not adjacent to an external wall.

(2)   Balances should not be placed next to a heater, radiator or oven, and must be protected from direct sunlight and from draughts. Balances of high accuracy are particularly sensitive to temperature changes.

Fig. 1.5   *Medium balance: semi-automatic, 7 kg × 0.5 g*

Fig. 1.6   *Fine balance: top pan automatic, 1200 g × 0.01 g*

(3)   A sensitive balance should have an independent support. A good arrangement consists
       of brick or concrete piers mounted on a solid floor carrying a concrete or slate slab 50
       mm thick — a paving slab is ideal (see Fig. 1.8). A layer of felt between the piers and slab
       will help to insulate the balance from vibration.

(4)   Sufficient space should be allowed for an operator to sit (on a stool of the correct height)
       or stand at the balance, with enough bench space for recording data as well as for weights
       and samples.

(5)   Wherever a balance is situated, it should stand on a firm and solid level surface.

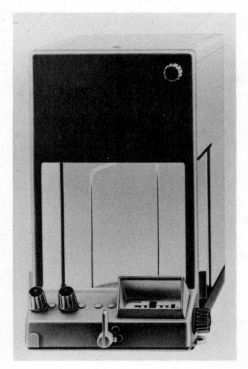

Fig. 1.7   *Analytical balance: semi-automatic, 160 g × 0.1 mg*

Fig. 1.8   *Chain-dial balance, 250 g × 1 mg*

(6)   The balance should be levelled with the adjusting feet (if fitted), and it should be ensured that it stands firm.

(7)   The balance should be protected from dust when not in use.

Recommendations for the general use of balances are given in Section 1.3.4.

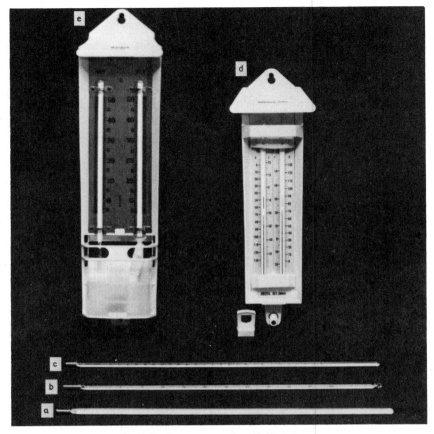

Fig. 1.9   *Thermometers* (*see Table* 1.5)

1.2.4   Ovens and Drying Equipment

CONVENTIONAL DRYING OVENS

Laboratory ovens provide a convenient means of drying material by heating to a predetermined and controlled temperature. Electrically heated ovens are normally used, and two sizes of oven are shown in Figs. 1.10 and 1.11. Ovens operating on town gas and bottled gas are also available, the latter being convenient for site use.

Before an oven is used, it is essential to calibrate the thermostat setting. The scale on the thermostat dial is not necessarily marked in °C but may be only a calibration number (often from 1 to 10, with intermediate divisions). The oven is calibrated by setting the thermostat to a certain mark and recording the temperature indicated by a thermometer when the temperature becomes steady. This is repeated for each major calibration mark on the scale, and a temperature–calibration mark curve (similar to Fig. 1.12) is plotted. For any required temperature the corresponding setting is read off the curve. The usual temperature setting for drying soil is 105–110°C, although for some soils a lower temperature (60–80°C) may be necessary (see Section 2.5.2). It is important to calibrate the oven in the surroundings in which it is to be used. If an oven is moved to a different part of the laboratory, a fresh calibration may be necessary. A full oven may give a different calibration from an empty oven, so the calibration should be done with 'average' contents in the oven.

A laboratory drying oven must be capable of maintaining a steady temperature which is substantially uniform over the whole of the drying space. Free circulation of air within the

Fig. 1.10   *Large-capacity electric drying oven*

Fig. 1.11   *Bench oven with moisture containers and accessories*

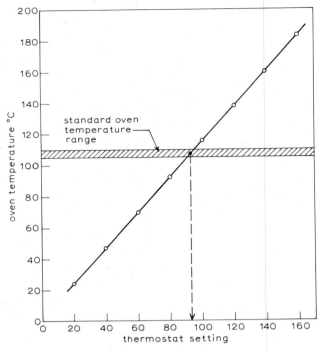

Fig. 1.12   Calibration of oven

oven is essential. In a small oven this can be achieved by convection alone, but in a large oven convection is assisted by means of a fan. A good oven should show a temperature fluctuation at any one point over a period of time within ±0.5°C, and a maximum variation throughout the drying space to within 4°C. Oven shelves should be a perforated plate or wire grid. The vent opening must be kept free both inside and outside, and the oven must never be so overloaded that the circulation of air is restricted.

Good insulation helps to maintain a uniform temperature and to economise on power consumption.

The thermostat unit must not be obstructed or knocked. Some ovens incorporate a secondary thermostat as a safety feature in case the main unit fails. Once the thermostat is set to the required temperature, it should be locked or held fast in that position, or at least marked clearly. Separate neon lamps indicate when the oven is switched on and when the heater is actually taking current.

Samples for drying should be placed in the oven in metal containers. Occasionally it is desirable to dry out a sample in a glass jar, in which case it should be handled with extreme care and placed on a high shelf away from the heating elements. Hot sample containers should always be handled with oven tongs and placed on an asbestos mat immediately on removal from the oven. Hot glassware should NEVER be placed on to a cold metal surface, as otherwise the glass is likely to crack. Containers should be transferred to a desiccator cabinet to cool with minimum delay.

The time required to achieve complete drying at the specified temperature may be determined by removing the sample from the oven at intervals, cooling, weighing and returning to the oven. By plotting the mass of the sample against time it can be seen at what point the mass becomes constant. Alternatively, a difference between successive weighings of less than 0.1% of the dry mass of soil used is an acceptable criterion for constant mass. Drying for 16–24 h (generally overnight) is usually sufficient, but a longer time may be necessary for a very large or very wet sample, or if the oven also contains a number of very wet samples.

Fig. 1.13   *Containers for drying soil*

Fig. 1.14   *Infra-red drying cabinet*

CONTAINERS FOR DRYING SOIL

Some typical containers used for drying soil in an oven are shown in Fig. 1.13. Containers used specifically for the determination of moisture content are discussed in Section 2.5.2.

OTHER DRYING METHODS

Four other types of equipment are available for drying soils:

(1)   Infra-red drying cabinet (Fig. 1.14). This is useful for the rapid drying of large samples of granular soil in preparation for particle size or compaction tests. However, it is not suitable for determining moisture content, especially if organic matter is present, because there is no means of temperature control.

(2)   Microwave oven. Microwave heating with forced air convection gives a very rapid method of drying, but without temperature control. This method also is not suitable for the determination of moisture content.

Fig. 1.15   *Warm air blower*

(3)   Warm air blower. A hand-held electric blower (Fig. 1.15) with a built-in heater unit providing a stream of air at a nozzle temperature of about 100°C is useful for rapidly reducing the moisture content of small quantities of soil. It is not suitable for drying soil for the determination of moisture content.

(4)   Electric hot-plate (shown in Fig. 4.28). Electric hot-plates fitted with a heat control unit are available in several sizes. They are useful for the rapid drying of non-cohesive soils spread out on trays, but an uneven distribution of heat requires that the soil be constantly watched and agitated. Hot-plates may be used for the determination of moisture content of some soils by the sand-bath method. They also provide a convenient source of controlled heat for chemical tests, and for many purposes can be used in place of bunsen burners.

1.2.5   Other Major Items of Equipment

DESICCATORS

After a soil sample has been removed from a hot oven, it is important to allow it to cool in dry air. If exposed to a damp atmosphere, the soil can pick up moisture as it cools. A desiccator is a sealed enclosure in which the air is kept dry by means of a desiccant, and may consist of a conventional laboratory glass desiccator (vacuum or non-vacuum type; Fig. 1.16) or a small cabinet (Fig. 1.17).

The most convenient desiccant is silica gel, in the form of crystals 4–6 mm in size. When freshly dried and active, the crystals are bright blue, but they lose their colour as they absorb moisture, and turn pink when saturated. They can be completely reactivated by drying in the oven 120°C for a few hours, but they must not be overheated.

The crystals should be spread about 10 mm deep on a small tray under the perforated floor of the desiccator. A desiccator cabinet consisting of two or more separate enclosures must have desiccant in each enclosure.

Usual practice is to keep two trayfuls of silica gel for each desiccator enclosure, one in the desiccator and one drying in the oven at 105–110°C. They are interchanged daily so that the desiccant in use is always active. A convenient arrangement is to position the desiccator between the oven and the balances most often used for moisture content measurements. The flow sequence is balance–oven–desiccator–balance.

Fig. 1.16 *Vacuum desiccator and protective cage*

Fig. 1.17 *Desiccator cabinet*

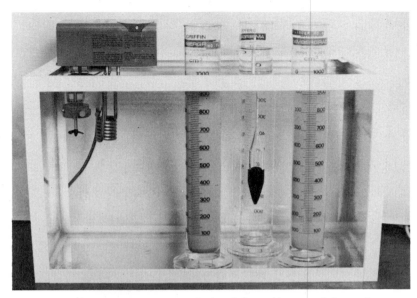

Fig. 1.18   *Constant-temperature bath with heater/thermostat/stirrer unit*

CONSTANT-TEMPERATURE BATH

For tests in which it is necessary to maintain a uniform temperature for any length of time (such as a sedimentation test, or a specific gravity test using density bottles), a constant-temperature water-bath is essential. A bath for sedimentation tests needs to be deep enough to surround the suspension in the cylinder, and should have clear glass sides. A suitable size is about $600 \times 300 \times 380$ mm deep, as shown in Fig. 1.18. A smaller bath can be used for specific gravity (SG) bottles or, alternatively, a perforated tray supported from the sides of a deeper tank can be provided for supporting SG bottles at the appropriate level.

An electrically operated unit consisting of a stirrer, heating coil and adjustable thermostat, together with indicator lamps, can be rested on or clamped to the sides of the bath. As with an oven, the thermostat setting should be calibrated against temperature in the environment in which the bath is to be used. The water level should be maintained at the correct level so that the heater and stirrer units remain fully submerged. Use of distilled or boiled water for topping up will minimise build-up ot 'scale' on the sides of the tank. A cover placed over the tank keeps the water clean and reduces heat and evaporation losses. Small plastics balls which float on the surface are available as an alternative means of insulation.

A suitable constant temperature for most work is 25°C. If 20°C is chosen, cooling of the water will be necessary in hot weather unless the laboratory is air-conditioned to maintain this temperature.

To inhibit algae growth, a water stabiliser as used in aquarium tanks may be added to each fresh batch of water placed in the tank.

VACUUM

A simple way of providing a moderate vacuum is by using a filter pump attached to a running water tap (Fig. 1.19). With a good mains water pressure a vacuum of about 15 torr (15 mmHg) can be obtained. (Atmospheric pressure is about 760 mmHg, or 760 torr.) This arrangement is adequate for a single item requiring a moderate degree of vacuum, but is not suitable for multiple uses and is wasteful of water.

A vacuum of 1 torr or less can be produced by means of an electrically driven vacuum pump (Fig. 1.20). A pump of this kind, if of sufficient capacity, can be connected to a vacuum

Fig. 1.19   Filter pump and vacuum filtration flask

Fig. 1.20   Electric vacuum pump and ancillary items

line fitted throughout the laboratory, from which connections are taken for a variety of purposes as required.

There are a few points to bear in mind if a system of this kind is to be used. A water-trap which can be drained off periodically should be installed near the pump to minimise the amount of condensation collecting in the pump. A similar trap is advisable at any outlet which is to be used extensively for vacuum filtration. Valves must be of a type suitable for use under vacuum, and should be included where they can isolate lengths of line when not in use. A vacuum gauge, reading in torr, should be fitted adjacent to the vacuum pump, and also at outlets where it is necessary to maintain a high vacuum such as one serving a vacuum desiccator used for de-airing density bottles (see Fig. 1.20).

Regular maintenance of a vacuum pump, in accordance with the manufacturer's instructions, is essential. The oil level should be topped up, and regular draining of water out of the reservoir chamber should not be overlooked. A properly maintained pump can be left running continuously for long periods. Before the pump is switched off, all connecting lines should first be isolated from it or opened to atmosphere.

Whenever a glass vessel such as a vacuum desiccator is to be connected to the vacuum line, it should first be covered by a metal cage designed for the purpose, as shown in Fig. 1.16, as a protective measure in case of implosion. A high vacuum should be applied gradually, not suddenly by rapid opening of the connecting tap. Only vessels designed to withstand external atmospheric pressure should be put under vacuum.

WATER PURIFICATION

In many soil laboratory tests use of distilled or de-ionised water is called for. Ordinary tap-water contains dissolved solids and bacteria, which may lead to ion-exchange or other reactions taking place with the minerals in the soil and could affect the test results.

Distilled water is traditionally produced by means of an electrically heated still, although gas-heated stills are also available. A convenient type is the Manesty with a 1.5 kW heating element, which can produce up to 1.7 litres/h (Fig. 1.21). Installation and operating instructions are provided by the manufacturer, but one vital precaution is to ensure that the water is always circulating when the power supply is switched on. The heating elements need to be descaled or replaced frequently in hard-water districts to maintain efficiency.

In Britain it is a legal requirement to notify the local Excise Officer of HM Customs and Excise when a still is installed. An Inspector may visit occasionally to confirm that only water is being distilled.

In recent years various types of de-ioniser have been introduced. The most convenient system is the renewable cartridge type, such as the one shown in Fig. 1.22. The unit is connected to the water supply and enables de-ionised water to be drawn off at the turn of the tap. An electrical resistance meter, requiring connection to the mains supply, provides a simple indication of the quality of the water being produced, and the cartridge is easily replaceable when necessary. De-ionised water is considered to be of a greater degree of purity than ordinary distilled water.

In this Manual, the term 'distilled water' also includes de-ionised water.

WAXPOT AND WAX

Coating with paraffin wax provides an effective and inexpensive means of sealing samples and of protecting them against moisture content changes when in storage. A low-melting-point wax (about 52–54 °C) should be used, to avoid damage to samples while being coated. Waxes with a higher melting point tend to be more brittle. Microcrystalline wax is preferable to ordinary paraffin wax because its shrinkage is less.

Wax should be used at a temperature only just above its melting point, and should not be overheated, as otherwise its sealing properties will be impaired and it is likely to become brittle. A thermostatically controlled electrically heated waxpot or wax-bath (Fig. 1.23) will maintain

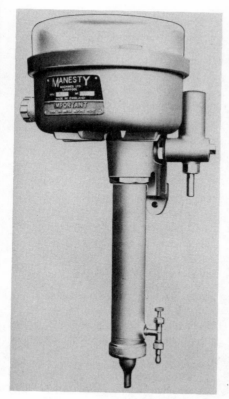

Fig. 1.21   *Electric water still*

the wax at the desired temperature, and is an essential requirement in a soil laboratory. Additional items needed are:

Ladle.
Paint brushes (say 10 mm and 50 mm).
Cutting rings, the same diameter as a U100 tube, for making wax discs.

An unprotected sample should not be immersed in molten wax. A thin coating of just molten wax should first be applied by brush, making sure that there are no cavities. If the sample is porous, it is first covered with a layer of waxed paper. When the wax has hardened, the sample may be dipped in the wax-bath two or three times, allowing the wax to set between each dipping.

For protecting block samples of cohesive material, muslin or cheesecloth is first wrapped around the sample to reinforce and retain the wax on the sides and at corners.

When a sample is sealed in a U100 tube, a wax disc is first inserted adjacent to the end of the sample. Then two or three coats of wax are brushed on to seal the gap between the disc and the tube. Finally a thin layer of molten wax is poured on top of the disc, followed by a second layer of about 20 mm thickness when the first has hardened. To seal a smaller tube, a disc of paper is inserted and brushed with wax to seal it in, and then two or three layers of molten wax are poured on.

SOIL MIXER

An electrically operated mixer such as a Hobart mixer (Fig. 1.24), working on the same principle as a food mixer, facilitates thorough mixing of water with soil. It is particularly useful

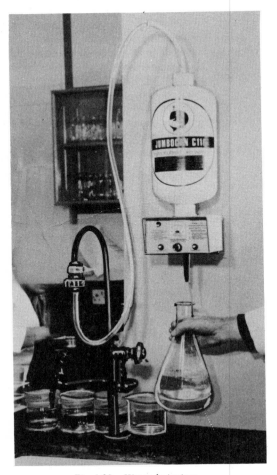

Fig. 1.22   Water de-ioniser

with non-cohesive soils. Clays may require a 'dough-hook' paddle, but machine mixing may not be practicable for clays of a firm to stiff consistency.

PRESSURE VESSEL FOR DE-AIRING WATER

Tap-water or distilled water may be made free of air by applying a vacuum to the water in a suitable vessel, or by boiling followed by cooling in a sealed container. With either method, the container must be strong enough to resist the full external atmospheric pressure.

RIFFLE BOX

A representative subdivision of a large sample of cohesionless soil may be obtained conveniently and rapidly by using a sample divider, or riffle box. Riffle boxes of several different sizes are available, ranging in capacity from 0.3 litre to 18 litres, the largest accepting gravel size particles up to 50 mm. A complete range of sizes is shown in Fig. 1.25, of which the three suggested as being the most useful for soil testing are listed in Table 1.6.

A riffle box consists of two separate containers beneath a row of slots; half the slots feed soil into one container, half into the other, arranged alternately. This ensures that each container receives an identical half of the original sample. Use of a third container quickens the process of successive subdivisions.

Fig. 1.23   *Electrically heated wax melting pot*

Fig. 1.24   *Soil sample mixer*

Fig. 1.25 *Riffle boxes*

Table 1.6. RIFFLE BOXES

| *Maximum particle size (mm)* | *Slot width (mm)* | *No. of slots* | *Capacity (approx.)* | |
			litres	*cm³*
50	64	8	18	18 000
20	30	10	4.4	4 400
5	7	12	0.3	300

SIEVE SHAKERS

A mechanical sieve shaker relieves the operator of physical effort, and ensures a uniform sieving procedure if properly used. There are two main types of shaker, each available in various sizes and with a built-in time switch.

The 'Inclyno' type (Fig. 1.26) imparts a jarring action at a frequency of about 5 Hz to the nest of sieves, which is clamped to the shaking table a few degrees out of vertical. The table slowly rotates so that the soil is spread evenly over each sieve mesh. This type is useful for gravelly soils, and can be fitted with a means of washing during shaking. It is rather noisy in operation, and the cam mechanism is subject to appreciable wear. Three sizes are available, to accommodate sieves of 200, 300 and 450 mm diameters.

The 'Endecott' type (Fig. 1.27) operates at a higher frequency, and imparts both a vertical and a rotary motion to the sieves. It is more suitable for dry sieving of sand and fine gravel sizes after washing out the fines, and is relatively quiet in operation. Two sizes are available, to accommodate 200 mm or 300 mm diameter sieves.

Fig. 1.26   Sieve shaker, Inclyno type

CENTRIFUGE

The centrifuge specified in BS1377:1975 (Tests 7(C) and 7(D)) for sedimentation tests is one capable of holding bottles of 250 ml capacity. This is a large and expensive machine, and is not likely to be available in a soil testing laboratory. However, by splitting the test sample into smaller portions a machine of more reasonable size and cost, accepting 50 ml centrifuge tubes, may be used instead. A typical centrifuge of this type is shown in Fig. 1.28.

A centrifuge is not an essential item, because filtration under vacuum, although slower, may be carried out instead.

1.2.6   Special Apparatus

Some items of equipment are required specifically for particular tests. These special items are listed below against the number of the chapter in which they are described more fully, so that a complete record of soil testing equipment may be presented within these pages.

Fig. 1.27   Sieve shaker, Endecott type

Chapter 2:   Liquid limit apparatus, Casagrande type
              Casagrande grooving tool and gauge
              Liquid limit cone penetrometer apparatus
              Metal cups for penetrometer test
              Shrinkage limit cell (Transport and Road Research Laboratory apparatus)
              Shrinkage limit containers ASTM apparatus
              Shrinkage limit mould
              Shrinkage limit prong plate
Chapter 3:   End-over-end shaker
              Magnetic stirrer
              Siphon can
              Submerged density attachments for balance
Chapter 4:   Sieves 450 mm diameter ⎫
              Sieves 300 mm diameter ⎬ all with lids and receivers
              Sieves 200 mm diameter ⎭
              Sieve brush
              High-speed stirrer
              Vibro-stirrer
Chapter 5:   BDH soil test kit
              Lovibond comparator
              pH meter

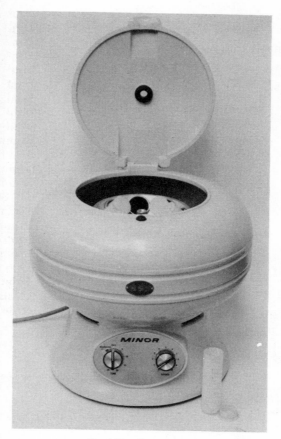

Fig. 1.28 *Centrifuge*

Chapter 6:   Compaction mould, collar, baseplate
Compaction rammer: 'ordinary' (2.5 kg)
'heavy' (4.5 kg)
CBR mould, collar, baseplate
Mould extractor
Vibrating hammer
Automatic compaction machine
Dietert apparatus and mould
Harvard Compaction Tamper

1.2.7 Glass and Ceramic Ware

GLASSWARE

Beakers, Pyrex, with watch glass covers	100 ml
	250 ml
Measuring cylinders, calibrated	25 ml
	100 ml
	250 ml
	1000 ml
one mark, no spout	1000 ml

Volumetric flasks	500 ml
	1000 ml
Conical flasks, wide-mouth	500 ml
	1000 ml
Pipettes, graduated	10 ml
	25 ml
	50 ml
Pipettes, bulb	25 ml
Burettes	50 ml
Weighing bottle	50 mm dia × 30 mm
	40 mm dia × 80 mm
Desiccator	200 mm diameter
(vacuum)	250 mm diameter
(cabinet)	300 × 300 × 300 mm
Funnels	50 mm diameter
	110 mm diameter
Filter flasks	1000 ml
Stirring rods	7 mm diameter × 200 mm
	3 mm diameter × 100 mm
Watch-glasses	50 mm and 75 mm diameter
Gas jar with cover	1 litre
Density bottle with capillary vent stopper	50 ml
Pycnometer jar with brass cone	1 kg
Ion-exchange column	
Constant-head device	
Suction syringe	
Sedimentation tubes	500 ml
Sampling pipette	
Soil hydrometer	
Tubing	
Glass plate	500 × 500 × 10 mm

CERAMIC WARE

Evaporating dish	150 mm diameter
Buchner funnel	110 mm diameter
Crucible, silica	35 ml
porcelain with lid	25 ml
Mortar and pestle	200 mm diameter
Rubber pestle	

1.2.8 Hardware

METALWARE

Moisture content containers	50 g
	150 g
	500 g
	3 kg
Tray, for quartering	1200 × 1200 × 75 mm
	900 × 900 × 75 mm
	600 × 600 × 60 mm
	300 × 300 × 40 mm
Spatula	100 × 20 mm
	200 × 25 mm

Chattaway spatula	150 × 3 mm
	130 × 10 mm
Tongs,	
oven	400 mm
crucible	200 mm
bow	200 mm
Bunsen burner	
Tripod stand	
Wire gauze	
Iron wire triangle	
Asbestos cement mat	
Stand	
burette	
retort	
funnel	
Bosshead and clamps	
Cheese grater	
Trays	
Trolley	
Diamond-tipped pencil	

PLASTICS WARE, ETC.

Beakers	250 ml
	600 ml
	1000 ml
Measuring cylinders	250 ml
	1000 ml
	2000 ml
Funnels	115 mm diameter
	200 mm diameter
Wash bottle	500 ml
Polythene bottle	500 ml
Aspirator bottle	4.5 litres
Buckets	9 litres
Dustbin	50 litres
Rubber tubing	
Rubber vacuum tubing	
Plastics tubing	

1.2.9   Small Tools

Scoop	
Hand shovel, flat blade	
Gardening trowel	
Pointing trowel	
Float, steel	
Shovel	
Trimming knife	
Wire saw	
Steel straight edge	300 mm
Ball-pein hammer	0.5 kg
	1 kg

Club hammer
Geological hammers $\frac{1}{2}$ kg
                                       1 kg

Hide mallet
Scraper
Wire brush
Soft hair brush
Sieve brush
Wax ladle
Wax brush
Test-tube brushes
ʹGlass cutter
Magnet
Steel rule
Calipers
   gauging
      external
      internal
   Vernier
Depth gauge
Pliers
   flat
   gas
   electrical
Pincers
Screwdrivers
Files, hand   ⎫ Smooth, and
      round   ⎬ second cut,
      triangular⎭ 150 mm or 250 mm
Engineers' vice

1.2.10   Chemicals and Consumables

REAGENTS

Ammonia (SG 0.880)
Barium chloride
Barium sulphate
Bromine
Ferric ammonium sulphate
Ferrous sulphate
Hydrochloric acid (SG 1.18)
Hydrogen peroxide
Nitric acid
Orthophosphoric acid, 85% (SG 1.70–1.75)
Potassium chloride
Potassium chromate
Potassium dichromate
Potassium thiocyanate
Silver nitrate
Sodium carbonate
Sodium hexametaphosphate ('Calgon')
Sodium hydroxide

Sulphuric acid (SG 1.84)
3,5,5-Trimethylhexan-1-ol

OTHER MATERIALS

Cationic exchange resin (Zeo-Karb 225 or Amberlite IT-120)
Mercury, redistilled
Silica gel granules
Kerosene
Leighton Buzzard silica sand, 425–600 μm
Plasticine or putty

INDICATORS

Litmus paper
   red
   blue
Methyl orange (screened)
Buffer solution
   pH 4.0 (potassium hydrogen phthalate)
   pH 9.2 (sodium tetraborate)
Bromothymol blue
Methyl red
Thymol blue
Soil indicator
Sodium diphenylaminesulphonate
Special indicator strips

MISCELLANEOUS

Paraffin wax (melting point 52–54 °C)
Microcrystalline wax (melting point 60–63 °C)
Silicone grease and vacuum grease
Labels
   tie on
   stick-on
Felt-tip markers
   spirit base
   water base
Glass sample jars
Plastics adhesive tape
Filter papers
   Whatman No. 541
   Whatman No. 44
   Whatman No. 50

1.2.11 Cleaning Materials

HARDWARE

Wire brush
Bottle brushes
Test-tube brushes
Soft hair brush
Washing-up brush

Viscose sponges
Nylon pan scourer
Tea towels
Hand towels
Dusters
Polishing cloths

CONSUMABLES

Soap
Teepol
Scouring powder
Woodwork polish
Metal polish
Cellulose cloths
Brillo pads
Steel wool
White spirit
Acetone
Alcohol
Ether
Slaked lime
Flowers of sulphur

1.3   TECHNIQUES

1.3.1   General

Good laboratory practice depends first and foremost upon the development of correct techniques in performing tests, observing data and recording observations. Careful plotting of graphs where necessary, accuracy in calculations and correct reporting of results are other skills which a technician must acquire.

Procedures for carrying out the tests are detailed in the respective chapters. The techniques given here are those which are relevant to laboratory testing generally and which are required for most tests on soil, such as the correct use of balances. Some of these procedures may appear to be elementary, but it is essential to acquire the right techniques at the outset.

During the course of laboratory testing, the need for certain small items recurs continually. It is, therefore, convenient to carry items such as the following in the pockets, usually of the laboratory coat or overall:

Pencils (well sharpened) — HB grade for writing; H grade for plotting graphs.
Ball-point pen.
Rubber eraser.
Notebook.
Felt-tip pen (waterproof).
Small dusting brush.
Steel rule (150 mm).
Small spatula.
Pocket knife.
Hand lens.
Clean cloth.

A small pocket slide-rule or calculator is useful for on-the-spot calculations. Laboratory test data sheets, and graph and calculation sheets are conveniently carried on a clip-board.

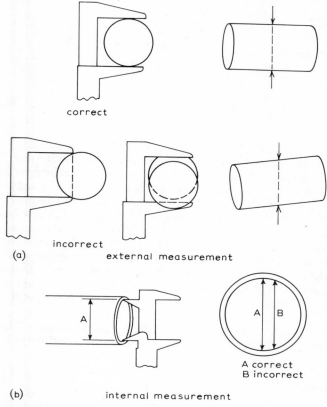

correct

incorrect

(a)   external measurement

(b)   internal measurement

A
A
B

A correct
B incorrect

Fig. 1.29   Use of measuring calipers

1.3.2   Use of Calipers

Calipers consist of a pair of hinged steel jaws which are used for measuring dimensions of solid objects where a scale rule cannot be applied directly. There are two types (illustrated at f and g in Fig. 1.1), for external and internal measurements, respectively.

When calipers are being used to measure the external diameter of a circular or cylindrical object, it is essential that a true diameter be measured, and that the measurement be made squarely and not on the skew (Fig. 1.29a). The true diameter is the *smallest* measurement attainable between the jaws. The jaws are closed until they just touch the object at the desired points of measurement, and should be brought up just tight, but not over-tightened. The distance between the jaws is then measured on a steel rule, to an accuracy of 0.5 mm.

The same principle applies when the internal diameter (of a sample tube, for instance) is being measured, as shown in Fig. 1.29(b). But here the true diameter is the *largest* measurement between the jaw extensions, provided that the measurement is made normal to the axis of the tube — that is, at a square end.

1.3.3   Using a Vernier Scale

A vernier scale is fitted to many types of measuring and surveying instruments. It is commonly used in soil testing as the basis of slide calipers, or vernier calipers (Fig. 1.1e). The vernier, invented by Pierre Vernier in the seventeenth century, enables measurements to be made by

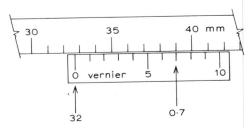

Fig. 1.30   *Reading a vernier scale*

direct reading to 0.1 mm without having to estimate fractions of a division. The scales can be manufactured quite simply without requiring high-precision machine tools.

Vernier calipers consist of a steel scale marked in millimetres with a fixed jaw at one end, and a sliding jaw carrying a scale which is 9 mm long and divided into 10 equal parts (Fig. 1.30). A vernier division is therefore 0.9 mm long — that is, 0.1 mm shorter than a 1 mm scale division. When the sliding jaw is brought up to the object being measured, the nearest scale division to left of the zero on the vernier indicates the number of whole millimetres being measured, and the vernier mark which exactly coincides with a scale mark gives the number of tenths of a millimetre to be added. In the example shown in Fig. 1.30 the vernier zero is between 32 and 33 mm, so the measurement is 32 mm plus a fraction. The seventh mark on the vernier lies directly opposite a scale mark (which is not read), so the fraction is 0.7 mm. The measurement is therefore 32.7 mm.

Slide calipers have jaws for both internal and external measurements. They should be kept clean and the sliding parts should be kept free from dust and soil particles.

1.3.4   Use of Balances

The following notes provide a general guide to the proper use of balances and procedures for weighing. Details will vary with each individual type of balance. Six main types are listed in Table 1.3. Notes on the setting up of balances were given in Section 1.2.3.

CHOICE OF BALANCE

To achieve sufficient accuracy, it is necessary before weighing to select the balance which is appropriate to the mass to be measured. This will usually be the balance with the highest sensitivity which has the required capacity. Under no circumstances should a balance be loaded beyond its stated capacity, otherwise damage may be caused to the mechanism. If in doubt, rough-weigh first on a coarse balance.

WEIGHING

(1)   Check that the pans are clean, and that the reading is zero when unloaded (after switching on the indicator light if fitted).
(2)   Set the tare adjustment to zero with the empty container on the pan, if this type of balance is used.
(3)   Allow the sample to cool in a desiccator if weighing dry from the oven.
(4)   Powdered substances (including soils) should never be placed directly on the pan of a sensitive balance, but on a watch-glass or in a weighing-bottle or moisture container. (This does not apply when scales fitted with a scoop-type pan designed for holding soil are used.)
(5)   Place sample and container on the pan gently and add loose weights (if appropriate) until balanced (see point 6). If a two-pan balance is being used, the object being weighed is

placed in the left-hand pan and loose weights are added to the right-hand pan, as in Fig. 1.8; but when a given mass of soil or other substance is to be weighed out, weights making up the given mass are first placed in the left-hand pan so that the material can be added to the right-hand pan.

(6)   When a balance which incorporates a 'rest' position to take the weight of the pans off the knife-edges is being used, the pans should be in this position whenever adjustments are made to the contents of the pans. Raising on to the knife-edges should be done steadily.

(7)   When the balance has steadied, read the indicator, and add to this the total value of any loose weights.

(8)   Write down this weight immediately in the appropriate space on the test sheet, while in front of the balance.

(9)   Check back to the indicated weight and loose weights, and confirm that the correct value has been written down.

(10) Remove any loose weights and in doing so recheck the total. Replace them in the box.

(11) Remove sample from pan, and clean the pan.

(12) Switch off the indicator lamp.

GENERAL CARE

(1)   Before use check that the balance is level and firm.

(2)   Always keep a balance clean and dust-free. If it has a cover, replace it after use.

(3)   Use the tweezers for handling small loose weights.

(4)   Loose weights must always be replaced in the weights box.

(5)   Never use a highly sensitive balance for weighing out corrosive liquids.

(6)   A regular maintenance contract with one of the specialist firms who provide this service will ensure that balances are kept in good working order. Separate contracts may be necessary with two different firms to cover analytical and fine balances, and the heavier balances and scales.

1.3.5   Recording Test Data

TEST FORMS

The first essential is to record a full identification of the sample to be tested on the appropriate test form, together with an engineering description. Any special testing instructions should also be recorded on the form. All entries should be plainly and clearly written. Readings should be accurately recorded at the time they are taken, directly into the appropriate space on the form. Errors should be crossed out (not altered) and fresh figures entered above, so that there can be no possibility of misunderstanding what has been written.

The use of odd pieces of paper for the recording of data should be strictly forbidden. Mistakes are easily made when transferring data from one piece of paper to another, and odd sheets often get lost. Incorrect calculations are another source of error. Figures and calculations should be double-checked, preferably by different methods or by someone else.

On completion of the test, the operator should sign the record sheet and enter the data of the test. An example of a laboratory test form is given in Fig. 2.15.

METHODS OF PRESENTING DATA

All laboratory tests involve the recording of numbers, which are observed in various ways during the test. Sometimes these observations form part of the data required from the test; and sometimes the observed data must be processed by calculation or other means to arrive at the required results. In either case the observed or calculated figures must be presented in a form in which they can be clearly understood. This can be done in two ways: either by drawing up a table or by plotting a graph.

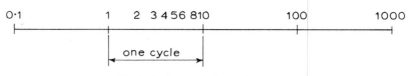

Fig. 1.31   *Logarithmic graph paper*

Tabulation of results is fairly straightforward and will be discussed under each test procedure. Graphs require the application of a few simple rules if the maximum benefit is to be derived from them.

1.3.6   Graphs

A graph is a means of representing a series of observed or calculated numerical data in diagrammatic form. The advantage of a graph over a table is that it links the data together and gives a general picture at a glance. It also enables values in between actual observed values to be estimated easily and reliably.

Most graphs are drawn on squared paper. One set of figures is represented along the horizontal axis (the abscissa) to a certain scale, and a second set along the vertical axis (the ordinate) to the same or a different scale. Each corresponding pair of observations or figures is plotted as a point, and a smooth curve or straight line is usually drawn through the points to represent the relationship between the two quantities. An example is shown in Fig. 2.15, in which penetration of the cone (mm) is plotted against moisture content (%).

TYPES OF GRAPH PAPER

Arithmetical graph paper is ruled up as a square grid. The sheets generally used are in 1 mm, 2 mm or 5 mm squares, sometimes with heavier rulings at 10 mm or 50 mm intervals.

Semi-logarithmic scales are used for some applications where a wide range of values is required. The ordinate is divided on an arithmetical scale, but the abscissa is divided into a number of cycles on a logarithmic scale. One cycle represents multiplication by a factor of 10. Thus the distances between 1 and 10, 10 and 100, 100 and 1000 units, etc., are all equal. Intermediate markings follow a fixed pattern, as shown in Fig. 1.31. A logarithmic scale enables very large numbers to be accommodated, while at the same time very small values can be plotted to the same degree of accuracy. Its main application is in particle size tests, for which are plotted sizes ranging from 0.002 mm to 200 mm or more (a range of five cycles — i.e. a ratio of 100 000).

Semi-logarithmic graph paper is supplied in various forms with up to seven logarithmic cycles across the width of the sheet.

CHOICE OF SCALES

The choice of suitable scales for the axes of a graph on an arithmetical grid is important, and must ensure: ease of plotting; accuracy of plotting; ease of reading off intermediate values.

The two basic rules to remember in choosing scales are, in order of importance:

(1)   Select a *convenient* scale in accordance with the recommendations given below.
(2)   Use the available area of the paper to best advantage.

It is of no advantage to cover the whole sheet if an inconvenient or non-standard scale has to be used to achieve this. The scale used should be one in which each major division represents a single unit of the quantity being plotted, or is related to it by a factor of 2, 5, 10; 20, 50, 100; etc. Do not use a multiple of 4, and *never* a multiple of 3. The rule is to use factors of 1, 2, 5 or their multiples of 10 (or of 1/10). Some examples of suitable and inconvenient scales using graph paper divided in mm and cm squares are given in Table 1.7.

Table 1.7.

Examples of recommended scales	Examples of inconvenient scales
1 cm = 1, 2, 5 units	1 cm = 4 units
10, 20, 50 units	25 units
0.1, 0.2, 0.5 units	3 units
1 mm = 0.001, 0.002, 0.005 units	1 mm = 0.004 units
	0.025 units

The selection of a scale for the arithmetical ordinate of a semi-logarithmic plot should follow the same rules as given above.

Whenever possible, use the same scale when comparing results of the same test on different samples. Otherwise, indicate very clearly that the scales are different.

1.3.7 Calculations

Calculations using laboratory test data may be made by the following methods:

Ordinary arithmetic.
Logarithmic tables (4-figure or 5-figure).
Slide rule.
Pocket calculator.
Programmable desk-top calculator.

The calculations given in this volume are set out in full, in the same way as they would be done in practice. In some instances, a slide rule gives sufficient accuracy, but in others either 5-figure logarithms or an electronic calculator are necessary.

All laboratory test calculations can be arranged for use in a programmable desk-top computer, so that the only figures seen are the laboratory readings and the final results. This practice is used today in many commercial testing laboratories, but until the procedures and the reasons behind them are fully understood, all stages of the calculations should be followed through in full and independently checked.

1.3.8 Reporting

Select the correct form on which to present the results. Test results should always be written legibly and must be reported in a clearly understandable and unambiguous manner.

Where there is no standard form, the information required for a complete record of the results consists of:

(1) Laboratory name and address.
(2) Nature of test.
(3) Sample identification data.
(4) Name of job and location number.
(5) Name of client (if known).
(6) Description of test sample.
(7) Test specification and any special instructions.
(8) Test results as required by the specification and special instructions.
(9) Particular reference to anything unusual that occurred during the test or in the results.
(10) Signature and date of test.

Local laboratory rules may require that every test sheet be signed by the person who carried

Table 1.8. RECOMMENDED ACCURACY OF REPORTING LABORATORY TEST RESULTS

Item	Accuracy	Unit
Density	0.01	mg/m³
Moisture content	<10%:0.1	%
	>10%:1	
Liquid limit ⎫		
Plastic limit ⎬	1	%
Plastic limit ⎭		
Specific gravity	0.01	—
Clay, silt, sand, gravel fraction	generally:1	%
	if <5%:0.1	%
Voids ratio	0.001	—

out the test, who should be responsible for ensuring that all information, including calculations and transfer of data, has been checked. The signed and dated forms with the completed results can then be passed to the supervisor.

The degree of accuracy to which results should be reported is indicated under 'Report results' at the end of each test description. These recommendations are summarised for reference in Table 1.8.

1.3.9 Laboratory Climate

Regular measurement of the climate in the laboratory is desirable and can be quite important for some tests, although those covered in this volume are not particularly sensitive to changes experienced under normal laboratory conditions. The items usually measured are:

> Laboratory temperature.
> Atmospheric pressure.
> Relative humidity.

TEMPERATURE

To record the overall temperature changes in the laboratory, a maximum/minimum thermometer is required. The usual type is shown in Fig. 1.9(d). The pointers indicating the maximum and minimum readings can be easily reset to the mercury surface by use of the small magnet which is provided. The bottom of each pointer indicates the maximum and minimum readings, and the mercury level on either side indicates the prevailing temperature.

Typically, five readings are taken and recorded each day, as follows.

At 9 a.m. (or start of the day) read overnight minimum temperature, the previous day's maximum and the prevailing temperature at the time. Set the pointers to the mercury surface.

At mid-day, and at 5 p.m. (or end of the day), read the temperature.

If a self-recording apparatus is available, all that is necessary is to renew the record chart each week, or at the appropriate intervals.

The thermometer or recording apparatus must be sited away from sources of heat, and shielded at all times from direct sunlight.

ATMOSPHERIC PRESSURE

The most satisfactory apparatus for atmospheric pressure is a Fortin barometer, which reads to 0.05 mmHg (0.05 torr) by means of a vernier on the scale. The instrument must be carefully installed. Alternatively, an aneroid barometer may be used.

Readings should be taken twice daily — morning and evening. Self-recording instruments are also obtainable.

RELATIVE HUMIDITY

A wet-and-dry bulb thermometer (Fig. 1.9e) is used to determine relative humidity. It is essential to keep the small reservoir topped up with *distilled* water, and to ensure that the wick is always moist. It should be replaced if it goes hard or non-absorbent.

From the room temperature (indicated by the dry bulb thermometer) and the difference between the readings of the two thermometers, the relative humidity (%) can be read off from tables supplied with the instrument.

1.3.10   Cleaning

It is important to keep glassware and ceramic ware absolutely clean and free from grease, especially when used for chemical tests. One of the simplest and most effective means of cleaning is to wash with warm soapy water, or water containing a little synthetic detergent (such as Teepol), and wiping with blotting paper or a soft cloth, or with a long-handled brush for burettes and the like. It is essential to rinse thoroughly afterwards to ensure complete removal of the cleaning agent. The items are then allowed to drain on a drying rack and wiped dry with a clean cloth.

1.4   CARE OF SAMPLES

1.4.1   Observations

Soil samples provide some of the most important evidence from a subsurface investigation. They are costly to obtain, and should be treated and handled with proper care.

Inspection and description of samples are usually carried out by a soils engineer or geologist, but observations made when samples are prepared and tested in the laboratory can be equally important. The operator should record as much information as possible about the samples, by means of descriptions of what is seen, felt and smelled. There will be closer contact with samples in the laboratory than anywhere else. For instance, an engineer may initially inspect an undisturbed tube sample only at each end, but it is not until the sample is extruded for test that the whole length of the sample can be observed. If there are significant variations from what has been seen at the ends, these should be brought to the notice of the engineer before any tests are started.

A systematic procedure for description of soils is given in Chapter 7. The following sections are concerned with the general handling and storage of samples.

1.4.2   Identification

Every sample should be clearly identified by a unique number, which will normally include the reference number and name of the location from which the sample was taken. This information should be recorded clearly at the outset on the sample description sheet. If there is any doubt about the sample numbering, this should be queried immediately.

The full identification of a sample will include the following details:

Name of location.
Location reference number.
Borehole or pit number, or similar reference.
Sample number.
Depth below surface (top and bottom).
Type of sample (disturbed or undisturbed).

Container type and number.
Date sampled.
Visual description of sample.
By whom described.
Date of description.
Signature.

The laboratory visual description may be subsequently modified by the Engineer in the light of additional data provided by the laboratory test results.

1.4.3   Opening a Sample Container

Before opening up a sample for inspection or for testing, examine the container to see whether it was properly sealed. Record any observations regarding inadequacy of sealing or packing, especially of undisturbed samples. Also note any damage to or deterioration of the sample container.

Sample containers must be opened carefully to avoid disturbance and loss of material or fluid. The protective wax coating applied to undisturbed samples should be gently prised off. Undisturbed samples must at all times be handled with care, and not dropped or bumped on the bench.

Samples should always be labelled on the container itself, not on the lid or end caps. Nevertheless the lid or caps should remain close to the sample container when removed, so that they do not get mixed up.

1.4.4   Resealing Samples

Samples should be resealed as soon as possible after description or removal of material for test, to prevent drying out.

When sufficient material from an undisturbed sample has been taken for testing, the portion which remains should be resealed and returned to the sample stores. The sealing procedure is described in Section 1.2.5 (p. 21).

Jar and bag samples should be resealed to prevent loss of moisture and material. After screwing on the lid of a glass jar, it may be sealed with adhesive plastics tape or by brushing on a coating of wax.

Polythene bags are best sealed by applying plastics adhesive tape over the opening, after removing as much air as possible. Twisting the neck of the bag and tying with wire or tape does not seal in the moisture, but this method may be adequate for coarse-grained soils in which the moisture content is not important. Never tie the neck of the polythene bag in a knot; it is very difficult to untie.

All samples must carry clear identification labels before being returned to stores. Remove or delete any old labels.

Material which has been used for a test should not be put back with the original material, but repacked separately in a suitable container with a label identifying the sample and showing clearly the type of test for which it has been used.

1.4.5   Storage of Samples

Soil samples should be stored under cover in a cool room which is protected from extremes of heat and cold. Ideally, the store should be maintained at a high humidity, but this may not be practicable except for a small room for special samples.

Disturbed samples in glass jars (honey type) can be conveniently stored in milk bottle crates.

Large disturbed samples in polythene bags ('bulk bag' samples) should not be piled one on top of another, but placed individually on shelves or racks. Pallet racking with the aid of a fork-lift truck is perhaps the most convenient method of storage, and enables the full height of a store shed to be used to maximum advantage. The layout of racks must allow sufficient manoeuvering space.

Undisturbed tube samples may be laid on racks designed for the purpose. However, tubes containing wet sandy or silty soil should be stored upright (suitably protected against being knocked over), to prevent possible slumping and segregation of water. The end caps of tube samples which are to be stored for any length of time should be sealed with tape or wax, in addition to the wax seal next to the sample itself. Any space between the wax and the end cap should be filled with packing material, such as sawdust. Sand or any other type of soil, and grass or straw, should not be used. These samples should be stored away from any possible source of heat, and not high up in the building, where warm air tends to collect.

Racks and shelves and all storage areas should be clearly numbered. It is essential to keep adequate records and to maintain an efficient filing system, so that any given sample can be readily located.

1.5 PREPARATION OF DISTURBED SAMPLES FOR TESTING

1.5.1 Scope

This section deals with the general principles involved in the preparation of disturbed samples for the classification and compaction tests described in Volume 1. It is based on Section 1.5 of BS1377:1975. The actual amount of material required for test is given with each test procedure, and depends upon the type of soil.

The preparation of undisturbed samples requires a different approach, which is touched on in Section 3.5 and will be covered in greater detail in Volume 2.

1.5.2 Selection for Test

The most important requirement for any test sample is that it be fully representative of the material from which it is taken. Taking samples on site is beyond the scope of this book, so the sample received in the laboratory has to be treated as being representative of the original material *in situ*.

For testing purposes a sample smaller than the received sample is usually required. The actual quantity of material depends upon the type of test and the nature of the soil. Coarse-grained soils require a larger sample, to be properly representative, than do fine-grained soils. The standard procedure for obtaining a small representative sample from a large quantity of granular soil is given in Section 1.5.5. This principle should always be followed. Taking a single random scoopful of material from a bagful or heap does not provide a true representative sample of the whole.

Even though the correct procedure has been followed, a check should be made by visual inspection that the selected test sample is indeed truly representative. For instance, if medium to coarse gravel is present, it can be observed whether the test sample contains an excess or deficiency of the larger particle sizes compared with the original material.

With cohesive soils (clays and silts) the riffling procedure is not possible. To obtain a representative disturbed sample of this type of material, a number of small samples should be selected from several locations and mixed together, rather than taking all the samples from one place. If the soil is laminated or non-homogeneous in any way, careful visual observation is needed to ensure that the test sample is properly representative, and it may be necessary to consider the sample as being of more than one type of material.

Fig. 1.32   *Sample preparation equipment*

1.5.3   Drying

Before a large disturbed sample of cohesionless soil can be properly subdivided, it must first be partially dried until it is in a state in which it can be crumbled. This should normally be done by air drying — that is, by leaving the soil spread out on trays in the laboratory, with free access to air, for a period of up to 3 days. The trays may be placed in a warm area, such as over an oven, but the temperature should not exceed 50 °C.

Air drying does not remove all the moisture in the soil, and if the air-dried moisture content is required, this may be found by taking representative specimens as described in Section 2.5.2, Moisture Content Test.

If it is known that oven drying will not affect the results of the tests to be carried out, the material may be dried in an oven at 105–110 °C. However, oven drying should not be used for that part of the sample on which plasticity tests are to be done (see Section 2.6.3). The same applies to material for pH tests. Organic soils and tropical soils should never be oven-dried before testing.

1.5.4   Disaggregation of Particles

Aggregations of particles must be broken down before cohesionless soil samples are prepared for test, but for most purposes crushing of individual particles must be avoided. This process is best done in a mortar, with a rubber pestle (Fig. 1.32). If a large quantity of material is to be prepared, it should be done in batches, to avoid the mortar being overloaded with soil.

If a quantity of material passing a certain size of sieve is required, the mortar should be emptied into the sieve at intervals and the material retained on the sieve returned to the mortar for further treatment. This avoids interference with the breakdown process due to the presence of fine material.

Disaggregation of particles may be considered to be complete when only individual particles are retained when sieved through a 2 mm sieve.

For some tests such as chemical tests, for which it is permissible to crush individual particles, a conventional stone pestle may be used with the mortar for grinding down to the required degree of fineness.

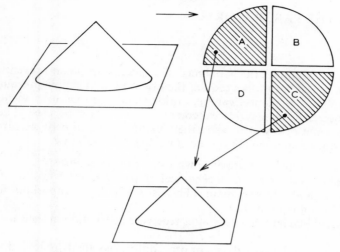

Fig. 1.33   *Cone-and-quartering*

1.5.5   Riffling and Quartering

Two methods may be used for subdividing a large sample of granular soil to obtain a representative test sample. These are:

(1)   Riffle box method
(2)   Cone and quartering method

Either process may be referred to as 'riffling'.

RIFFLE BOX

Three suggested sizes of riffle box which will cover most requirements are given in Section 1.2.5 (p. 22) and are shown in Fig. 1.25. The material to be divided must first be thoroughly mixed, then poured evenly into the riffle box from a scoop or shovel. It should be distributed along all or most of the slots, not confined to two or three slots near the middle. Each receiving container should then receive an identical sample, one of which is rejected and the other remixed and poured back into the riffle-box after an empty receiver has been inserted. This process is repeated as many times as necessary until the required reduction has been achieved, the material from alternate sides after successive pourings being retained.

CONE AND QUARTER

This procedure is slower and requires more effort than the use of the riffle box, but is equally reliable if the mixing is done thoroughly so that segregation of particle sizes does not occur.
   The initial material is mixed on a tray, and formed into a circular conical heap. Any coarse particles around the base of the cone should be evenly distributed. With a cruciform sample splitter, or a straight edge, the heap is divided into four equal portions as shown in Fig. 1.33. Two diagonally opposite portions (A and C) are separated out and mixed thoroughly together and formed into a smaller heap for quartering as before. The portions B and D are rejected and returned to the original sample container. The above process is repeated as many times as necessary until a small enough representative sample is obtained.

1.6 SAFETY IN THE LABORATORY

1.6.1 Basic Rules

Safety in the laboratory implies the avoidance of accidents. Accident prevention is largely a matter of common-sense, but it does require the observance of simple precautions, together with tidiness, cleanliness, vigilance, careful working and cautious movements at all times. Short-cuts and potentially dangerous expedients should never be used. The effects of an accident can be harmful or disastrous to others besides those directly involved.

The most important requirements can be summarised as follows:

(1)   Study and observe the laboratory's own rules and procedures.
(2)   Before handling electrical or mechanical apparatus, refer to the manufacturer's instructions — and observe them. (This is particularly important with a new or unfamiliar piece of equipment.)
(3)   Chemicals should be handled only after receiving instruction in their use and potential hazards.
(4)   If in doubt, or if the safety of a device or procedure looks at all dubious, do not hesitate to consult the person in charge.

The precautions which follow under various headings are given in summary form as a general guide to safe practice and conduct. These points may be amplified by the laboratory's own rules and practice.

1.6.2 General Conduct and Handling

Keep laboratory benches and working areas tidy.

Waste materials and soil should be disposed of in proper containers, which should be emptied regularly.

Cleanliness in a soil laboratory is difficult to maintain, but accumulations of dust and soil trimmings and old samples should be minimised by regular tidying up and cleaning.

NEVER RUN in a laboratory building (except in dire emergency).

A laboratory is not the place for practical jokes, which can easily misfire, with disastrous consequences.

Always wear appropriate protective clothing (see Section 1.6.11).

Winchester bottles and the like should not be lifted by the neck.

Never attempt to lift any item which is too heavy or too bulky to handle unaided. Obtain assistance.

Before picking up a heavy or awkward item, make sure that there is a clear and safe space in which to put it down.

Large or heavy objects should be transported any distance on a suitable trolley. This applies particularly to glassware, such as a tray loaded with glass weighing bottles.

A trolley should be pulled from in front, not pushed from behind, when passing by doorways.

Hot objects should be placed on a suitable heat-proof mat, not directly on to the bench.

Use oven gloves and tongs for handling hot items.

1.6.3 Fire

The local fire rules and drill must be learnt and understood.

Smoking is one of the greatest fire hazards, and should preferably be prohibited in a laboratory.

Paper-work and wastepaper containers should be kept away from naked flames and hot-plates.

Some vapours from chemical reactions are highly inflammable. The same applies to the use of certain adhesives which give off inflammable vapours. Smoking should certainly be banned in areas where these are used.

Glass reagent bottles and other containers should not be left in direct sunlight, because of the lens effect, which can cause localised intense heat.

Electrical overheating of motors and heaters, and electric sparking, are other potential fire hazards.

1.6.4 Electricity

Connection of electric items to a conventional socket outlet requires the usual precaution that the wires be correctly connected to the plug. The old British and current International colour codes are as follows:

Terminal	International Standard	Obsolete British Standard
Live	Brown	Red
Neutral	Blue	Black
Earth	Green and yellow	Green

Ensure that the capacity of the fuse fitted is adequate, but not excessive, for the maximum current rating of the equipment.

Many electric motors require connection to a 3-phase supply by a competent electrician.

Suitable leads and cables should be used, and should be secured neatly on the bench. Keep cables well away from sources of heat such as hot-plates.

Never touch electrical apparatus or switches with wet hands.

Switch off before removing a plug from its socket, and see that the switch is off before inserting a plug.

Avoid the use of multiple outlet adaptors on a single socket.

If in doubt, or if any modification is necessary, obtain the services of a competent electrician.

Check the condition of leads and cables regularly. If the outside covering is worn or damaged, the cable should be replaced.

1.6.5 Gas

Check flexible gas tubing regularly. Worn or punctured tubing should be replaced — do not attempt to make repairs.

If a leak is suspected, extinguish any naked flames, open the windows, and notify the Gas Board immediately. Do not attempt any emergency repairs.

Never search for a leak with a naked flame.

Learn the correct way to use a bunsen-burner or other gas appliances.

1.6.6 Drains

Sinks and gullies at which wet sieving (Section 4.6.4) is carried out, or where washing of equipment with adhering soil takes place, should be fitted with removable silt traps. The traps must be cleaned out at regular intervals, normally weekly, but every day if large numbers of wet sievings are being done. If not trapped in this way, silt and clay can very quickly build up in gullies and cause blockage to drains.

Waste soil and sample remains should never be disposed of by washing down a drain. Proper receptacles should be provided for this purpose.

1.6.7 Mercury

Handling of mercury can be hazardous unless the appropriate precautionary measures are taken. Mercury vapour is poisonous, and if sufficient liquid mercury is exposed in an enclosed room at normal temperature (18–22 °C), the concentration of mercury vapour can rise to more than 100 times greater than the accepted maximum allowable concentration of $100\,\mu g/m^3$. However, normal ventilation will prevent the concentration rising as high as that, but it is important to control the surface area of mercury exposed to the air. Further information on the hazards of mercury will be found in Technical Data Note 21 by HM Factories Inspectorate (1976).

PREVENTATIVE MEASURES

To avoid the scattering of mercury and build-up of vapour during handling, the following routine precautions should always be taken so far as they are applicable:

(1) Good general ventilation (preferably by mechanical means) should be provided.
(2) Storage should be in airtight or water-sealed vessels.
(3) Containers of mercury should be kept on trays (plastics types are very suitable) which will catch and retain any spillage which may result from the breakage of glass containers or from other causes.
(4) Mercury should be handled over trays sloping into a gulley leading to a water-sealed trap, or over a perforated tray above a bath containing a sufficient depth of water.
(5) Mercury should be handled in an enclosed system if possible, or in a fume-cupboard or an enclosure connected to an exhaust system which can maintain an air speed of about 0.5 m/s over all openings.
(6) The exhaust from vacuum pumps should be vented outside the workroom at a safe height.
(7) Care should be taken to ensure that no mercury can lodge on any heated surfaces, such as local lighting units, radiators.
(8) Contact with the skin should be avoided, protective clothing (laboratory coat, rubber gloves) being used where appropriate.

A method for the determination of mercury vapour in the air is given in booklet No. 13 in the series 'Methods for the Detection of Toxic Substances in Air' published by HMSO.

SPILLAGE

If any spillage is not immediately cleaned up, it will, during normal traffic over the area concerned, be broken up into tiny droplets offering a large surface area to the air. The droplets will lodge in any rough or irregular surface and in any cracks, and will be exceedingly difficult to remove.

This danger is not always appreciated until an area has already become contaminated, and cleaning will then be of vital importance.

Spillages should first be cleaned up as far as practicable by mechanical means, such as by vacuum probe (a flexible tube connected to a conical vacuum flask which is connected to a vacuum line). The affected areas should then be treated with a wash composed of equal parts of slaked lime and flowers of sulphur mixed with sufficient water to form a thin paste. This yellow wash, which has the appearance of distemper, should be liberally applied and allowed to dry on the floors, the lower parts of the walls, work-benches and any other contaminated surfaces. Twenty-four hours later the wash should be removed with clean water and the surfaces again allowed to dry.

This method of cleaning ensures that any droplets of mercury which cannot be mechanically removed are converted into mercury sulphide, so eliminating the danger of vaporisation. It also ensures by visual observation of the removal of the yellow wash that the cleaning is effectively performed.

1.6.8 Chemicals

Most chemicals are potentially hazardous, and must be handled with caution. Good technique, orderliness and cleanliness are essential.

Chemicals which emanate fumes should be used only if a properly ventilated fume-hood is available. If there is one — use it.

Do not sniff gases given off by chemical substances or evolved by a reaction.

Do not look into a test tube or flask while heating or mixing chemicals.

Always add acid to water — NEVER water to an acid.

Do not use the mouth to suck corrosive or toxic or volatile liquids into a pipette.

Any spillages should be cleared up immediately. Acids or alkalis must first be neutralised.

Acid or alkali (including ammonia) on the skin should be washed off immediately with plenty of clean water.

Unlabelled containers of chemicals are useless and potentially dangerous. They should be disposed of, but by the proper means.

Chemicals likely to be used in a soil laboratory which are particularly hazardous are as follows.

ACIDS AND ALKALIS

Neutralise spilled acids with sodium carbonate.

Neutralise spilled caustic alkali with ammonium chloride.

Strong solutions of these substances should be kept readily available for emergency use.

AMMONIA

Ammonia affects the skin and causes painful burns. Drops on the skin should be washed off immediately with plenty of clean water. Its effect on the eyes is immediate and disastrous, unless the eye is held open while being washed with clean water for at least 15 min.

Always open a bottle of ammonia in a fume-cupboard behind glass. Alternatively wear an eye-shield as a protection against the inevitable slight spray. Special care is needed with 0.880 SG ammonia; 25% ammonia (0.91 SG) is less dangerous.

BARIUM CHLORIDE

Extremely poisonous. Wash hands immediately after using. Keep well away from teacups and food.

BROMINE

Bromine is extremely poisonous and must always be handled in a fume-cupboard. Bromine has a high density and bottles could fracture if not handled carefully. Both the liquid and the vapour can cause burns on the skin; as an emergency measure, wash with petrol.

Bromine water (1.2% by volume of bromine) once made up is not so hazardous but should be clearly labelled and stored in a cool dark cupboard.

MERCURY

See Section 1.6.7.

NITRIC ACID

Strong nitric acid can ignite organic material, and spillages must not be mopped up with a rag or sawdust unless first neutralised.

SODIUM HYDROXIDE

Sodium hydroxide is a strong caustic alkali and should be handled with great care. Emergency precautions are as for ammonia.

SULPHURIC ACID

The addition of water to strong sulphuric acid generates heat rapidly and causes sputtering. Always add acid to water, NEVER water to acid.

1.6.9   Noise

Exceptionally noisy procedures, such as the use of a vibrating hammer, should be kept away from the main working area and carried out in an acoustically insulated room. Ear-muffs should be worn when operating very noisy equipment such as a vibrating hammer.

Intermittent noise of high intensity is not the only possible source of irritation and discomfort in a laboratory. A persistent noise at a much lower level, such as that from a continuously running vacuum pump, can be very troublesome in a laboratory working area. Items such as this should be housed elsewhere, or enclosed in an acoustic surround (making provision for air and exhaust inlet and outlet).

1.6.10   Miscellaneous

GLASSWARE

Any glassware which is damaged in any way, or even scratched, should be disposed of immediately.

If breakage occurs, gather up all fragments of glass immediately and place them in a stout box. (Someone else will have to dispose of the refuse.) Ensure that no glass splinters are left in sinks. A piece of Plasticine is effective in picking up small fragments for disposal.

Take care when inserting bungs in cylinders, or when pushing glass tubes through bungs. Alcohol may be used as a lubricating medium.

Use protective gloves when cutting glass tubes and the like. Smooth off-cut ends in a flame. Glass cutting and shaping should be done only by someone who has had sufficient experience.

All glass vessels should be properly supported when in use.

Do not store glass containers in strong sunlight, or on a high shelf, or adjacent to heaters or hot water supplies.

TOOLS

Do not use a metal file unless it is properly fitted with a handle.

Use the correct size spanner on nuts and bolts, and preferably not an adjustable spanner.

Always wear eye protection when using a grinding wheel.

Machinery with exposed gears, belt drives and other moving parts should be fitted with guards.

VACUUM

Examine glassware which is to be evacuated (filter flasks, desiccators, etc.). If the slightest flaw is seen, do not use it but dispose of it.

Always use a protective screen over glassware subjected to vacuum.

Filtration flasks must be of thick-walled glass and designed for use under vacuum. Never use an ordinary flask and two-hole stopper for filtering by suction.

Open vacuum valves slowly.

Lubricate glass vacuum cocks with silicone vacuum grease, and do not force them if stuck.
Ensure that rubber bungs are large enough to avoid being sucked in.
Do not stop the vacuum pump until all connections have been opened to atmosphere.

1.6.11 Protective Clothing

Normal laboratory practice requires the wearing of a white laboratory coat at all times.
Additional protection is necessary for certain operations, as referred to above. Items of
protective clothing which should be available in the laboratory are as follows:

Overalls.
Boiler suit.
Eye-shield.
Dust mask.
Ear muffs.
Hard hat.
Gumboots.
Toe-tector shoes or boots.
Rubber gloves.
Asbestos gloves.
Industrial gloves.

BIBLIOGRAPHY

*BS1377:1967, 'Methods of testing soils for civil engineering purposes'. British Standards Institution, London
*BS1377:1975, 'Methods of test for soils for civil engineering purposes'. British Standards Institution, London
**BS Code of Practice CP2001 (1957), 'Site investigations' (British Standards Institution, London); and draft revision
document No. 76/11937
The Concise Oxford English Dictionary of Current English, 5th edition. Clarendon Press, Oxford
Guide to Safety in Laboratories. Department of Health and Social Security
HM Factories Inspectorate, Health and Safety Executive (June, 1976). Technical Data, Note 21 (rev.), 'Mercury'.
HMSO, London
Methods for the Detection of Toxic Substances in Air. No. 13, 'Mercury and Compounds of Mercury'. HMSO, London
Nixon, I. K. and Child, G. H. (1975). *Civil Engineer's Reference Book*, 3rd edition (ed. L. S. Blake), Chapter 11, 'Site
investigation'. Butterworths, London
Safety in Laboratories. University of Oxford Committee on Safety
The Shorter Oxford English Dictionary, 3rd edition. Clarendon Press, Oxford

*   These British Standards are quoted frequently in the succeeding chapters, and are not listed elsewhere.
Extracts from BS1377:1975 are reproduced by permission of BSI, 2 Park Street, London W1A 2BS, from whom
complete copies can be obtained.
** It should be noted that draft revision documents are Drafts for Public Comment and are not British Standards.

Chapter 2

Moisture content and index tests

2.1 INTRODUCTION

2.1.1 Scope

This chapter deals with the moisture content (or water content) of soils, and the way in which the amount of water in soils can influence their behaviour. Measurement of moisture content, both in the natural state and under certain defined test conditions, can provide an extremely useful method of classifying cohesive soils and of assessing their engineering properties. These specified conditions have gained world-wide recognition, and are referred to as the index properties, or consistency limits. The tests to determine them are known as Index Tests or Limit Tests.

2.1.2 Types of Test

The moisture content of a soil is the characteristic which is most frequently determined, and applies to all types of soil. The other index properties considered here relate only to cohesive soils, often including silts, and are referred to as their plasticity characteristics.

The tests to be described are for the determination of:

Moisture content.
Liquid limit (LL).
Plastic limit (PL).
Shrinkage limit (SL).
Sticky limit.
Linear shrinkage.
Puddle characteristics.
Swelling capability.

The liquid limit, plastic limit and shrinkage limit are known collectively as the Atterberg Limits, after the Swedish scientist Dr A. Atterberg, who first defined them for the classification of agricultural soils in 1911. Originally they were determined by means of simple hand tests using an evaporating dish. The procedures were defined more precisely for engineering purposes by Professor A. Casagrande in 1932. The mechanical device he designed for determining the liquid limit is still known as the Casagrande apparatus, although more recently a cone penetrometer apparatus has been developed for routine use. The tests for determining the liquid and plastic limits are specified in BS 1377:1975, and are by far the most widely used of the index tests.

The tests grouped under the heading of Empirical Index Tests (Section 2.8) include simple traditional tests to assess the suitability of clay for use as a puddle material. A more recent simple test to indicate the possible swelling characteristics of clays, and the sticky limit test, are also included.

2.2 DEFINITIONS

MOISTURE CONTENT (*w*) The mass of water which can be removed from the soil by heating at 105 °C, expressed as a percentage of the dry mass. (Also referred to as WATER CONTENT.)

NATURAL MOISTURE CONTENT The moisture content of natural undisturbed soil *in situ*.

OVEN DRYING Drying a soil to constant mass in an oven controlled to within certain limits of temperature, as in the standard moisture content test. STANDARD oven drying is at a temperature of 105–110 °C.

AIR DRYING Allowing a soil to lose most of its moisture by exposure to a warm dry atmosphere.

DESICCATION The process of drying out.

LIQUID LIMIT (LL) The moisture content at which soil passes from the plastic to the liquid state, as determined by the liquid limit test.

PLASTIC LIMIT (PL) The moisture content at which a soil passes from the plastic state to the solid state, and becomes too dry to be in a plastic condition, as determined by the plastic limit test.

PLASTICITY INDEX (PI) The numerical difference between liquid limit and plastic limit.

NON-PLASTIC A soil with a plasticity index of zero, or on which the plastic limit cannot be determined by the standard procedure.

RELATIVE CONSISTENCY (*C*$_r$) The ratio of the difference between liquid limit and moisture content to the plasticity index.

LIQUIDITY INDEX (LI) The ratio of the difference between moisture content and plastic limit to the plasticity index.

SHRINKAGE LIMIT (SL) The moisture content at which a soil on being dried ceases to shrink.

SHRINKAGE RATIO (*R*) The ratio of the change in volume to the corresponding change in moisture content above the shrinkage limit.

SHRINKAGE RANGE The difference between the field moisture content, or the moisture content at which a clay is placed, and the shrinkage limit.

LINEAR SHRINKAGE (LS) The change in length of a bar sample of soil when dried from about its liquid limit, expressed as a percentage of the initial length.

STICKY LIMIT The moisture content below which a cohesive soil does not stick to metal tools.

2.3 THEORY

2.3.1 Moisture in Soils

Naturally occurring soils nearly always contain water as part of their structure. The moisture content of a soil is assumed to be the amount of water within the pore space between the soil grains which is removable by oven drying at 105–110 °C, expressed as a percentage of the mass of dry soil. By 'dry' is meant the result of oven drying at that temperature to constant mass, usually for a period of about 12–24 h. In non-cohesive granular soils this procedure removes all water present.

There are several ways in which water is held in cohesive soils, which contain clay minerals existing as plate-like particles of less than 2 μm across (see Section 7.2). The shape and very small size of these particles, and their chemical composition, enable them to combine with or hold on to water by several complex means. A simplified illustration of the 'zones' of water surrounding a clay particle is obtained by considering the following five categories of water, indicated diagrammatically in Fig. 2.1.

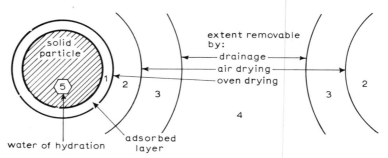

Fig. 2.1   *Representation of categories of water surrounding clay particles*

(1)   Adsorbed water, held on the surface of the particle by powerful forces of electrical attraction and virtually in a solid state. This layer is of very small thickness, perhaps of the order of 0.005 μm. This water cannot be removed by oven drying at 110 °C, and may, therefore, be considered to be part of the solid soil grain.

(2)   Water which is not so tightly held and can be removed by oven drying, but not by air drying.

(3)   Capillary water, held by surface tension, generally removable by air drying.

(4)   Gravitational water, which can move in the voids between soil grains and is removable by drainage.

(5)   Chemically combined water, in the form of water of hydration within the crystal structure. Except for gypsum, and some tropical clays, this water is not generally removable by oven drying.

For the purpose of routine soil testing, moisture content relates only to the water which is removable by oven drying at 105–110 °C. The water of category (1) above is not taken into account in the determination of moisture content and will not be referred to again in that context. The possible presence of water of category (5) is one reason for avoiding oven drying of tropical soils (Section 2.5.2).

Moisture content is usually expressed as a percentage, always on the basis of the *oven-dry* mass of soil. If the mass of water removed by drying at 105 °C is denoted by m_w, and the mass of dried soil by m_D, the moisture content, w, is given by the equation:

$$w \; (\%) = \frac{m_w}{m_D} \times 100$$

2.3.2   The Atterberg Limits

The condition of a clay soil can be altered by changing the moisture content; the softening of clay by the addition of water is a well-known example. For every clay soil there is a range of moisture contents within which the clay is of a plastic consistency, and the Atterberg limits provide a means of measuring and describing the plasticity range in numerical terms.

If sufficient water is mixed with a clay, it can be made into a slurry, which behaves as a viscous liquid. This is known as the 'liquid' state. If the moisture content is gradually reduced by allowing it to dry out slowly, the clay eventually begins to hold together and to offer some resistance to deformation; this is the 'plastic' state. With further loss of water the clay shrinks and the stiffness increases until there is little plasticity left, and the clay becomes brittle; this is the 'semi-solid' state. As drying continues, the clay continues to shrink in proportion to the amount of water lost, until it reaches the minimum volume attainable by this process. Beyond that point further drying results in no further decrease in volume, and this is called the 'solid' state.

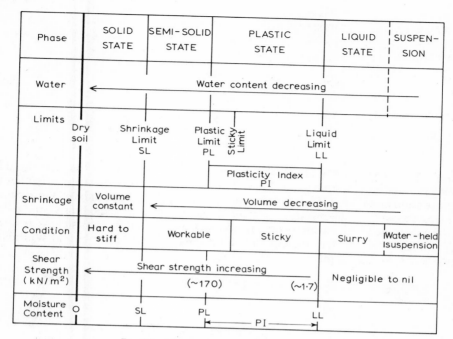

Fig. 2.2   *Phases of soil and the Atterberg limits*

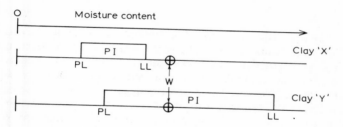

Fig. 2.3   *Consistencies of two clays*

These four states, or phases, are shown diagrammatically in Fig. 2.2. The change from one phase to the next is not observable as a precise boundary, but takes place as a gradual transition. Nevertheless three arbitrary but specific boundaries have been established empirically, as indicated in Fig. 2.2, and are universally recognised. The moisture contents at these boundaries are known as the

Liquid limit (LL)
Plastic limit (PL)   The Atterberg limits or consistency limits
Shrinkage limit (SL)

The moisture content range between the PL and LL is known as the plasticity index (PI), and is a measure of the plasticity of the clay. Cohesionless soils have no plasticity phase, so their PI is zero.

2.3.3   Consistency of Clays

Moisture content by itself is not sufficient to define the state of consistency of a clay soil. This can be done only by relating its moisture content to its liquid and plastic limits. For example, two different clays can have the same moisture content, denoted by *w* in Fig. 2.3, yet show quite

Table 2.1. RELATIVE CONSISTENCY AND LIQUIDITY INDEX

Moisture content range w	Relative consistency C_r	Liquidity index LI
Below PL	>1	negative
At PL	1	0
Between PL and LI	1 to 0	0 to 1
At LL	0	1
Above LL	negative	>1

different characteristics. For clay X the moisture content w is greater than the liquid limit, so the clay is in the 'liquid' state (i.e. a slurry). For clay Y the same moisture content lies between the plastic and liquid limit. This clay is in the 'plastic' state, and would be of a firm consistency.

The relationship of the moisture content to the liquid and plastic limits can be expressed numerically in two ways, using parameters known as the relative consistency, denoted by C_r (Terzaghi and Peck, 1948) or the liquidity index, denoted by LI (Lambe and Whitman, 1969). They are determined as follows:

$$C_r = \frac{LL - w}{LL - PL} = \frac{LL - w}{PI}$$

$$LI = \frac{w - PL}{LL - PL} = \frac{w - PL}{PI}$$

Note that $LI = 1 - C_r$.

Values of relative consistency and liquidity index throughout the moisture content range depicted in Fig. 2.2 are summarised in Table 2.1. These two parameters must not be confused.

The liquidity index is used more generally than the relative consistency. Between PL and LL it gives an easily visualised indication of the position in the plasticity range at which the soil lies. Below the plastic limit, the value of LI is negative.

At the liquid limit (LI = 1), a slowly drying clay slurry first begins to show a small but definite shear strength, the value of which is about $1.7 \, kN/m^2$ (Wood and Wroth, 1976). As the moisture content decreases (and LI approaches zero), the shear strength increases considerably, and at the plastic limit (LI = 0) the shear strength may be 100 or more times greater than at the liquid limit. (Skempton and Northey, 1953; Wood and Wroth, 1978).

2.3.4 Activity of Clays

The Atterberg limits are related to the combined effects of two intrinsic properties of a clay, namely particle size and mineral composition. For a particular clay mixed with coarser material, Skempton (1953) showed that the plasticity index depended on the clay fraction (the percentage of particles finer than $2 \, \mu m$), and that the ratio plasticity index/clay fraction was constant. For different clays different ratios were obtained, but the ratio was more or less constant for each clay type. The relationship between PI and clay fraction for a particular clay was termed the colloidal activity, or simply 'activity', where

$$activity = \frac{PI}{clay \; fraction}$$

On this basis, clays can be classified into four groups as shown in Table 2.2.

Approximate values of the activity of some clay minerals are shown alongside their liquid limits in Table 2.3.

Table 2.2. ACTIVITY OF CLAYS

Description	Activity
Inactive clays	<0.75
Normal clays	0.75–1.25
Active clays	1.25–2
Highly active clays	>2
(e.g. bentonite)	(6 or more)

Table 2.3. TYPICAL RANGES OF INDEX PROPERTIES OF SOME COMMON CLAY MINERALS

Clay mineral	Liquid limit range	PI range	Activity (approx.)
Kaolinite	40–60	10–25	0.4
Illite	80–120	50–70	0.9
Sodium montmorillonite	700	650	7
Other montmorillonites	300–650	200–550	1.5
Granular soils	20 or less	0	0

2.3.5 Flow Curve

The 'flow curve' for a cohesive soil is derived from the Casagrande liquid limit test (Section 2.6.6). Values of moisture content, w, are plotted against log N, where N is the number of standard blows required to close the groove. The resulting curve is known as the 'flow curve', and is virtually a straight line over the range covered by a typical test. (See Fig. 2.23.) A plot of log w against log N gives a straight line, inclined at an angle B to the horizontal axis. The slope of this line, tan B, is, according to the British Standard, equal to 0.092 for most British soils. This is the basis of the table of factors for the one-point liquid limit test (Table 2.5) described in Section 2.6.7.

The equation to the flow line is

$$LL = w \left(\frac{N}{25}\right)^{\tan B}$$

In this equation tan B is an exponent, not a multiplier, a point which is not clear in Note 5 of Test 2 (C) in BS 1377:1975.

2.3.6 Shrinkage Characteristics

SHRINKAGE LIMIT

When the water content of a fine-grained soil is reduced below the plastic limit, shrinkage of the soil mass continues until the shrinkage limit is reached. At that point the solid particles are in close contact and the water contained in the soil is just sufficient to fill the voids between them. Further reduction of water content cannot bring the particles closer together, so there is no further decrease in volume of the soil mass. Air enters the voids and the soil takes on a lighter colour. Below the shrinkage limit the soil is considered to be a solid in which the particles remain in contact and in an arrangement which gives a dense state of packing.

Clays are more susceptible to shrinkage than silts and sands. In most cohesive soils the

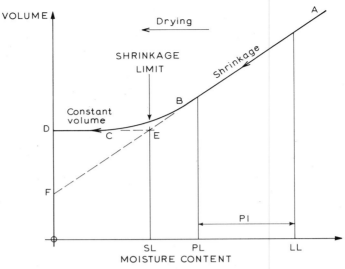

Fig. 2.4   *Shrinkage curve for clay soil*

shrinkage limit is appreciably below the plastic limit. In silts the two limits are close together and may be difficult to measure.

The shrinkage of a cohesive soil is illustrated graphically in Fig. 2.4, in which the change in volume of the soil mass is plotted against its moisture content as the soil is allowed to dry. The portion AB of the curve is linear, and shows that the decrease in volume is directly proportional to the loss of water. To the left of point C there is no further decrease in volume as the soil dries. The portion BC represents the transition zone between these two conditions. The point E is found from the intersection of the two straight lines AB and DC, and this enables the shrinkage limit value to be determined as shown.

The straight line AB is produced back to meet the volume axis at F, which represents the state of the soil if shrinkage could continue without the formation of air voids — that is, by the solid particles merging together to eventually form a single lump. The volume OF therefore represents the volume of the dry solids in the soil mass. If expressed in ml, it is theoretically equal to the mass of dry soil (g) divided by the specific gravity of the soil grains.

SHRINKAGE RATIO

The 'shrinkage ratio' is the ratio of the change in volume, expressed as a percentage of the final dry volume, to the change in moisture content, above the shrinkage limit. In Fig. 2.5 the shrinkage ratio is equal to the slope of the line AE. (Points A, E, D, F correspond to those in Fig. 2.4.) The heights of the vertical columns represent diagrammatically the total volumes of soil at various points on the graph, and their subdivisions indicate the volumes of solid, water (V_w), and (below SL) air, contained within the soil.

Volume of dry soil $= V_0$ (ml)
Mass of dry soil $= m_0$ (g)
Mass of water in soil $= \rho_w V_w$

Moisture content $w = \dfrac{\rho_w V_w}{m_0} \times 100\%$

Above the shrinkage limit (SL) water occupies the whole of the voids between particles. The change in volume on drying from point 1 to point 2 (both above SL) is therefore equal to the volume of water lost between points 1 and 2.

Change in volume $= V_1 - V_2$

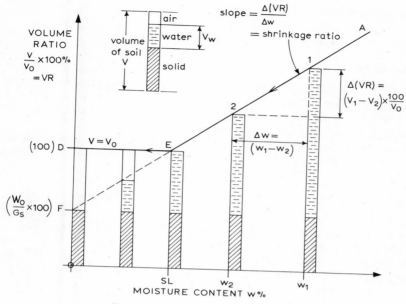

Fig. 2.5 *Derivation of shrinkage ratio*

Volume change (ΔV) as percentage of final dry volume:

$$\Delta V = \frac{V_1 - V_2}{V_0} \times 100\%$$

Change in moisture content (Δw):

$$\Delta w = \frac{\rho_w V_1}{m_0} - \frac{\rho_w V_2}{m_0} \times 100\%$$

$$= \frac{\rho_w}{m_0}(V_1 - V_2) \times 100\%$$

By definition, shrinkage ratio $SR = \Delta V / \Delta w$ (above SL). Therefore,

$$SR = \left[\frac{V_1 - V_2}{V_0} \times 100 \right] \div \left[\frac{\rho_w}{m_0}(V_1 - V_2) \times 100 \right] = \frac{m_0}{\rho_w V_0}$$

In SI units, $\rho_w = 1\ \text{g/cm}^3$. Therefore,

$$SR = \frac{m_0}{V_0}$$

Thus, the shrinkage ratio is equal to the ratio of the mass (g) to the volume (ml) of the oven-dried pat of soil at the end of the test described in Section 2.7.3.

SHRINKAGE RANGE

The 'shrinkage range' is the difference between the moisture content at which a clay is placed

(m_{pl}) and the moisture content at the shrinkage limit:

$$\text{shrinkage range} = m_{pl} - \text{SL} \ (\%)$$

LINEAR SHRINKAGE

The measurement of linear (one-dimensional) shrinkage of fine-grained soils is a different procedure from the measurement of volumetric shrinkage referred to above. The linear shrinkage is found by determining the change in length of a semi-cylindrical bar sample of soil when it dries out, starting from near the liquid limit.

If the original length when made up at about the LL is denoted by L_0, and the dried length by L_D, the change in length is equal to $L_0 - L_D$, and the linear shrinkage, LS, is given by:

$$\text{LS} = \frac{L_0 - L_D}{L_0} \times 100\%$$

In addition to indicating the amount of shrinkage, this test can provide an approximate estimate of the plasticity index for soils in which the liquid and plastic limits are difficult to determine. Examples are soils of low clay content, or soils from which it is difficult to obtain reproducible results, such as those with a high mica content. In these instances the linear shrinkage test can give more consistent results. The plasticity index is given approximately by the relationship:

$$\text{PI} = 2.13 \times \text{LS}$$

which was based on experience with British soils. This equation was given in BS 1377:1967, but it does not appear in the 1975 Standard, although the test remains unaltered.

2.4   APPLICATIONS

2.4.1   Moisture Content

The reasons for carrying out moisture content tests on soils fall into three categories:

(1)   To determine the moisture content of the soil *in situ*, using undisturbed or disturbed samples.
(2)   To determine the plasticity and shrinkage limits of fine-grained soils, for which moisture content is used as the index.
(3)   To measure the moisture content of samples used for laboratory testing, usually both before and after test. This is normally done on all test samples as a routine procedure.

2.4.2   Classification

The liquid and plastic limits provide the most useful way of identifying and classifying the fine-grained cohesive soils. Particle size tests provide quantitative data on the range of sizes of particles and the amount of clay present, but say nothing about the type of clay. Clay particles are too small to be examined visually (except under an electron microscope), but the Atterberg limits enable clay soils to be classified physically, and the probable type of clay minerals to be assessed.

Classification is usually accomplished by means of the plasticity chart (also referred to as the A-line chart). This is a graphical plot of the liquid limit (LL) as ordinate against plasticity index (PI) as abscissa. The standard chart is shown in Fig. 2.6.

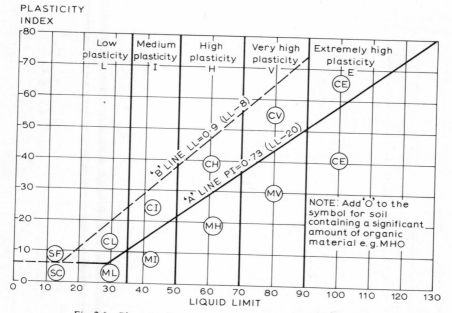

Fig. 2.6 Plasticity chart (reproduced from draft revision of CP2001)

When the values of LL and PI for inorganic clays are plotted on this chart, most of the points lie just above the line marked 'A-line', and in a narrow band parallel to it. The A-line is defined by the relationship

$$PI = 0.73 \, (LL - 20)$$

This line was derived from experimental evidence and does not represent a well-defined boundary between soil types, but it does form a useful reference datum.

The dashed line labelled 'B-line' is a tentative upper limit for all soils, which has also been drawn from experimental data. It is defined by the relationship

$$PI = 0.9 \, (LL - 8)$$

The British practice this chart is divided into five zones, giving the following categories for clays:

(1) Clays of low plasticity (CL), less than 35 liquid limit.
(2) Clays of medium plasticity (CI), liquid limit from 35 to 50.
(3) Clays of high plasticity (CH), liquid limit from 50 to 70.
(4) Clays of very high plasticity (CV), liquid limit from 70 to 90.
(5) Clays of extremely high plasticity (CE), liquid limit exceeding 90.

The letters in parentheses are the standard symbols by which each group is known.

Silts when plotted on this chart generally fall below the A-line. They are divided into five categories similar to those for clays. The group symbols are ML, MI, MH, MV, ME.

Clays containing appreciable amounts of organic matter also plot below the A-line. Organic clays and silts include the letter 'O' after the group symbol (for example, CHO, MIO).

In the USA, the plasticity chart is divided into three groups:

Low plasticity (CL), liquid limit less than 30.
Intermediate plasticity (CI), liquid limit from 30 to 50.
High plasticity (CH), liquid limit exceeding 50.

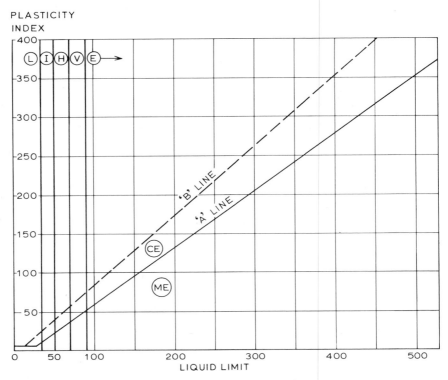

Fig. 2.7   *Extended plasticity chart*

An extended version of the A-line chart, embracing liquid limits up to 500, is given in Fig. 2.7.

Liquid limit values for some of the commoner clay minerals are typically within the ranges given in Table 2.3. Approximate values of activity (PI/clay fraction) are also included.

2.4.3   Engineering Properties

PLASTICITY

The Atterberg limits may be used to correlate soil strata occurring in different areas of a site, or to investigate in detail the variations of soil properties which occur within a limited zone. Results of limits tests can also be applied to the selection of soils for use as compacted fill in various types of earthworks construction.

In general, clays of high plasticity are likely to have a lower permeability, to be more compressible and to consolidate over a longer period of time under load than clays of low plasticity. High-plasticity clays are more difficult to compact when used as fill material.

While the liquid and plastic limits indicate the *type* of clay in a cohesive soil, the *condition* of the clay is dependent upon its moisture content in relation to those limits, as expressed by the liquidity index (Section 2.3.3). Those engineering properties which govern shear strength and compressibility are largely dependent upon this relationship.

For many straightforward applications it is possible to obtain sufficient understanding of the nature of clay soil from Atterberg limits and moisture content tests, and little else, if the geological history of the soil is also known. Where it is necessary to obtain additional information by further testing, the results obtained from limits tests carried out as a first stage facilitate the selection of samples for more complex tests later.

SHRINKAGE AND STICKINESS

The shrinkage limit of most British soils lies within the range 10–15% moisture content. For some tropical and lateric soils the range may be much wider, and for these soils the shrinkage limit is perhaps more useful as a criterion for classification.

The shrinkage limit and shrinkage ratio of a clay soil, considered in relation to the field moisture content, can indicate whether it is likely to shrink on drying (such as by exposure to atmosphere, movement of ground water or loss of moisture taken up by vegetation or trees); and if so, by how much. These values are particularly useful in connection with the placing of puddle clay in reservoir embankments or canal linings. To prevent excessive cracking if some drying out of the clay is likely to occur, the shrinkage range can be limited to, say, 20% by controlling the placement moisture content in relation to the shrinkage limit.

The sticky limit can provide useful information to a contractor prior to an earthworks contract. If the sticky limit is lower than the natural moisture content of a clay to be excavated, or is lower than the moisture content at which it is to be placed, the handling of the clay may prove to be difficult, owing to its tendency to stick to construction tools and equipment.

2.5   MOISTURE CONTENT

2.5.1   Test Methods

The standard method for determining moisture content of soils is the oven-drying method, and this is the procedure recommended for a soils laboratory. A subsidiary method using a heated sand-bath is also described for use on site where an oven is not available or would be impracticable. Another site method used extensively for moisture control of fill is the carbide method using a 'Speedy' moisture tester. This is not included here, but detailed instructions are provided by the manufacturer with the equipment.

Another method, in which methylated spirit is mixed with the soil and ignited, is given in the British Standard, where it is referred to as the Alcohol Method. This procedure is not recommended as a laboratory method, because of the risk of fire, and is not described here.

2.5.2   Oven Drying (BS 1377:1975, Test 1(A))

A drying temperature of 105–110 °C is specified as the standard procedure, and this should be used as a general rule. However, this temperature may be too high for certain types of soil. For peats and soils containing organic matter a drying temperature of 60 °C is to be preferred, to prevent oxidation of the organic content. For soils containing gypsum the water of crystallisation may be lost at temperatures above 100 °C, so a temperature not exceeding 80 °C should be used. The same can apply to some soils from tropical regions. These lower temperatures may require longer drying times.

Throughout this book the apparatus and procedure detailed in this section are referred to as the 'standard moisture content apparatus and procedure'.

APPARATUS

(1)   Thermostatically controlled drying oven, capable of maintaining a temperature of 105–110 °C, and adjustable to a lower temperature (down to 60 °C) if necessary. (See general notes on Ovens in Chapter 1, Section 1.2.4.)
(2)   Desiccator container or cabinet, containing anhydrous silica gel.
(3)   Balance, of a capacity and accuracy appropriate to the size of specimens to be tested. (See under 'Selection of sample', below.) An automatic top-pan balance reading to 0.01 g is suitable for all except coarse-grained soils.

Table 2.4.   MOISTURE CONTENT CONTAINERS

	Size of container			
Type of soil	*diameter* *(mm)*	*height* *(mm)*	*Capacity*	*Typical mass* *(approx., g)*
Clay and silt	65	20	75 g	12
Medium-grained	90	20	90 g	18
	150	15	500 g	60
Stony	300 × 220 (tray)	65	3 kg	300

(4)   Numbered glass weighing bottles or non-corrodible containers. For work requiring high accuracy, glass weighing bottles with ground-glass stoppers (both of which are numbered) are preferable. For routine work aluminium containers (usually referred to as moisture tins) or trays are used. Container and lid should each be stamped with the reference number. Suitable dimensions are given in Table 2.4.

Typical weighing bottles and containers are shown in Fig. 1.13.

All containers must be washed clean, dried thoroughly before use and weighed carefully. The mass of glass weighing bottles can be permanently recorded, although they should be check-weighed periodically. Aluminium containers should be weighed before each use.

(5)   Trimming knife, spatula, scoop; other small tools as appropriate.

(6)   Moisture content printed forms, or printed forms for other tests with provision for moisture content test data.

PROCEDURAL STAGE

(1)   Weigh container.
(2)   Select soil sample.
(3)   Weigh wet.   Sequence: Balance
(4)   Dry in oven.   Oven
(5)   Cool in desiccator.   Desiccator
(6)   Weigh dry.   Balance
(7)   Calculate.
(8)   Report results.

The layout of balance, oven and desiccator should allow for convenient operation of this repetitive sequence.

TEST PROCEDURE

(1)   *Weighing container*

If necessary, clean and dry the container and lid. Make sure that both are marked with the same number or reference letter. Weigh to the appropriate degree of accuracy, usually to within 0.02% of the mass of specimen to be used, as follows:

Mass of specimen up to:	50 g	500 g	3 kg
Weigh to nearest:	0.01 g	0.1 g	1 g

The correct use of balances is described in Section 1.3.4.
Enter the mass in the space provided on the test-form.

(2) *Selection of sample*

The test sample must be selected so that it is properly representative of the soil sample from which it is taken. The nature of the soil material must also be taken into account. Normal practice should be to make three, or at least two, separate moisture content determinations and to average the results. However, if only a very small quantity of soil is available, it is better to use all the material for one measurement.

For the measurement of natural moisture content, the approximate mass of specimen required for different soil types is as follows:

Homogeneous clays and silts	30 g
Medium-grained soils	300 g
Coarse-grained (stony) soils	3 kg

In laminated clays where it is required to measure the moisture content of individual thin layers, it may be necessary to use much less than 30 g, and greater accuracy in weighing may then be desirable. With very stony soils it is sometimes preferable to measure the moisture content of the finer fraction only (for instance, the fraction passing a 6.3 mm sieve), and then to adjust this measured value by proportioning up from the percentage finer than that size, to obtain the moisture content of the material as a whole.

The test specimen should be taken from near the middle of a mass of soil — not from the outside, where the soil may have partially dried by exposure to the atmosphere. The specimens should be crumbled and placed loosely in the containers so that the soil will dry right through. Replace the lids immediately to prevent loss of moisture before weighing.

Observe and record the type of sample, the soil type, and its condition, at this stage. Include comments on any apparent inadequacy of sealing of the sample in its container, possibly drying out or segregation of water. Always record any unusual features which relate to the condition of the sample being tested.

(3) *Wet weighing*

Each container with wet soil should be weighed as soon as practicable after taking the specimen. If weighing is likely to be delayed, the lid must be fitted tightly and the containers left in a cool place.

The container and soil is weighed to the appropriate degree of accuracy referred to in stage (1), and the mass is entered in the space on the test-sheet.

(4) *Oven drying*

Remove the lid from the container and place both on a shelf in the oven. The thermostat control should already be set to maintain the required temperature, normally 105–110 °C unless a lower temperature is desirable for the reasons referred to above. To avoid overheating of the soil, it is essential that air be able to circulate around the container, which should not be placed on the floor of the oven or close to the heating elements. The proper use of ovens is outlined in Section 1.2.4.

Drying in the oven should continue until the specimen has dried out to reach a constant mass. (See Section 1.2.4 and stage (6), below.)

(5) *Cooling in desiccator*

Remove the specimen container and lid from the oven and place both separately in a desiccator cabinet until cool. Weighing must not be done while hot, because warm air currents can lead to inaccuracies with a sensitive balance. The soil must not be allowed to cool in the open, because dry soil can absorb moisture from the atmosphere.

(6)   *Dry weighing*

When cool, replace the lid on the container and weigh. Enter on the moisture content sheet as the mass of container plus dry soil.

If it is uncertain whether or not the soil has been completely dried, repeat stages (4), (5) and (6) at intervals of 2–4 h until a constant mass is observed. Constant mass is achieved when the difference between successive weighings is less than 0.1% of the dry mass of soil used. Repeated weighings should always be made if an oven temperature lower than 105 °C has to be used.

(7)   *Calculation*

The moisture content of a soil is expressed as a percentage of its *dry* mass:

$$\text{moisture content } w = \frac{\text{loss of moisture}}{\text{dry mass}} \times 100\%$$

Let $m_1 = $ mass of container; $m_2 = $ mass of container and wet soil; $m_3 = $ mass of container and dry soil. Then

$$w = \frac{m_2 - m_3}{m_3 - m_1} \times 100\%$$

The value of w for each individual moisture content specimen is calculated to the nearest 0.01%. If two or three separate measurements have been made on one sample, the average value is then calculated.

The specimen should not be disposed of until the moisture contents have been calculated and verified as being sensible values. If there is any doubt, check the dry weighings, or repeat the whole test if necessary.

(8)   *Results*

The final moisture content w is reported to two significant figures. That is:

If less than 10%, report to nearest 0.1% (e.g. 7.6%).
If 10% or greater, report to nearest 1% (e.g. 13%).
If an exact 0.5%, report the nearest even number. For example,
| | 13.50% would be reported as 14%; |
| 16.50% would be reported as 16%; |
| but | 16.51% would be reported as 17%; |
| 16.49% would be reported as 16%. |

Remarks on the results sheet should include whether the test was to determine natural moisture content or whether it was done in connection with another test; and the oven temperature, if different from the standard value.

Typical calculations are given in Fig. 2.8.

2.5.3   Sand-bath method (BS 1377:1975, Test 1(B))

This method is intended as a site test for moisture content, where a drying oven is not available. It is sometimes used in a main laboratory as a rapid method for granular soils.

This procedure should not be used for soils containing gypsum, calcareous matter or organic matter.

Moisture Content

| Location: | Bracknell | | Loc. No. | 2456 |

| Sample Nos. | 17/7 |

| Relevant test | Natural Moisture | Operator | C.B.A. | Date started | 3.1.78 |

Sample No. & Ref.			17/7 A	B	C
Container No.			6	17	32
Wet soil & Container	g	m_2	52·68	61·39	58·42
Dry soil & Container	g	m_3	47·17	54·31	52·01
Container	g	m_1	15.53	16·22	15·75
Dry soil	g	m_3-m_1	31·64	38·09	36·26
Moisture loss	g	m_2-m_3	5.51	7·08	6·41
MOISTURE CONTENT	%	w	17.41	18.59	17·68
AVERAGE MOISTURE	%			17.89	

Moisture content $= w = \dfrac{m_2 - m_3}{m_3 - m_1} \times 100\%$

Example (Container no. 6)

$$w = \frac{52.68 - 47.17}{47.17 - 15.53} \times 100$$

$$= \frac{5.51}{31.64} \times 100 = 17.41\%$$

Average value $= \dfrac{17.42 + 18.59 + 17.68}{3} = 17.89$

Reported moisture content: 18%

Fig. 2.8   *Moisture content test results and calculations*

APPARATUS

(1)   Sand-bath containing clean sand to a depth of at least 25 mm.
(2)   Moisture content containers for fine soils, as used for oven drying.
(3)   For coarser soils heat-resistant trays 200–250 mm square and 50–70 mm deep, the size depending on the quantity of soil required for test.
(4)   Heating equipment, such as a bottled gas burner or paraffin pressure stove, or electric hot-plate if mains electricity is available.
(5)   Scoop, spatula, appropriate small tools.

PROCEDURAL STAGES

(1)   Weigh container or tray.
(2)   Select soil sample.
(3)   Weigh wet.
(4)   Dry on sand-bath.
(5)   Cool.
(6)   Weigh dry.
(7)   Calculate.
(8)   Report results.

Stages (1)–(3)

Similar to the oven drying method.

If a tray is used as the specimen container, it should first be cleaned and weighed, the wet soil should be spread out evenly on it and the trayful then weighed immediately.

(4) *Drying on sand-bath*

Place the sample container or tray on the sand-bath, and heat on the stove. Do not overheat. Small pieces of white paper mixed with the soil will act as an indicator and turn brown if overheated. The soil should be frequently turned with a spatula during heating, to assist evaporation of moisture.

The operiod of drying will vary with the type of soil, size of sample and prevailing conditions. Check weighings should be made to determine the minimum drying time necessary. If the loss in mass after heating for a further period of 15 min does not exceed the following, the soil may be considered to be dry:

Fine-grained soils	0.1 g
Medium-grained soils	0.5 g
Coarse-grained soils	5 g

(5) *Cooling*

Since this is a less precise test than the oven-drying method, cooling in a desiccator is not essential, but a container should have the lid fitted when cooling. A tray can be dry-weighed as soon as it is cool enough to handle.

Stages (6) *and* (7)

As for oven drying.

(8) *Results*

The moisture content is reported to the nearest whole number, and the method used is reported as the sand-bath method.

2.6 LIQUID AND PLASTIC LIMITS

2.6.1 Types of Test

The following tests for the determination of the Atterberg limits are given in BS1377:1975:

Test 2 (A) Liquid limit (cone penetrometer).
Test 2(B) Liquid limit (Casagrande apparatus).
Test 2 (C) Liquid limit (one-point method).
Test 3 Plastic limit.

These tests are described in this section.

The method devised by Casagrande (1932), using the apparatus known by his name, was for 40 years the universally recognised standard method for the determination of the liquid limit of clay soils. This has now been superseded in Great Britain by the cone penetrometer test, although the Casagrande procedure is retained in the British Standard as a subsidiary method. The plastic limit test, in which the soil is rolled into a thread until it crumbles, remains unaltered in principle, although some details of procedure have been changed.

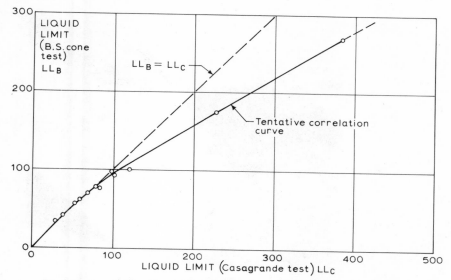

Fig. 2.9   *Correlation of liquid limit results from two test methods (from BS1377:1975)*

2.6.2   Comparison of Methods

Results obtained by the cone method have been proved to be more consistent, and less liable to experimental and personal errors, than those obtained by the Casagrande method (Sherwood and Ryley, 1968). For liquid limit values of up to 100% there is little difference between the results obtained by each method. Above 100% the cone method tends to give slightly lower values (Littleton and Familo, 1977). The relationship between results obtained from the Casagrande and cone methods, based on evidence available in 1978, is shown graphically in Fig. 2.9.

Both tests are carried out on remoulded soil, the fraction passing a 425 μm sieve being used. The cone method may not be any quicker to perform, but it is fundamentally more satisfactory, because the mechanics of the test depend directly on the static shear strength of the soil. The Casagrande procedure, on the other hand, introduces a dynamic component which is not related to shear strength in the same way for all soils. The definition of the liquid limit is dependent on the point at which the soil begins to acquire a recognisable shear strength (about 1.7 kN/m^2), so a test which is based on this property is to be preferred to any other empirical method.

The Casagrande one-point method is useful as a 'rapid' test, or when only a very small amount of soil is available. A 'one-point cone test' has been suggested by Clayton and Jukes (1978), and is given in Section 2.6.8.

2.6.3   Sample Preparation

EFFECT OF DRYING

Wherever possible the soil used for Atterberg limits tests should not be dried prior to testing. Drying out, even at the controlled temperature of 105–110°C, can alter the soil in various ways, such as: causing a chemical change, especially in organic soils; causing particles to subdivide; causing particles to agglomerate; driving off some of the adsorbed water, which may not be readily recoverable within a reasonable time on rewetting. These changes can result in changes in the index properties, but their effect is small for most inorganic British soils. The effect is greater for organic clays, which tend to give lower liquid and plastic limits if oven-dried. Many

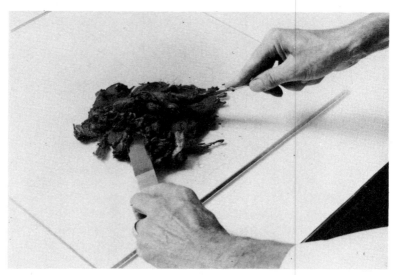

Fig. 2.10   Mixing soil into water for liquid limit test

tropical soils are also likely to be affected by drying and should not be oven-dried before testing.

Air drying may be used as a compromise when it is necessary to sieve the soil before testing (see below). The soil is spread out and left, usually for a few days, exposed to the atmosphere in a warm area of the laboratory (such as on top of an oven), but it must not be allowed to get hot. Air drying will not remove all the moisture, but the air-dried moisture content can be ascertained, if necessary, by weighing and oven drying a small representative portion of the whole.

Whichever procedure is followed, the condition of the soil as tested must be reported with the test results: for example, natural condition; air-dried; oven-dried; whether sieved or not.

SIEVING BEFORE TEST

The liquid and plastic limits tests must be carried out on material passing a 425 μm sieve. If the soil is first dried, either in an oven or by air drying, it is pestled if necessary (Section 1.5.4), and then sieved by hand through a 425 μm BS sieve nested on a receiving pan. About 250 g of sieved dry soil is needed for the two tests.

However, if the soil is predominantly clay, it is preferable, and more convenient, to take the soil in its natural state (without drying) and to remove any coarse particles by hand during the mixing process. Organic soils, and most tropical soils, should always be tested in their natural condition. Whenever possible this should be the standard procedure for all soil types.

IMPORTANCE OF MIXING

Whichever procedure is used for determining the liquid limit, an essential factor is the thorough mixing of the soil with water, and proper maturing. There is no short-cut for these processes, which demand patience and skill on the part of the operator. Mixing water into soil on a glass plate is shown in Fig. 2.10.

WATER

In carrying out these tests, tap-water must not be used for mixing the soil paste. Always use distilled water, or de-ionised water. Otherwise there is the possibility that ion exchange will take place between the soil and impurities in the water, which could affect the plasticity of the soil.

Fig. 2.11   Apparatus for cone penetrometer liquid limit test

2.6.4   Liquid Limit — Cone Penetrometer Method (BS 1377:1975, Test 2(A))

This is the British Standard preferred method for determining the liquid limit of soils. It is based on the measurement of penetration into the soil of a standardised cone of specified mass. At the liquid limit the cone penetration is 20 mm. The method was developed at TRRL from various cone tests which have been in use in other countries, and was adopted by the BSI with a few modifications. It requires the same apparatus as is used for bituminous materials testing, (BS 4691:1974), but fitted with a special cone.

APPARATUS

(1)   Penetrometer apparatus complying with the requirements of BS 4691, and illustrated in Fig. 2.11 (which also shows most of the items listed below.)
(2)   Cone for the penetrometer, the main features of which are as follows (see also Fig. 2.12.):
      stainless steel or duralumin
      smooth, polished surface
      length approximately 35 mm
      cone angle 30°
      sharp point
      mass of cone and sliding shaft 80.00 g
(3)   Sharpness gauge for cone, consisting of a small steel plate 1.750 mm thick with a 1.50 mm diameter hole bored through.
(4)   Flat glass plate, about 500 mm square and 10 mm thick, with bevelled edges and rounded corners.
(5)   Metal cups, of brass or aluminium alloy, 55 m diameter and 40 mm deep. The rim must be parallel to the base, which must be flat.
(6)   Wash bottle containing distilled or de-ionised water.

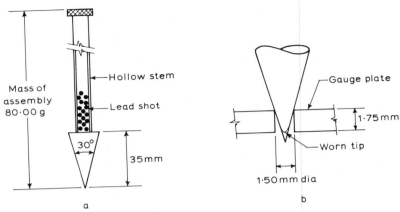

Fig. 2.12   Details of cone for penetrometer

(7)   Metal straight-edge.
(8)   Palette knives or spatulas:
        two 200 mm long × 30 mm
        one, 150 mm long × 25 mm
        one 100 mm long × 18 mm
        one square-ended, 150 mm long × 25 mm
(9)   Standard moisture content apparatus (Section 2.5.2).

PROCEDURAL STAGES

(1)   Select and prepare sample.
(2)   Mix with water.
(3)   Mature.
(4)   Check apparatus.
(5)   Remix.
(6)   Place in cup.
(7)   Adjust cone.
(8)   Measure cone penetration.
(9)   Repeat penetration.
(10) Measure moisture content.
(11) Remix with extra water.
(12) Calculate and plot graph.
(13) Report result.

PROCEDURE

(1)   *Selection and preparation of sample*

A sample of 200–250 g of soil passing the 425 μm BS sieve is required. Wherever possible this should be the natural soil with coarse particles removed by hand. Alternatively, air-dried soil can be used, which is then passed through a 425 μm sieve (see Section 2.6.3).
    Natural soil should be cut up into small pieces or shredded by using a cheese grater.

(2)   *Mixing with water*

Place the soil on the glass plate and mix thoroughly with a little distilled or de-ionised water, using two palette knives or spatulas. If the plastic limit test is also to be done, set aside a small portion in a sealed bag or container before adding too much water, and while the soil is still firm (see Section 2.6.5). To the remainder of the soil add more water, a little at a time, and work into a thick homogeneous paste. If the soil has not been passed through the 425 μm sieve, pick out any coarse particles by hand, or by using tweezers.

(3) Maturing

Place the paste into an airtight container, and seal. A small polythene bag is suitable if it is adequately sealed with adhesive tape. Leave standing for a curing period of 24 h, or overnight, in a cool place on the bench, to allow water to permeate through the soil mass.

For soils of a low clay content, such as very silty soils, the curing period may be omitted and the test carried out immediately after thorough mixing.

(4) Checking apparatus

The cone penetrometer apparatus must comply with Clauses 2.2.1.2 (3) and (4) of BS 1377:1975. The main points to check are:

(a) The cone designed specially for testing soils must be fitted.
(b) Mass of cone and stem 80.00 ± 0.05 g. This is *most important*. The stem is hollow, so that lead-shot can be inserted to bring the cone and stem assembly to the specified mass.
(c) Sharpness of the cone point can be checked by pushing the tip into the hole of the sharpness gauge plate. If the point cannot be felt when brushed lightly with the tip of the finger, the cone should be replaced.
(d) The cone must fall freely when the release button is pushed, and the sliding shaft must be clean and dry.
(e) The apparatus must stand on a firm level bench.

(5) Remixing

Remove the soil from the container after maturing. Remix with the spatulas for at least 10 min. Some soils, especially heavy clays, may need a longer mixing time, up to 40 min. It is essential to obtain a uniform distribution of water throughout the sample.

Keep the soil together near the middle of the glass plate, to minimise drying out due to exposure to air. Cover with a damp cloth or polythene when not mixing.

Thorough mixing and kneading is hard work, but it is the most important feature of the Atterberg tests (see Section 2.6.3), and must never be overlooked.

(6) Placing in cup

(a) Press the soil paste against the side of the cup, to avoid trapping air. (b) Press more paste well into the bottom of the cup, without creating an air-pocket. (c) Fill the middle and press well down. The small spatula is convenient for these operations. The top surface is finally smoothed off level with the rim.

(7) Adjustment of cone

Lower the cone and its support carefully, without exerting any force on the stem, so that the tip just touches the surface of the soil. Moving the cup slightly sideways will just mark the surface. The reading of the dial gauge is noted and recorded to the nearest 0.1 mm (R_1). See that the cone is over the centre of the cup.

(8) Measuring cone penetration

The cone is allowed to fall by pressing the button, which must be held in the pressed position for 5 s, timed with a seconds timer or watch (see Fig. 2.13). The apparatus must remain steady and must not be jerked.

After 5 s release the button so as to lock the cone in place. The dial gauge is lowered to this

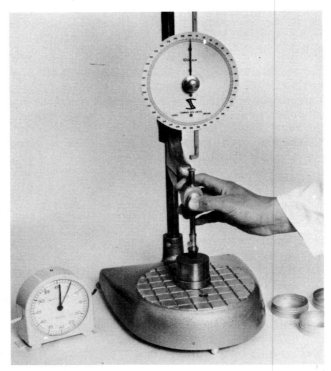

Fig. 2.13   Cone penetrometer test immediately after penetration

position and the reading is recorded to the nearest 0.1 mm (R_2). The difference between R_1 and R_2 is recorded as the cone penetration.

(9) Repeat penetration

Lift out the cone and clean it carefully. Add a little more wet soil to the cup, without entrapping air, smooth off, and repeat stages (7) and (8). If the second cone penetration differs from the first by less than 0.5 mm, the average value is recorded, and the moisture content is measured (stage 10).

If the second penetration is between 0.5 and 1 mm different from the first, a third test is carried out, and provided the overall range does not exceed 1 mm, the average of the three penetrations is recorded and the moisture content is measured (stage 10).

If the overall range exceeds 1 mm, the soil is removed from the cup and remixed, and the test (stages 5–8) is repeated.

(10) Moisture content measurement

Take a moisture content sample of about 10 g from the area penetrated by the cone, using the tip of a small spatula. This is placed in a numbered moisture content container, which is weighed, oven dried and weighed as in the standard moisture content procedure (Section 2.5.2). The moisture content sample should not be 'smeared' into the container, but dropped cleanly in by tapping the spatula on another spatula held close to the container, as shown in Fig. 2.14.

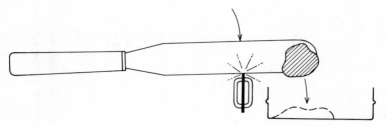

Fig. 2.14 Transferring moisture sample to container

(11) *Remixing*

The soil remaining in the cup is remixed with the rest of the sample on the glass plate together with a little more distilled water, until a uniform softer consistency is obtained. The cup is wiped clean and dried, and stages (6)–(10) are repeated at least three more times (making four in all), with further increments of distilled water.

A range of penetration values from about 15 mm to 25 mm should be covered.

(12) *Calculation and plotting*

The moisture content of the soil from each penetration reading is calculated from the wet and dry weighings as in the moisture content test.

Each cone penetration (mm) is plotted as ordinate, against the corresponding moisture content (%) as abscissa, both to linear scales, on a graph as shown in Fig. 2.15, which also shows typical data. The best straight line fitting these points is drawn.

(13) *Results*

From the graph the moisture content corresponding to a cone penetration of 20 mm is read off. The result is quoted to the nearest whole number as the liquid limit (cone test).

The percentage of material passing the 425 μm sieve is reported, together with whether the sample was tested in its natural state, or how dried.

The plastic limit and plasticity index are usually reported with the liquid limit.

2.6.5 Plastic Limit (BS 1377:1975, Test 3)

This test is to determine the lowest moisture content at which the soil is plastic. It can be carried out only on soils with some cohesion, on the fraction passing a 425 μm sieve. The test may be carried out either on soil in its natural state or on air-dried soil which has been remixed with water. The test is usually carried out in conjunction with the liquid limit test.

APPARATUS

(1)   The most important piece of apparatus for this test is the hand of the operator, which should be clean and free from grease.
(2)   Glass plate and small tools used for the liquid limit test.
(3)   A short length (say 100 mm) of 3 mm diameter metal rod (useful but not essential).
(4)   Standard moisture content apparatus (Section 2.5.2).

PROCEDURAL STAGES

(1)   Prepare sample.
(2)   Roll into ball.
(3)   Roll into threads.

Atterberg Limits - Cone Test

Location	BRACKNELL				Loc. No.	2456
Soil description	Fim blue-grey silty CLAY				Sample No,	6/5
Sample type	Undisturbed	Operator	A.B.S.		Date started	4.1.78

PLASTIC LIMIT

Test number		1	2	3	4	5	Average
Container no.		C1	G5	B7			
Wet soil & container	g	14.99	15.06	17.62			
Dry soil & container	g	13.48	13.60	15.58			
Container	g	7.94	7.99	7.97			
Dry soil	g	5.54	5.61	7.61			
Moisture loss	g	1.51	1.46	2.04			
MOISTURE CONTENT	g	27.26	26.02	26.81			26.70

LIQUID LIMIT

Test number		1		2		3		4	
Cone penetration	mm	15.5	15.1	19.0	19.0	22.0	21.8	25.4	25.2
Average penetration	mm	15.3		19.0		21.9		25.3	
Container no.		20		56		59		62	
Wet soil & container	g	46.78		57.20		63.60		71.72	
Dry soil & container	g	32.51		38.31		41.64		45.78	
Container	g	8.31		8.35		8.26		8.29	
Dry soil	g	24.20		29.96		33.38		37.49	
Moisture loss	g	14.27		18.89		21.96		25.94	
MOISTURE CONTENT	%	58.97		63.05		65.79		69.19	

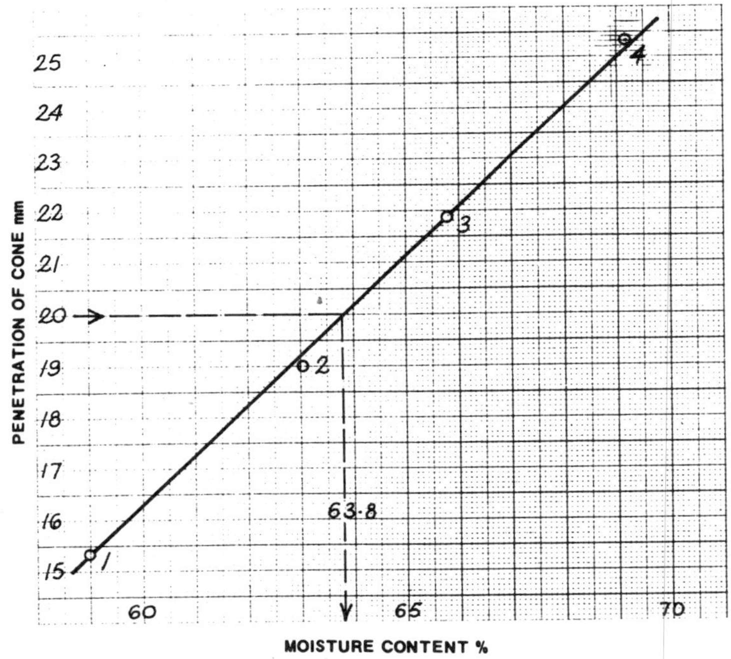

PREPARATION

As received ✓
~~Air dried~~
~~Oven dried~~
~~Pestled~~
~~Passed through~~
~~425 μm sieve~~

RESULTS

LL	64
PL	27
PI	37

63.8

PENETRATION OF CONE mm

MOISTURE CONTENT %

Fig. 2.15   *Liquid limit (cone test) and plastic limit results and graph*

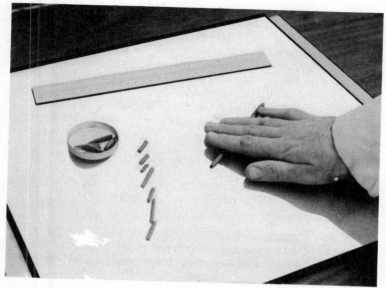

Fig. 2.16   Plastic limit test

(4)   Measure moisture content.
(5)   Repeat tests.
(6)   Calculate.
(7)   Report results.

PROCEDURE

(1)   *Preparation of sample*

About 20 g of soil as used for the liquid limit is required. This may be obtained by air drying and sieving through a 425 μm sieve, or by taking the natural soil and removing coarse particles by hand (see Section 2.6.3).

After thoroughly mixing, it is convenient to set aside a little of the soil for the PL test before proceeding with the LL test. If it is initially too wet, the sample should be allowed to partially dry in air on the glass plate until it is the right consistency.

(2)   *Rolling into a ball*

When the soil is plastic enough, it is well kneaded and then shaped into a ball. Mould the ball between the fingers and roll between the palms of the hands so that the warmth of the hands slowly dries it. When slight cracks begin to appear on the surface, divide the ball into two portions each of about 10 g. Further divide each into four equal parts, but keep each set of four parts together.

(3)   *Rolling into threads*

One of the parts is formed into a thread about 6 mm diameter, using the first finger and thumb of each hand. The thread must be intact and homogeneous. Using a steady pressure, roll the thread between the fingers of one hand and the surface of the glass plate (see Fig. 2.16). The pressure should reduce the diameter of the thread from 6 mm to about 3 mm after between five and ten back-and-forth movements of the hand. Some heavy clays may need more than this because this type of soil tends to become harder near the plastic limit. It is important to maintain a uniform rolling pressure throughout; do not reduce pressure as the thread approaches 3 mm diameter.

Mould the soil between the fingers again to dry it further. Form it into a thread and roll out again as before. Repeat this procedure until the thread crumbles when it has been rolled to 3 mm diameter. The metal rod serves as a reference for gauging this diameter. By 'crumbling' is meant shearing both longitudinally and transversally as it is rolled.

Crumbling must be the result of the decreasing moisture content only, and not due to mechanical breakdown caused by excessive pressure, or oblique rolling or detachment of an excessive length beyond the width of the hand.

The first crumbling point is the plastic limit. It may be possible to gather the pieces together after crumbling, to reform a thread and to continue rolling under pressure, but this should not be done.

(4) *Moisture content measurement*

As soon as the crumbling stage is reached, gather the crumbled threads and place them into a weighed moisture content container. Replace the lid immediately.

(5) *Repeat tests*

Repeat stage (3) for the other three pieces of soil, and place in the same container. Weigh the container and soil as soon as possible, dry in the oven overnight, cool and weigh dry, as in the standard moisture content procedure (Section 2.5.2).

Repeat stages (3)–(5) on the other set of four portions of the soil, using a second moisture content container.

(6) *Calculations*

Calculate the moisture content of the soil in each of the two containers. Take the average of the two results. If they differ by more than 0.5% moisture content, the test should be repeated.

Typical test data are included in Fig. 2.15.

(7) *Results*

The average moisture content referred to above is expressed to the nearest whole number and reported as the plastic limit (PL) of the soil. The treatment of the soil (e.g. natural, air-dried, oven-dried) is reported, and so is the percentage of material passing the 425 μm sieve if it was sieved. The result is usually reported on the same sheet as the liquid limit test.

(8) *Plasticity Index* (referred to as Test 4 in BS 1377:1975)

The difference between the liquid limit and plastic limit is calculated to give the plasticity index (PI) of the soil:

$$PI = LL - PL$$

This value is also reported to the nearest whole number. If it is not possible to perform the plastic limit test, the soil is reported as non-plastic (NP). This also applies if the plastic limit is equal to or greater than the liquid limit; the latter can occur in some soils with a high mica content (Tubey and Webster, 1978).

2.6.6 Liquid Limit — Casagrande Method (BS 1377:1975, Test 2 (B))

This test procedure has been retained in the British Standard as a subsidiary method, but the cone penetrometer method is now the standard preferred method. Results obtainable from the two types of test are discussed in Section 2.6.2.

Fig. 2.17 *Casagrande liquid limit apparatus and tools*

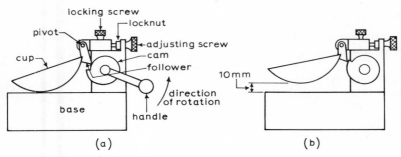

Fig. 2.18 *Principle of Casagrande apparatus*

APPARATUS

(1) A mechanical device (the Casagrande apparatus), shown in Fig. 2.17. The principle of its action is shown in Fig. 2.18. The cup must fall freely, without too much side-play. The height to which the cup is lifted must be exactly 10 mm above the base. This can be checked with the spacer gauge (a steel block 10 mm thick, or a block on the handle of the grooving tool; Fig. 2.19), which should just pass between cup and base when the cup is at its maximum height. The adjusting screw provides a simple means of adjustment. The lock-nut must be tightened after adjustment, and the maximum height rechecked with the gauge. The hardness of the rubber base has a considerable effect on the test results. The material and construction must conform to the requirements of the relevant British Standards. Details are given in BS 903:1969, Part A26 and 1963, Part A8.

(2) Grooving tool as illustrated in Fig. 2.19. It must be kept clean and dry, and the dimensions of the cutting profile must be checked periodically. The V-groove profile wears with use, and the essential dimensions must not differ by more than 0.25 mm from those specified. When necessary the tool must be reground to the correct profile.

(3) Flat glass plate, with bevelled edges and rounded corners, 10 mm thick and about 500 mm square.

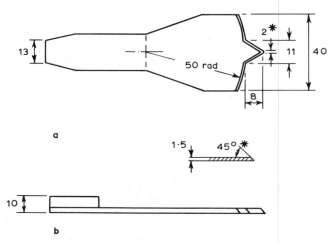

Fig. 2.19   *Grooving tool for Casagrande apparatus (dimensions in mm) (from BS1377:1975)*

(4)   Wash bottle containing distilled or de-ionised water.
(5)   Two palette knives (spatulas), blades 200 mm long and 30 mm wide.
(6)   Spatula with blade about 150 mm and 25 mm wide.
(7)   Standard moisture content apparatus (Section 2.5.2).

PROCEDURAL STAGES

(1)   Select and prepare sample.
(2)   Mix with water.
(3)   Mature.
(4)   Adjust apparatus.
(5)   Remix.
(6)   Place in bowl.
(7)   Cut groove.
(8)   Apply blows.
(9)   Repeat run.
(10)   Measure moisture content.
(11)   Repeat tests.
(12)   Calculate.
(13)   Report results.

PROCEDURE

Stages (1)–(3)

As for the cone penetrometer test (Section 2.6.4).

(4)   *Adjustment of apparatus*

The Casagrande apparatus must be clean, and the bowl must be dry and oil-free. Check that the bowl moves freely but without too much side-play, and that the drop is 10 mm; adjust if necessary. If a blow counter is fitted, turn it to zero. Check that the grooving tool conforms to the correct profile. (See itesm (1) and (2) under 'Apparatus', above.)
    The machine should be placed on a firm, solid part of the bench, so that it will not wobble. The position should also be convenient for turning the handle steadily and at the correct speed (two turns per second). Practise against a seconds timer with the cup empty, to get accustomed to the correct rhythm.

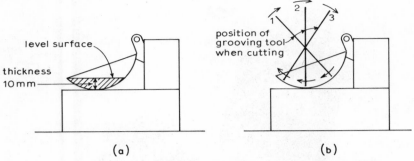

Fig. 2.20   Soil placed in Casagrande bowl, and use of grooving tool

A motorised version of the machine is available, which is designed to apply the bumps at the correct speed.

(5)   Remixing

As for the cone penetrometer test (Section 2.6.4).
The same remarks regarding the vital importance of thorough mixing apply to this test.

(6)   Placing in bowl

See that the bowl of the apparatus is resting on the base, and is not supported by the cam. Place a portion of the mixed soil in the cup, pressing it from the middle outwards to prevent trapping any air. The smaller spatula is best for this. The surface of the soil paste should be smoothed off level and parallel to the base, giving a depth at the greatest thickness of 10 mm (see Fig. 2.20a).

(7)   Cutting groove

A groove is cut through the sample from back to front, dividing it into two equal halves. Starting near the hinge, draw the grooving tool from the hinge towards the front in a continuous movement with a circular motion, keeping the tool normal to the surface of the cup (see Fig. 2.20b). The chamfered edge of the tool faces the direction of movement. The tip should lightly scrape the inside of the bowl, but do not press hard. The completed groove is shown in Fig. 2.21.

It is sometimes difficult to cut a smooth groove in soils of low plasticity with this tool. An alternative tool is available, known as the ASTM grooving tool (shown on the right in Fig. 2.17). If this is inserted as a wedge into the soil paste in the bowl, it will separate the soil into two halves which slide on the surface of the bowl. During the test the soil then tends to slide back again on this surface, instead of flowing as it should do. This objection can be overcome to a certain extent by pressing the soil around the tool, and forming the groove as the soil is placed instead of cutting it afterwards.

If a smooth groove cannot be cut with the standard tool, this fact must be recorded. If a groove is formed by any other means (such as by using the ASTM tool), the method must be reported.

(8)   Application of blows

Turn the crank handle of the machine at a steady rate of two revolutions per second, so that the bowl is lifted and dropped. Use a seconds timer if necessary to obtain the correct speed. If a revolution counter is not fitted, count the number of bumps, counting aloud if necessary. Continue turning until the groove is closed along a distance of 13 mm. The back end of the standard grooving tool serves as a length gauge. The groove is closed when the two parts of the soil come into contact at the bottom of the groove (Fig. 2.22). Record the number of blows

Fig. 2.21   Groove before applying bumps

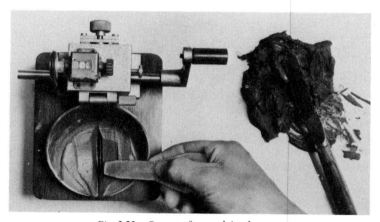

Fig. 2.22   Groove after applying bumps

required to reach this condition. If there is a gap between two points of contact, continue until there is a length of continuous contact of 13 mm, and record the number of blows. If this exceeds 50 blows, remove the soil, mix in a little more water, and repeat stages (6)–(8).

Closing of the groove should occur because of plastic flow of the soil, and not as a result of sliding on the surface of the bowl. If sliding occurs, this should be recorded and the result discarded. The test is repeated until flow does occur. If sliding persists after adding and mixing more water, the test is not applicable and it should be reported that the liquid limit could not be obtained.

(9)   Repeat run

Add a little more soil from the mixture on the glass plate, about as much as was removed in forming the groove. Mix with the soil in the cup. Repeat stages (6)–(8) until two consecutive runs give the same number of blows for closure. Record the number immediately.

Ensure that the soil does not dry out between repeat tests, because the number of blows will increase with decreasing moisture content.

(10)   *Moisture content measurement*

Take a small quantity (about 10 g) of soil from the zone of the groove where the two portions have flowed together, using the small spatula. Place it in a moisture content container, replace the lid and determine the moisture content as in the moisture content test (Section 2.5.2). (See stage (10) of Section 2.6.4.)

(11)   *Repeat tests*

Remove the soil from the bowl and mix it in with the remaining soil paste on the glass plate. Add a little more water, and mix in thoroughly. Clean the bowl of the apparatus and wipe it dry. Reset the revolution counter, if fitted, to zero. Clean and dry the grooving tool and spatula.

Repeat stages (6)–(10) at least three more times (four determinations in all), adding a little more water each time. The moisture contents should be such that the number of blows is roughly evenly spaced over the range from about 50 to 10. There should preferably be at least two blow-counts each side of 25. If necessary, do an additional repeat test. The test must start at the drier condition (about 50 blows) and proceed towards the wetter condition (10 blows).

Whenever the soil is left on the glass plate, it should be covered with a damp cloth to prevent drying out.

(12)   *Calculations*

Calculate the moisture content for each blow count, as in the moisture content test. Using a semi-logarithmic chart, plot the moisture content as ordinate (linear scale) against the corresponding number of blows as abscissa (logarithmic scale). Draw the best straight line fitting the plotted points. This is called the 'flow curve' (Fig. 2.23).

Draw the ordinate representing 25 blows, and where it intersects the flow curve draw the horizontal line to the moisture content axis. Read off this value of moisture content and record it on the horizontal line to the nearest 0.1%, as in Fig. 2.23.

(13)   *Results*

The moisture content read off from the flow curve is reported to the nearest whole number as the liquid limit (LL) of the soil.

The method of test (using the Casagrande apparatus) is reported. Also report the percentage of material passing the 425 μm sieve (if the sample was sieved), and whether the soil was air-dried or tested in its natural state.

The plastic limit (PL) is usually reported at the same time.

2.6.7   Liquid Limit — Casagrande One-point Method (BS 1377:1975, Test 2 (C))

This method provides a quick means of determining the liquid limit of a soil, because only one moisture content measurement is needed. However, the result is likely to be less reliable than the four-point Casagrande method or the cone penetrometer method.

It is a useful method when only a very limited quantity of soil is available.

APPARATUS

Similar to that listed in Section 2.6.6.

PROCEDURAL STAGES

(1)   Select and prepare sample.
(2)   Mix with water.
(3)   Mature.

Atterberg Limits Tests

Location	North Bromwich			Loc. No.	3210
Soil description	Grey clay with brown mottling			Sample No.	4/9
Sample type	Jar	Operator	F. B. J.	Date started	15. 6. 78

PLASTIC LIMIT

Test Number		1	2	3	4	5	Average
Container no.							
Wet soil & container	g						
Dry soil & container	g						
Container	g						
Dry soil	g						
Moisture loss	g						
MOISTURE CONTENT	%						

LIQUID LIMIT

Test Number		1	2	3	4	5	6
Number of bumps		49	40	29	22	16	
Container no.		64	95	57	74	82	
Wet soil & container (m_2)	g	19.47	19.77	21.01	22.72	19.26	
Dry soil & container (m_3)	g	14.78	14.82	15.39	16.31	14.38	
Container (m_1)	g	8.51	8.43	8.38	8.49	8.64	
Dry soil $(m_3 - m_1)$	g	6.27	6.39	7.01	7.82	5.74	
Moisture loss $(m_2 - m_3)$	g	4.69	4.95	5.62	6.41	4.88	
MOISTURE CONTENT W	%	74.8	77.5	80.2	81.9	85.1	

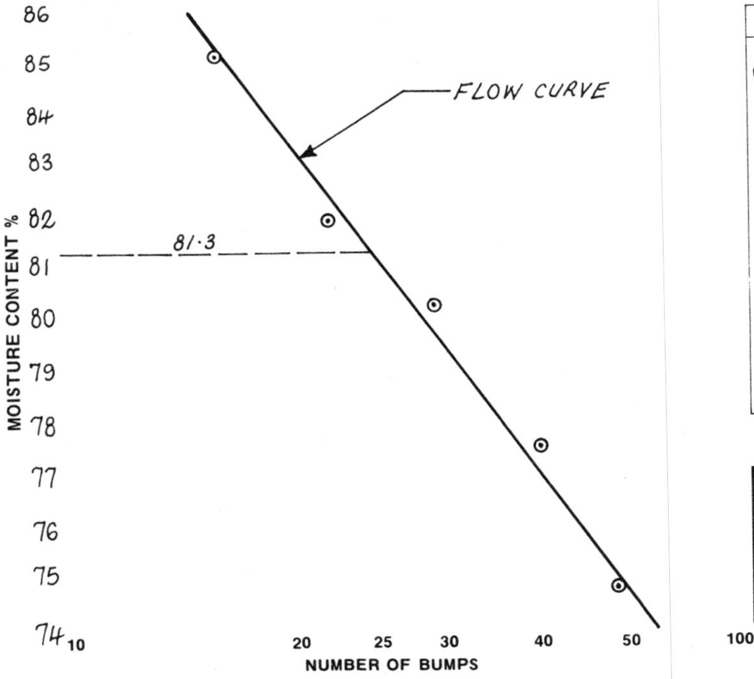

FLOW CURVE

MOISTURE CONTENT %

81.3

NUMBER OF BUMPS

PREPARATION

As received

Air dried

Oven dried

Pestled

Passed through
425 μm sieve

RESULTS

LL	81
PL	29
PI	52

Fig. 2.23 *Liquid limit (Casagrande test) results and graph*

(4)   Adjust apparatus.
(5)   Remix.
(6)   Place in bowl.
(7)   Cut groove.
(8)   Apply blows.
(9)   Repeat run.
(10)   Moisture content.
(11)   Calculate.
(12)   Report results.

PROCEDURE

(1)   *Selection and preparation of sample*

About 50 g of soil is required for the test; otherwise the procedure is as described in Section 2.6.6.

Stages (2)–(7)

As in Section 2.6.6.
   Even though a small sample is used, the mixing process is just as important as for the four-point test. The period of maturing should not be omitted for clay soils.

(8)   *Application of blows*

This is carried out as in Section 2.6.6 but the moisture content should be such that the number of blows for closure of the groove is between 15 and 35.

(9)   *Repeat run*

Add a little more soil to the bowl, mix in and make an immediate repeat run. This should give the same number of blows for closure. If not, repeat until two consecutive runs do give the same number.

(10)   *Moisture content measurement*

Since only one moisture content determination is necessary, most of the soil in the bowl after the second run can be used for this purpose. It is placed in a moisture container and treated as in the moisture content test (Section 2.5.2). Clean out and dry the bowl.

(11)   *Calculation*

The moisture content of the soil from the bowl is calculated in the usual way, and expressed to the nearest 0.1%. This moisture content is multiplied by a factor, given in Table 2.5, corresponding to the number of blows, to give the liquid limit. The theoretical basis of these factors is discussed in Section 2.3.5.

Table 2.5. FACTORS FOR CASAGRANDE ONE-POINT LIQUID LIMIT TEST (from BS1377:1975)

Number of blows	Factor	Number of blows	Factor	Number of blows	Factor	Number of blows	Factor
15	0.95	21	0.98	26	1.00	31	1.02
16	0.96	22	0.99	27	1.01	32	1.02
17	0.96	23	0.99	28	1.01	33	1.02
18	0.97	24	0.99	29	1.01	34	1.03
19	0.97	25	1.00	30	1.02	35	1.03
20	0.98						

Table 2.6. SUGGESTED FACTORS FOR CONE PENETRATION ONE-POINT LIQUID LIMIT TEST (from Clayton and Jukes, 1978)

Penetration (mm)	Soil of high plasticity	Soil of intermediate plasticity	Soil of low plasticity
15	1.098	1.094	1.057
16	1.075	1.076	1.052
17	1.055	1.058	1.042
18	1.036	1.039	1.030
19	1.018	1.020	1.015
20	1.001	1.001	1.000
21	0.984	0.984	0.984
22	0.967	0.968	0.971
23	0.949	0.954	0.961
24	0.929	0.943	0.955
25	0.909	0.934	0.954
Measured moisture content range	above 50%	35% to 50%	below 35%

The moisture content of the soil should be as close as possible to the liquid limit, so that the number of blows required is close to 25. This will minimise the error due to variation of the slope of the flow line from the mean value used for the derivation of the conversion factors.

If the liquid limit exceess about 120%, the variation is too large for this procedure to be reliable. One of the four-point tests should then be carried out instead.

(12) *Report results*

The liquid limit (LL) calculated as above is reported to the nearest whole number. The method is reported as the one-point method using the Casagrande apparatus. The percentage passing the 425 μm sieve, and whether natural or air-dried soil was used, are also reported.

The plastic limit test may also be carried out on the same sample.

2.6.8 Liquid Limit — One-point Cone Penetrometer Method

This method was suggested by Clayton and Jukes (1978) as a possible less elaborate routine method of assessing the liquid limit of a soil than the four-point cone penetrometer test described in Section 2.6.4. It is based on a statistical analysis of their experimental data, and on similar work carried out by Soil Mechanics Limited. The principle is analogous to the one-point Casagrande test described in Section 2.6.7.

The apparatus is the same as that described in Section 2.6.4. The procedure is identical with stages (1)–(10) of that section, except that a smaller quantity (about 100 g) of soil is required. Proper maturing, and thorough mixing with water, are just as important as in the standard test. The moisture content of the soil should be adjusted so that a cone penetration of between 15 mm and 25 mm is obtained.

The moisture content, determined as in stage (10) of Section 2.6.4, is expressed to the nearest 0.1% and is then multiplied by a factor obtained from Table 2.6 to obtain the liquid limit of the soil. The factor to be used for a given penetration depends upon whether the soil is of low, medium or high plasticity. These categories are explained in Section 2.4.2, but for this purpose high plasticity refers to all soils having a liquid limit exceeding 50%.

The measured moisture content indicates to which group the soil belongs, as indicated at the bottom of Table 2.6. The factor to be used is read off from the appropriate column opposite the measured penetration.

The liquid limit (LL) calculated in this way is reported to the nearest whole number, and the method is reported as the suggested one-point cone penetrometer test. The percentage passing the 425 μm sieve, and whether natural or air-dried soil was used, are also reported.

The plastic limit test may also be carried out on the same sample.

2.7 SHRINKAGE TESTS

2.7.1 Types of Test

Tests to measure two aspects of the shrinkage properties of soils discussed in Section 2.3.6, are considered here. These are the shrinkage limit and linear shrinkage. Two tests for the determination of the shrinkage limit and the BS test for linear shrinkage are described.

No test for measurement of the shrinkage limit of soils is given in BS 1377:1975. Both methods presented here depend upon the accurate measurement of volume of a soil specimen as it dries out, by immersion in mercury. The first method was developed at the Transport and Road Research Laboratory (TRRL), and the second was adapted from ASTM Standards by the US Bureau of Reclamation (USBR). Either method can be used with undisturbed or remoulded specimens. The TRRL method was developed for the use of standard 38 mm or 51 mm (2 in) diameter undisturbed samples. This method is the more elaborate, and would appear to be more versatile and to offer greater accuracy. In practice, however, in a laboratory where this test is not often performed, corrosion effects at a small electrical contact can render this method unreliable. The USBR method requires simple apparatus, but involves the use of exposed mercury. Adequate ventilation is essential, together with precautions to guard against spillage and availability of treatment facilities if spillage does occur (see Chapter 1, Section 1.6.7).

The linear shrinkage test indicates only the amount of shrinkage, and is included in BS 1377:1975.

2.7.2 Shrinkage Limit — TRRL Method

This test is described in *Soil Mechanics for Road Engineers* (TRRL, 1952), Sections 3.96–3.108.

APPARATUS

(1) Shrinkage limit apparatus with micrometer device, soil cage, dry cell battery and electric lamp, shown in Fig. 2.24.
(2) Mercury, sufficient to fill the cell of the apparatus to the required level.
(3) Mould of known volume, for preparation of remoulded specimen.
(4) Trimming and measuring tools, for preparation and measurement of undisturbed specimens.
(5) Balance, 200 g capacity, 0.01 g accuracy.
(6) Tongs, small brush, watch glass and other small tools.
(7) Standard moisture content apparatus (Section 2.5.2).
(8) Large tray containing a small depth of water, to retain any spilled mercury.

PROCEDURAL STAGES

(1) Prepare specimen.
(2) Measure and weigh.
(3) Adjust apparatus.
(4) Measure volume and mass.
(5) Allow partial drying.

Fig. 2.24   Shrinkage limit test — TRRL apparatus

(6)   Repeat volume measurements.
(7)   Oven dry and weigh.
(8)   Calculate and plot.
(9)   Report results.

TEST PROCEDURE

(1)   *Preparation of specimen*

A cylindrical specimen with a height equal to twice the diameter is convenient for this test. The size will depend upon the dimensions of the mercury container, but 38 mm diameter and 75 mm long is a typical specimen size.

An undisturbed specimen may be obtained by carefully pushing a thin-walled cutting tube into the main sample, if it is in a relatively large container. A sample in a 100 mm diameter sampling tube should be jacked out and the 38 mm diameter cutting tube pushed in as it emerges. The sample is then pushed out of the cutting tube and trimmed to length. Alternatively, an undisturbed specimen may be prepared by hand trimming, or with the use of a soil lathe.

For a remoulded specimen, the soil is mixed with water if necessary to a stiff plastic consistency (wetter than the plastic limit), and compacted without entrapping pockets of air into a cutting tube or split mould. It is then extruded or removed, and trimmed to length.

(2)   *Initial measurements*

The volume of the specimen may be determined initially from its linear measurements, as described in Chapter 3, Section 3.5.2. This is useful as a check on the volume-measuring apparatus, but is not essential.

The specimen should be weighed to an accuracy of 0.01 g and this mass recorded as the initial mass.

(3)   *Adjustment of apparatus*

Set the apparatus on the tray on a firm bench and use the base-adjusting screws to level it, checking with the built-in bubble. Add mercury to the required depth, and recheck for level (the mercury is probably almost as heavy as the apparatus which contains it). The apparatus should remain undisturbed in the same position until the tests in hand are completed, which may cover a period of several days.

Check that the batteries are fresh and that the indicator lamp lights up when the circuit is closed. A test switch is usually provided for this purpose. Lower the cage into the mercury tank until it is completely immersed. Rotate the cage two or three times to remove trapped air bubbles. Adjust the micrometer until the platinum contact just touches the surface of the mercury. This is indicated by the bulb lighting up. Record the reading of the micrometer as the zero reading, R_0. Raise the cage out of the tank.

(4)   *Measurements of volume and mass*

Place the specimen in the cage, and lower into the mercury until the whole is completely immersed. Air bubbles must be dislodged by rotating or agitating the cage. Reset the micrometer so that electrical contact with the mercury surface is just made, and the bulb just lights up. Record the micrometer reading (R).

Raise the cage, remove the specimen and carefully brush off any droplets of mercury back into the tank. Weigh the specimen immediately, to 0.01 g (*W*).

Wash hands after handling mercury. Cover up the apparatus until it is needed again.

(5)   *Partial drying*

Place the specimen on a watch-glass and leave it standing exposed to the air at room temperature for about 2 h so as to partially dry.

(6)   *Repeat measurements*

Repeat stages (4) and (5) so as to obtain a series of readings of volume and corresponding mass. Continue until three successive readings show no change in volume with reducing mass. At this stage the soil will appear lighter in colour. This may require several days, but the process must not be hurried. If the specimen dries too quickly, cracks may develop. If necessary, cover the specimen overnight with a damp cloth under a piece of polythene sheet so that it does not lose too much moisture between readings. If fine shrinkage cracks develop, it is possible that mercury will not be able to penetrate them fully, which will result in a false reading of volume. Other cracks may become filled with mercury which may remain entrapped, giving a false mass.

(7)   *Oven drying and weighing*

When it is clear that no further shrinkage is taking place, dry the specimen overnight, or to constant mass, in an oven at 105–110 °C. The oven and its surroundings must be adequately ventilated so as to get rid of any mercury vapour (Section 1.6.7). Weigh to 0.01 g after cooling (W_D). Measure its volume in the mercury tank (V_D).

(8)   *Calculation and plotting*

The volume of the specimen at any stage is calculated from the height to which the mercury is displaced in the tank. Let D = diameter of tank, R_0 = zero reading of micrometer (mm) without specimen, R = micrometer reading (mm) at any intermediate stage, V = volume at that stage (cm^3), W = mass at that stage (g), W_D = dry mass of soil. Then

$$V = (R - R_0) \times \frac{\pi D^2}{4} \times \frac{1}{1000} \text{ cm}^3$$

If the area of cross-section of the tank is 50 cm^2 (i.e. a diameter of about 80 mm), the volume is given by

$$V = (R - R_0) \times 5 \text{ cm}^3$$

Some older cells were fitted with a micrometer calibrated in inches, and gave a direct reading of volume in cubic inches. Multiply by 16.39 to convert to cubic centimetres.

Calculate the unit volume (U) per 100 g of dry soil at each stage from the equation

$$U = \frac{100 \times V}{W_D} \text{ cm}^3$$

Calculate the moisture content (w) of the specimen at each stage from the expression

$$w = \frac{W - W_D}{W_D} \times 100\%$$

Plot a graph of U (as ordinate) against w. A typical curve, together with test data and calculations, is shown in Fig. 2.25. The unit volume U_D in the fully dry state is plotted at zero moisture content, where

$$U_D = \frac{100 \times V_D}{W_D} \text{ cm}^3$$

(9)   Results

A smooth curve may be drawn through the plotted points. But the graph consists essentially of two straight portions. Draw the horizontal abscissa through U_D, and produce the inclined straight line portion back so that these two lines intersect at E, as in Fig. 2.25.

The moisture content corresponding to this point of intersection is read off, and is reported as the shrinkage limit (SL) to the nearest 1%.

It is preferable to carry out two tests simultaneously, and to report the average result as the shrinkage limit.

2.7.3   Shrinkage Limit — ASTM Method

This test is described in the USBR Earth Manual under 'Test Designation E-7, Part C'. It relates to remoulded soil, starting at wetter than the liquid limit. It requires only one measurement of volume of the soil specimen, which is done by weighing displaced mercury. Shrinkage limit is calculated directly from initial moisture content, dried mass and volume. A shrinkage curve is not produced, but intermediate volume measurements could be made if a curve is required. This would be necessary if the test were used as an alternative method for measuring the shrinkage limit of an undisturbed soil sample.

APPARATUS

(1)   Shrinkage dish, porcelain, about 42 mm diameter and 12 mm deep.
(2)   Glass cup, about 57 mm diameter and 38 mm deep, with rim ground flat.

Shrinkage Limit

Location: Hemel Hempstead					Loc. No. 4402		

T.R.R.L. method

Location:			Hemel Hempstead			Loc. No. 4402	
Soil description		Light brown silty clay				Sample No. 5/2	
Sample type	Jar		Operator	G. H. W.		Date started 8. 11. 78	
Time	Zero reading R_o units	Volume reading R units	$R - R_o$ $=V'$	$5 \times V'$ $=V$ cm^3	Wet weight of Sample g	Moisture Content %	Volume of 100 g dry soil $=100\frac{V}{W}$ cm^3
	3.42	20.66	17.24	86.20	149.65	35.8	78.2
	2.18	18.95	16.77	83.85	147.12	33.5	76.1
	6.21	21.95	15.74	78.70	142.27	29.1	71.4
	6.27	21.17	14.90	74.50	137.64	24.9	67.6
	6.30	19.23	13.93	69.65	131.14	19.0	63.2
	6.36	29.91	13.55	67.75	122.43	11.1	61.5
	5.42	18.93	13.51	67.55	117.58	6.7	61.3
	6.03	19.43	13.40	67.00		0	60.8
			Final dry weight W	110.20			

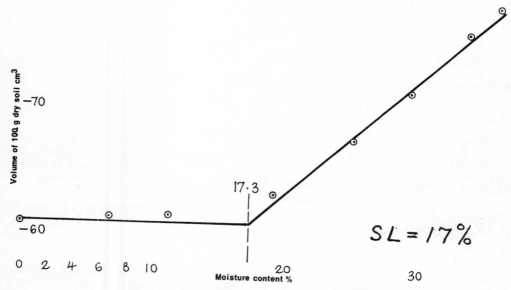

Fig. 2.25   Shrinkage limit test results and graph (TRRL method)

(3) Prong plate, glass, fitted with three steel prongs as shown in Fig. 2.26. Dimensions are not critical but the plate should be large enough to completely cover the glass cup.

(4) Glass plate, large enough to cover the shrinkage dish.

(5) Two porcelain evaporating dishes 150 mm diameter.

(6) Measuring cylinder 25 ml.

Fig. 2.26   Shrinkage limit test — ASTM apparatus

(7)   Mercury, rather more than will fill the glass cup.
(8)   Straight edge, spatula, small tools.
(9)   Balance, 200 g capacity, reading to 0.01 g.
(10) Standard moisture content apparatus (Section 2.5.2).
(11) Large tray containing a small amount of water, to retain any spilled mercury.

PROCEDURAL STAGES

(1)   Prepare specimen.
(2)   Measure and prepare apparatus.
(3)   Form soil pat.
(4)   Weigh.
(5)   Dry.
(6)   Weigh dry.
(7)   Measure volume.
(8)   Calculate.
(9)   Report result.

PROCEDURE

(1) *Preparation of specimen*

About 30 g of air-dried soil passing the 425 µm sieve is required. Place the soil in an evaporating dish and thoroughly mix with distilled water to make into a readily workable paste. Air bubbles must not be included. The moisture content should be somewhat greater than the liquid limit. The consistency should be such as to require about 10 blows of the Casagrande liquid limit apparatus to close the groove or to give about 25–28 mm penetration of the cone penetrometer.

(2) *Preparation of apparatus*

Clean and dry the shrinkage dish, and weigh it to 0.01 g (W). Its internal volume is determined by measuring the volume of mercury held. Place the dish in an evaporating dish and fill it to overflowing with mercury. The evaporating dish will catch the overflow. Press the small glass plate firmly over the top of the shrinkage dish so that excess mercury is displaced, but avoid trapping any air. Remove the glass plate carefully and transfer the mercury to the 25 ml measuring cylinder. Record the volume of mercury in ml, which is the volume of the shrinkage dish (V).

Coat the inside of the shrinkage dish with a thin layer of Vaseline or similar grease. This is to prevent soil sticking to the dish.

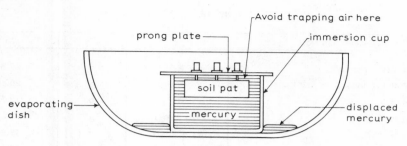

Fig. 2.27   *Immersion of sample in shrinkage limit test*

(3)   *Forming soil pat*

Add the wetted soil paste to the shrinkage dish so as to about one-third fill it. Avoid trapping air. Tap the dish on the bench surface to cause the soil to flow to the edges of the dish. This should also release any small air bubbles present. The bench should be padded with a few layers of blotting paper or similar material.

Add a second amount of soil, about the same as the first, and repeat the tapping operation until all entrapped air has been released. Add more soil, and continue tapping, so that the dish is completely filled with an excess standing out. Strike off the excess with a straight edge, and clean off adhering soil from the outside.

(4)   *Weighing*

Immediately after the above, weigh the soil and dish to 0.01 g (W_1). Calculate the mass of wet soil (W_2) from

$$W_2 = W_1 - W$$

(5)   *Drying*

Leave the soil in the dish to dry in the air for a few hours, or overnight, until its colour changes from dark to light. Place it in oven at 105–110 °C, and dry to constant mass.

(6)   *Weighing dry*

Cool in a desiccator, and weigh the dry soil and container to 0.01 g (W_3). Calculate the mass of dry soil W_0 from

$$W_0 = W_3 - W$$

(7)   *Measurement of volume*

Remove the dried soil-pat carefully from the shrinkage dish. It should be intact if it was given long enough drying in air before transferring to the oven.

Place the glass cup in a clean evaporating dish standing on the large tray. Fill the cup to overflowing with mercury, and remove the excess by pressing the glass prong plate firmly on top of the cup. Avoid trapping air under the glass plate. Carefully remove the prong plate, and brush off any mercury drops adhering to the glass cup. Place the cup into another clean evaporating dish without spilling any mercury.

Place the soil-pat on the surface of the mercury (it will float). Press the three prongs of the prong plate carefully on the sample so as to force it under the mercury (Fig. 2.27). Avoid trapping any air. Press the plate firmly on to the dish. Displaced mercury will be held in the evaporating dish. Brush off any droplets of mercury adhering to the cup into the dish. Transfer all the displaced mercury to the measuring cylinder and record its volume (V_0). This is equal to the volume of the dry soil-pat.

SHRINKAGE LIMIT

Shrinkage Dish No.			3
Mass of dish + wet soil	W_1	g	35.84
Mass of dish + dry soil	W_3	g	26.31
Mass of dish	W	g	12.73
Mass of water	$(W_1 - W_3)$	g	9.53
Mass of dry soil	$(W_3 - W) = W_0$	g	13.58
Mass of wet soil	$(W_1 - W) = W_2$	g	23.11
Moisture content $\dfrac{(W_1 - W_3) \times 100}{W_0}$	w	%	70.2
Volume of dish	V	cm³	14.81
Volume of dry soil	V_0	cm³	7.23
Volume change $(V - V_0)$	ΔV	ml	7.58
Unit Volume Change $\dfrac{\Delta V \times 100}{W_0}$	ΔU	%	55.8
Shrinkage Limit $(w - \Delta U)$	SL	%	14.4
Shrinkage Ratio $\dfrac{W_0}{V_0}$	SR		1.88

Fig. 2.28   Shrinkage limit test results (ASTM method)

(8) *Calculations*

Calculate the moisture content of the initial wet soil-pat, w, from the equation

$$w = \left(\frac{W_2 - W_0}{W_0} \right) \times 100\%$$

The shrinkage limit SL can then be calculated from the equation

$$SL = w - \left(\frac{V - V_0}{W_0} \right) \times 100\%$$

where V = volume of wet soil pat (ml); V_0 = volume of dry soil pat (ml); W_0 = mass of dry soil-pat (g).

The shrinkage ratio, R, can be calculated from

$$R = \frac{W_0}{V_0} \quad \text{(see Section 2.3.6)}$$

Typical results and calculations are given in Fig. 2.28.

(9) *Results*

The shrinkage limit is reported to the nearest whole number. The test procedure is reported as USBR Earth Manual Test E-7, Part C, adapted from ASTM Designation D 427-39.

The shrinkage ratio is reported to two places of decimals.

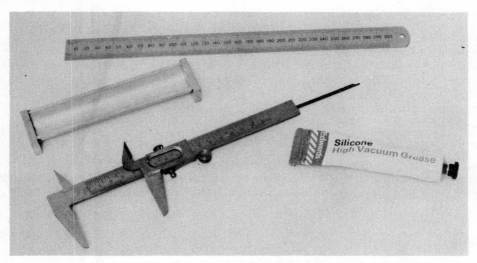

Fig. 2.29   Linear shrinkage apparatus

2.7.4   Linear Shrinkage (BS 1377:1975, Test 5)

This test gives the percentage linear shrinkage of a soil. It can be used for soil of low plasticity, including silts, as well as for clays.

APPARATUS

(1)   Mould as illustrated in Fig. 2.29, of brass or other non-corrodible metal, 140 mm long and 25 mm diameter.
(2)   Flat glass plate (as for LL test).
(3)   Palette knives.
(4)   Silicone grease.
(5)   Vernier calipers measuring up to 150 mm and reading to 0.1 mm.
(6)   Standard moisture content apparatus (Section 2.5.2).

PROCEDURAL STAGES

(1)   Prepare mould.
(2)   Prepare soil sample.
(3)   Place in mould.
(4)   Dry.
(5)   Measure length.
(6)   Calculate.
(7)   Report result.

PROCEDURE

(1)   *Preparation of mould*

Clean and dry the mould. Apply a thin film of silicone grease to the inner surfaces to prevent soil from sticking.

(2) *Preparation of sample*

About 150 g of air-dried soil passing the 425 μm sieve is required. The proportion of the original sample passing the 425 μm sieve is recorded.

Place the soil on the glass plate and mix thoroughly with distilled water, as for the liquid limit test. Continue mixing until it becomes a smooth homogeneous paste at about the liquid limit. This is not critical, but it may be checked by using the cone penetrometer, which should give a penetration of about 20 mm.

(3) *Placing in mould*

Place the paste in the mould, avoiding the trapping of air as far as possible, so that the mould is slightly overfilled. Tap it gently on the bench to remove any air pockets. Level off along the top edge of the mould with a palette knife or straight-edge. Wipe off any soil adhering to the rim of the mould.

(4) *Drying*

Leave the mould exposed to the air but in a draught-free position, so that the soil can dry slowly. When the soil has shrunk away from the walls of the mould it can be transferred to an oven set at 60–65 °C. When shrinkage has virtually ceased, increase the drying temperature to 105–110 °C to complete the drying.

(5) *Measurement of length*

Allow the mould and soil to cool in a desiccator. Measure the length of the bar of soil with the vernier calipers, making two or three readings and taking the average (L_D).

If the specimen has curved during drying, remove it carefully from the mould and measure the lengths of the top and bottom surfaces. Take the mean of these two lengths as the dry length L_D.

If the specimen has fractured in one place, the two portions can be fitted together before measuring the length. If it has cracked badly, and the length is difficult to measure, repeat the test, using a slower drying rate or leaving the mould longer in air before transferring to the oven.

(6) *Calculations*

Calculate the linear shrinkage (LS) as a percentage of the original length of the specimen from the equation

$$LS = \left(1 - \frac{L_D}{L_0}\right) \times 100\%$$

where L_0 = original length (140 mm if a standard mould is used); L_D = length of dry specimen. If the plasticity index (PI) is required from this test, it is calculated from the equation

$$PI = 2.13 \times LS$$

(7) *Results*

The linear shrinkage of the soil is reported to the nearest whole number. The percentage of soil passing tthe 425 μm sieve is also reported, together with the previous treatment of soil.

The PI, if calculated, is reported to the nearest whole number, with a statement that it was calculated from the linear shrinkage.

2.8 EMPIRICAL INDEX TESTS

2.8.1 Types of Test

Six simple empirical tests on clays are described in this section. They are known as

Puddle-clay index tests: pinch test
                                      tenacity test
                                        elongation test
                                        soaking test
Free swell test
Sticky limit test

The first four are traditional tests for assessing the suitability of a clay for use as puddle-clay in applications such as earth dams, reservoirs and canal bank linings. They were referred to by Bishop (1946) and Glossop (1946), and were adapted for laboratory use in 1954 by I. K. Nixon. While they do not provide precise numerical results, they do give an indication of the properties of the clay which may be used as a guide to the selection of more elaborate tests. These tests may also be useful in situations where no laboratory testing facilities are available.

The free swell test is taken from a procedure described by Gibbs and Holtz (1956). Its purpose is to indicate the possible expansive characteristics of a clay, whether in its natural state *in situ* or as a fill after being compacted.

The sticky limit test is that referred to by Terzaghi and Peck (1948).

Very little apparatus is required for these tests, other than standard oven-drying equipment for moisture content measurements.

2.8.2 Puddle Tests

PREPARATION OF SAMPLE

The clay is tested at the moisture content considered to be suitable for puddling. If this is not known, first determine the liquid and plastic limit of the natural (undried) soil. Calculate the moisture content required to give a liquidity index (LI) of about 40% (see Section 2.3.3). Use either air-dried soil or undried soil at a known moisture content, and take a quantity equivalent to a dry mass of about 1 kg. If undried soil is used, calculate the amount of additional water needed to bring the moisture content up to the required value. Add the water, mix it well in and allow to mature in an airtight container for a few hours or preferably overnight.

Whether the clay is tested 'as received' or after remoulding, the actual moisture content should be measured on a small representative sample immediately before carrying out these tests.

PINCH TEST

(1) Knead the soil well in the hands and form a ball of clay about 75 mm diameter. It must not be fissured at this stage.
(2) Squeeze the ball flat on a glass plate or between the hands (see Fig. 2.30) until it forms a disc 25 mm thick.
(3) Sketch and describe any cracks, if formed. If there are no cracks, state 'no cracks', and the clay has passed the test.

If the test is not possible because the sample is too friable, record that fact.

TENACITY TEST

(1) Roll a cylinder 300 mm long and 25 mm diameter from the clay sample.

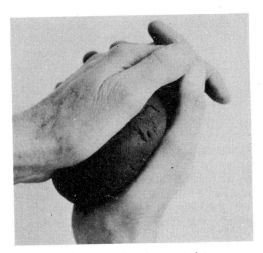

Fig. 2.30   Puddle clay — pinch test

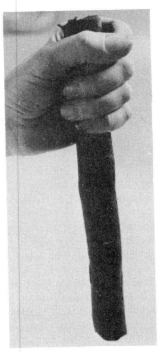

Fig. 2.31   Puddle clay — tenacity test

(2)   Hold it up vertically from one end so that 200 mm is unsupported, for a period of 15 s (Fig. 2.31).
(3)   If the clay supports its own weight, record this fact; it has passed the test. If necking or stretching occurs, record the details. Also record whether any cracks are visible.

ELONGATION TEST

(1)   Roll another cylinder 300 mm long and 25 mm diameter.
(2)   Grip each end firmly in the hands, leaving 100 mm unsupported (Fig. 2.32).
(3)   Holding the cylinder horizontal, stretch it gradually by pulling on both hands until it breaks.
(4)   Record the length of neck formed at failure and the type of break. The longer the neck, the more suitable the clay (see Fig. 2.33). If a break occurs after little or no stretching, this should be recorded.

SOAKING TEST

(1)   Make up a ball of clay 50 mm diameter, without fissures.
(2)   Place it in a 600 ml beaker and cover it with water. Note the time of immersion.
(3)   Record the state of the sample at the following intervals after immersion: $\frac{1}{2}$, 1, 2, 4, 8, 24 h; 2, 4 days. Include details of cracking, flaking, breaking away of pieces and final disintegration. If no change takes place, state 'nil'. A suitable clay should not disintegrate (see Fig. 2.34).

   Alternatively, if several clays are to be tested, record the times after immersion at which the following events occur: cracking; spalling; flaking; splitting; severe splitting; collapse or disintegration. If these events are marked at equal intervals along the X-axis of a graph (starting with 'unchanged' at zero), and if time is plotted to a suitable scale on the Y-axis, the

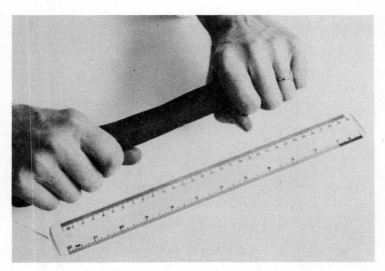

Fig. 2.32   Puddle clay — elongation test

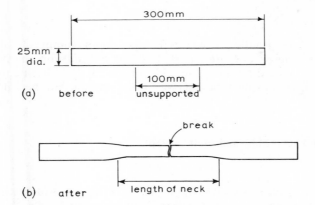

300mm

25mm
dia.

100mm

(a)   before   unsupported

break

(b)   after   length of neck

Fig. 2.33   Puddle clay — extension of elongation test specimen

Fig. 2.34   Puddle clay — ball soaking test on two specimens

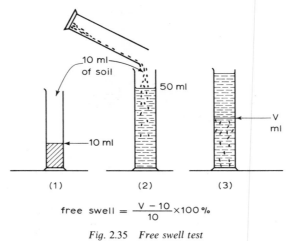

$$\text{free swell} = \frac{V - 10}{10} \times 100 \%$$

Fig. 2.35   Free swell test

behaviour of a clay can be represented graphically. Several different clays can be compared at a glance by this method.

REPORT RESULTS

The above four tests are usually carried out together on one sample of clay. Recorded details, and sketches where appropriate, are reported as a set of results on one sheet for a particular clay. After each test should be stated 'suitable' or 'unsuitable', depending on the results.

   The moisture content at which the tests were done, and the Atterberg limits, should also be reported.

2.8.3   Free Swell Test (Gibbs and Holtz, 1956)

CHARACTERISTICS

Free swell is defined as the increase in volume of the soil from a loose dry powder form when it is poured into water, expressed as a percentage of the original volume.

   Soils with free swell values less than 50% are not likely to show expansive properties. Values of 100% or more are associated with clays which could swell considerably when wetted, especially under light loadings. High-swelling soils such as bentonite might have free swell values up to 2000%.

PREPARATION OF SAMPLE

About 50 g of soil is oven dried, and passed through a 425 μm sieve. Place the dried soil loosely in a dry 25 ml cylinder up to the 10 ml mark. The powder should not be compacted or shaken down.

PROCEDURE

(1)   Place 50 ml of distilled water in a 50 ml glass measuring cylinder.
(2)   Pour the dry soil powder slowly and steadily into the water. The word used in the original text is 'drizzle', and this is a very apt description.
(3)   Allow the main part of the solid particles to come to rest; this will take from a few minutes to half an hour. The finest particles may remain in suspension for much longer but these can be ignored. (See Fig. 3.35.)
(4)   Read off and record the volume of settled solids (V ml).

CALCULATION AND REPORT

'Free swell' is defined as the change in volume of the dry soil expressed as a percentage of its original volume. It is calculated from the equation

$$\text{free swell} = \frac{V - 10}{10} \times 100\%$$

if the original dry loose volume was 10 ml.
   The result is reported to the nearest whole number.

2.8.4   Sticky Limit

This is a simple test on a clay to determine the lowest moisture content at which the clay adheres to metal tools. This method is based on a procedure outlined by Terzaghi and Peck.
   Use a pat of clay which has been matured at a moisture content within the plastic range, such that it is 'sticky' — that is, the clay sticks to a clean dry spatula blade or a plated grooving tool. Allow the clay to dry gradually by exposure to the atmosphere, and at intervals draw the tool lightly over the surface of the clay-pat. When the tool no longer picks up any clay, measure the moisture content. Add a little water to the clay so that it becomes sticky again, and repeat the process once or twice more.
   If the measured moisture contents are in reasonable agreement (an overall range of 2%), calculate the average moisture content to the nearest 1% and report it as the sticky limit of the clay.

BIBLIOGRAPHY

American Society for Testing and Materials (1974). Part 11. Test Designation D427-61, 'Standard method of test for shrinkage factors of soils'. ASTM, Philadelphia
Atterberg, A. (1911). 'Über die physikalische Bodenuntersuchung und über die Plastizität der Tone'. *Internationale Mitteilungen für Bodenkunde, Vol.* 1.
Bishop, A. W. (1946). 'The leakage of a clay core wall'. *Trans. Inst. Water Engineers*, Vol. 51, pp. 97–116
BS 903: 'Methods of testing vulcanized rubber'. Part A8: 1963, 'Determination of rebound resilience'. British Standards Institution, London
BS 903: Part A26: 1969, 'Determination of hardness'. British Standards Institution, London
BS 4691:1974, 'Method of determination of penetration of bituminous materials'. British Standards Institution, London
Casagrande, A. (1932). 'Research on the Atterberg limits of soils'. *Public Roads*, Vol. 13, No. 8
Casagrande, A. (1947). 'Classification and identification of soils'. *Proc. Am. Soc. Civ. Eng.*, Vol. 73, No. 6, Part 1
Casagrande, A. (1958). 'Notes on the design of the liquid limit device'. *Géotechnique*, Vol. 8, No. 2
Clayton, C. R. I. and Jukes, A. W. (1978). 'A one point cone penetrometer liquid limit test. *Géotechnique*, Vol. 28, No. 4, p. 469, December 1978
Gibbs, H. J. and Holtz, W. G. (1956). 'Engineering properties of expansive clays'. *Trans. Am. Soc. Civ. Eng.*, Vol. 121, No. 1, Paper 2814
Glossop, R. (1946). Discussion on 'The leakage of a clay core wall'. *Trans. Inst. Water Engineers*, Vol. 51, p. 124
Hansbo, S. (1957). 'A new approach to the determination of the shear strength of clay by the fall-cone test'. *Proc. Royal Swedish Geotechnical Institute, Stockholm*, No. 14
Karlsson, R. (1961). 'Suggested improvements in the liquid limit test, with reference to flow properties of remoulded clays'. *Proc. 5th Int. Conf. S.M. & F.E., Paris, July* 1961, Vol. I
Lambe, T. W. and Whitman, R. V. (1969). *Soil Mechanics*. Wiley, New York
Littleton, I. and Farmilo, M. (1977). 'Some observations on liquid limit values, with reference to penetration and Casagrande tests'. *Ground Engineering*, Vol. 10, No. 4
Norman, L. E. J. (1958). 'A comparison of values of liquid limit determined with apparatus having bases of different hardness'. *Géotechnique*, Vol. 8, No. 2
Norman, L. E. J. (1959). 'The one-point method of determining the liquid limit of a soil'. *Géotechnique*, Vol. 9, No. 1
Nixon, I. K. (1956). Internal communication. Soil Mechanics Limited

Sherwood, P. T. (1970). 'The reproducibility of the results of soil classification and compaction tests'. TRRL Report LR 339, Transport and Road Research Laboratory, Crowthorne, Berks.

Sherwood, P. T. and Ryley, D. M. (1968). 'An examination of cone-penetrometer methods for determining the liquid limit of soils'. TRRL Report LR 233, Transport and Road Research Laboratory, Crowthorne, Berks.

Skempton, A. W. (1953). 'The colloidal "activity" of clays'. *Proc. 3rd Int. Conf. Soil Mech.*, Zurich, Vol. 1, Session 1/14

Skempton, A. W. and Northey, R. D. (1953). 'The sensitivity of clays'. *Géotechnique*, Vol. 3, No. 1

Terzaghi, K. and Peck, R. B. (1948). *Soil Mechanics in Engineering Practice.* Wiley, New York

Transport and Road Research Laboratory (1952). *Soil Mechanics for Road Engineers*, Chapter 3, HMSO, London

Tubey, L. W. and Webster, D. C. (1978). 'The effects of mica on the roadmaking properties of materials'. TRRL Supplementary Report No. 408, Transport and Road Research Laboratory, Crowthorne, Berks.

US Department of the Interior, Bureau of Reclamation (1974), *Earth Manual*, 2nd edition. Test designation E.7, Part C. US Government Printing Office, Washington, DC

Wood, D. M. and Wroth, C. P. (1976). 'The correlation of some basic engineering properties of soils'. *Proc. Int. Conf. on Behaviour of Offshore Structures, Trondheim*, Vol. 2

Wood, D. M. and Wroth, C. P. (1978). 'The use of the cone penetrometer to determine the plastic limit of soils'. *Ground Engineering*, Vol. 11, No. 3, April 1978

Chapter 3

Density and specific gravity

3.1 INTRODUCTION

3.1.1 Scope

In this chapter are described three different laboratory tests for the measurement of density, and three for the measurement of specific gravity of soil particles. All these tests are to be found in BS 1377:1975 and 1967. In addition, tests for the determination of the 'limiting densities' of granular soils are included. These tests do not appear in the British Standard, but they are based on experience or accepted practice.

The section on theory (Section 3.3) includes illustrations of the meaning of voids ratio, porosity and degree of saturation, and their relationship to density and moisture content.

The measurement of density *in situ* is beyond the scope of laboratory work and is not included here.

3.1.2 Terminology

The concept of 'density' is well known, and refers to mass per unit volume. The term 'specific gravity' is also widely used, and is sometimes confused with density. It is essential to appreciate the difference between these two terms, because they have different meanings when applied to soils. Both density and specific gravity can be qualified in several different ways, each one of which has a particular application. These terms are discussed in this chapter.

In British practice the various terms using 'density' relate to the mass per unit volume (or unit mass of the soil in bulk). The accepted SI term (see Appendix) is 'mass density', and is denoted by the symbol ρ. It is expressed as a mass per unit volume (e.g. kg/m^3 or Mg/m^3), which should not be confused with the 'weight density', denoted by γ. Weight density relates to the gravitational force (the weight) acting on the mass, and is expressed as force per unit volume (e.g. N/m^3 or kN/m^3). Weight (force) is equal to mass multiplied by the local acceleration due to gravity, so, if we consider a unit volume,

$$\gamma = \rho \times g$$

The average terrestrial value of g is about 9.81 m/s^2, which for many practical purposes can be written as (10) m/s^2.

The weight density expressed in kilonewtons per cubic metre (kN/m^3), is then given by

$$\gamma = 9.81 \times \rho \quad kN/m^3$$

or approximately

$$\gamma = (10) \times \rho \quad kN/m^3$$

Throughout this book, unless stated otherwise, the term 'density' refers to mass density, ρ. For soils it is expressed in Mg/m^3, which is the same magnitude as g/cm^3.

The various terms using 'specific gravity' refer to the actual particles which make up the soil mass. Specific gravity is the ratio of the mass of particles to the mass of water they displace and is therefore dimensionless — i.e. a pure number without units. In some European countries the specific gravity is called 'relative density', which in Britain has quite a different meaning (see Section 3.3).

The term 'specific gravity' (or SG) throughout this book refers to the particles as they occur naturally. This is sometimes called the 'apparent SG' (see Section 3.3.4).

3.2 DEFINITIONS

DENSITY   Mass per unit volume.

BULK DENSITY   Mass of bulk soil, including solid particles, water and air, contained in a unit volume.

DRY DENSITY   Mass of dry soil, after drying at 105 °C, contained in a unit volume of undried soil.

LIMITING DENSITIES   The dry densities corresponding to the extreme states of packing (loose and dense) at which the grains of soil can be placed. Relevant only for granular soils.

MAXIMUM DENSITY   The dry density at the densest practicable state of packing of the particles. (Maximum dry density has a different meaning in connection with the compaction of soil. See Chapter 6.)

MINIMUM DENSITY   The dry density at the loosest state of packing of dry particles which can be sustained.

SPECIFIC GRAVITY (abbreviated to SG)   The ratio between the mass of dry solids and the mass of distilled water displaced by the dry soil particles. (In some countries this is known as *relative density*.)

VOIDS   The spaces between solid particles of soil. They may contain gas (usually air) or water, or both.

VOIDS RATIO   The ratio between the volume of voids (water and air) and the volume of solid particles in a mass of soil.

POROSITY   The volume of voids (water and air) expressed as a percentage of the total (bulk) volume of a mass of soil.

DEGREE OF SATURATION   The volume of water contained in the void space between soil particles expressed as a percentage of the total voids.

SATURATION   A soil is fully saturated when all the voids are completely filled with water.

CRITICAL DENSITY   The dry density corresponding to the critical voids ratio of a granular soil.

CRITICAL VOIDS RATIO   The voids ratio of a granular soil at which it will neither expand nor contract when subjected to shear strains.

UNIT MASS   Same as DENSITY.

UNIT WEIGHT   Weight (force) per unit volume, equal to mass per unit volume multiplied by local acceleration due to gravity.

ABSOLUTE SPECIFIC GRAVITY   The specific gravity of the mineral constituents present in the soil. It is measured by pulverising the soil to silt size or finer, so that all impermeable voids in the coarser grains are exposed. (Mainly applicable to rocks.)

APPARENT SPECIFIC GRAVITY   The specific gravity of the soil particles as they occur naturally, and referred to simply as specific gravity. (See definition under SPECIFIC GRAVITY.)

*BULK SPECIFIC GRAVITY (SATURATED, SURFACE DRY)   The specific gravity with the permeable or surface voids of the particles filled with water.

*BULK SPECIFIC GRAVITY (WET, SURFACE DRY)   The specific gravity when the permeable voids are not entirely filled with water.

*BULK SPECIFIC GRAVITY (OVEN DRY)   The minimum specific gravity of particles, in which the permeable and impermeable voids associated with the particles are included.

*These terms are used mainly for concrete aggregates and not for soils

3.3 THEORY

3.3.1 Mass, Volume and Density

In the laboratory mass is measured in grams. Linear dimensions are measured in millimetres, and a volume calculated from such measurements will be in cubic millimetres (mm³). It is convenient to convert to cubic centimetres (cm³) at the outset by dividing by 1000, because $1 \, cm^3 = 1000 \, mm^3$. Thus, for a rectangular prism, the volume V is given by

$$V = \frac{LBH}{1000} \ cm^3$$

and for a right cylinder

$$V = \frac{\pi D^2 H}{4000} \ cm^3$$

where L, B, H, D relate, respectively, to length, breadth, height and diameter, and are measured in millimetres.

If the mass of a soil sample is in grams, and its volume is $V \, cm^3$, the density ρ is given by

$$\rho = \frac{m}{V} \ g/cm^3$$

$$= \frac{m}{V} \ Mg/cm^3$$

because $1 \, Mg = 10^6 \, g$ and $1 \, m^3 = 10^6 \, cm^3$.

The practical unit of density is Mg/m^3, and is obtained directly from laboratory measurements if we convert mm³ to cm³.

If the volume V is obtained by direct measurement, such as by water displacement, it is measured in millilitres (ml). For practical purposes $1 \, ml = 1 \, cm^3$, so the above relationship still applies.

The density of pure water is at its maximum value of $1.000 \, g/cm^3$ (or Mg/m^3) at a temperature of $4 \, °C$. The density of water at other temperatures may be obtained from the graph in Fig. 3.1, or from Table 4.10 in Section 4.7.3 (Kay and Laby, 1973). At $20 \, °C$ it is equal to $0.998 \, 20 \, Mg/m^3$.

For most soil-testing purposes (except the hydrometer sedimentation test; Section 4.8.3), the density of pure water may be taken to be $1.00 \, Mg/m^3$.

3.3.2 Voids Ratio and Porosity

The amount of void space within a soil has an important effect on its characteristics. Two expressions are used to provide a measure of the void space — namely 'voids ratio' and

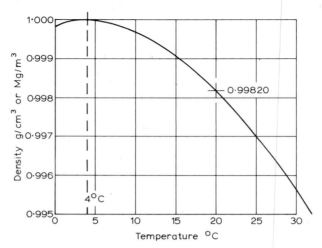

Fig. 3.1   Density of pure water

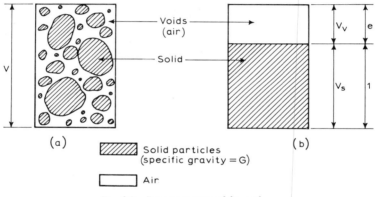

Fig. 3.2   Representation of dry soil

'porosity'. The relationships between density, moisture content, specific gravity, voids ratio and porosity are developed below for soil in four conditions:

(1)   dry
(2)   fully saturated
(3)   partially saturated
(4)   submerged

Equations are summarised at the end of this Section for easy reference, but it is better to understand the principles than to attempt to memorise all the formulae.

DRY SOIL

A dry soil consists of solid particles separated by air spaces (voids), as shown diagrammatically in Fig. 3.2(a). The soil occupies a total volume V and its mass is m. The specific gravity of the solid particles is G_s.

If we imagine that all the solid particles are fused together into a solid lump they will occupy a volume V_s which is less than V. The difference is the volume of voids V_v, as indicated in Fig. 3.2(b).

The voids ratio, e, is the ratio of volume of voids to volume of solids, or

$$e = \frac{V_v}{V_s} \tag{3.1}$$

This is a pure number which is usually expressed as a decimal (e.g. 0.35). It can be greater than 1.

Porosity, n, is the ratio of volume of voids to the total volume, or

$$n = \frac{V_v}{V} = \frac{V_v}{V_v + V_s} \tag{3.2}$$

This is often expressed as a percentage, and must be less than 100%.

The relationship between voids ratio and porosity is given by the equations

$$n = \frac{1}{1+e} \quad \text{and} \quad e = \frac{n}{1-n} \tag{3.3}$$

The mass of the volume of soil is equal to the volume of solids multiplied by their density, the voids having zero mass. If the density of water is denoted by ρ_w, the density of the solid particles is $G_s \rho_w$. Therefore,

$$\text{mass of the soil} = V_s G_s \rho_w$$

The volume of this soil is $(V_s + V_v)$.

The density (which in this case is the dry density, ρ_D) is equal to mass divided by volume, or

$$\rho_D = \frac{V_s G_s \rho_w}{V_s + V_v}$$

$$= \frac{G_s \rho_w}{1 + \dfrac{V_v}{V_s}}$$

i.e.

$$\rho_D = \frac{G_s}{1+e} \rho_w \tag{3.4}$$

SATURATED SOIL

In a completely saturated soil the void spaces are completely filled with water, as in Fig. 3.3(a). If we again imagine the solids fused together, the remaining voids will be filled with water as in Fig. 3.3(b).

Voids ratio and porosity are defined in the same way as before, i.e.

$$e = \frac{V_w}{V_s} \tag{3.5}$$

$$n = \frac{V_w}{V} = \frac{V_w}{V_w + V_s} \tag{3.6}$$

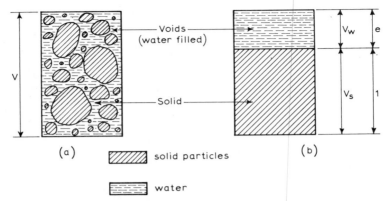

Fig. 3.3   *Representation of fully saturated soil*

The total mass is now made up of two parts:

$$\text{solid, } m_s = V_s G_s \rho_w$$

$$\text{water, } m_w = V_w \rho_w$$

The volume is

$$V = V_s + V_w$$

The density is now the saturated density, ρ_{sat}, and is given by

$$\rho_{\text{sat}} = \frac{(V_s G_s \rho_w) + (V_w \rho_w)}{V_s + V_w}$$

$$= \frac{\rho_w \left(G_s + \dfrac{V_w}{V_s} \right)}{1 + \dfrac{V_w}{V_s}}$$

Therefore

$$\rho_{\text{sat}} = \frac{G_s + e}{1 + e} \rho_w \tag{3.7}$$

PARTIALLY SATURATED SOIL

Many natural soils contain both air and water in the voids — that is, they are partially saturated. This condition can be represented diagrammatically as in Fig. 3.4. The symbols on the left of the diagram are those used previously. On the right-hand side the volume of solids, V_s, is reduced to unity, so that the volume of voids is denoted by e. This notation simplifies calculations involving voids ratio changes or degree of saturation. It is much easier to remember this diagram, and to work from the first principles, than to remember these equations.

The proportion of the voids occupied by water is denoted by s, so that

$$s = \frac{V_w}{V_v} = \frac{V_w}{V_a + V_w} \tag{3.8}$$

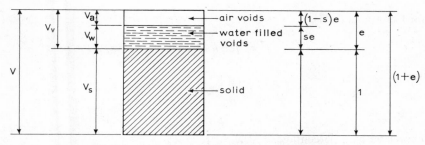

Fig. 3.4 Representation of partially saturated soil

The ratio s is known as the degree of saturation. It is sometimes expressed as a percentage, and is then denoted by S. In fully saturated soil $s = 1$ /$S = 100\%$), and the volume of air voids, V_a, is zero. In dry soil $s = 0$ and $V_a = V_v$.

The mass of the volume $(1 + e)$ is made up of:

solid: $1 \times (G_s \rho_w)$
water: $se\rho_w$
air: 0
total mass $= G_s \rho_w + se\rho_w$
The density is the wet density, or bulk density, ρ, and

$$\rho = \frac{G_s + se}{1 + e} \rho_w \tag{3.9}$$

The moisture content, w, is the ratio of mass of water to mass of solid (Section 2.3.1), or

$$w = \frac{se\rho_w}{G_s \rho_w} = \frac{se}{G_s} \tag{3.10}$$

If expressed as percentages, this can be written

$$S = \frac{G_s}{e} w\% \tag{3.11}$$

The dry density ρ_D is the mass of dry soil per unit volume; i.e.

$$\rho_D = \frac{G_s \rho_w}{1 + e}$$

which is the same equation as (3.4)

From Equation (3.10)

$$se = wG_s$$

Substituting in Equation (3.9),

$$\rho = \frac{G_s + wG}{1 + e} \rho_w$$

$$= \frac{G_s \rho_w}{1 + e} (1 + w)$$

and from Equations (3.11),

$$\rho = \rho_D(1 + w)$$

Therefore

$$\rho_D = \frac{1}{1 + w} \rho$$

If w is expressed as a percentage,

$$\rho_D = \frac{100}{100 + w\%} \rho \tag{3.12}$$

Equation (3.12) should be remembered, because it is often needed for calculating dry density from bulk density and moisture content.

SUBMERGED SOIL

If partially saturated soil is completely submerged and the air in the voids remains entrapped, the apparent (submerged) density is denoted by ρ' and is derived as follows.

total mass $= (G_s + se)\rho_w$
volume of water displaced $= (1 + e)$

Therefore

mass of water displaced $= (1 + e)\rho_w$

By the principle of Archimedes (Section 3.3.3),

$$\text{apparent mass} = (G_s + se)\rho_w - (1 + e)\rho_w$$

$$\text{Submerged density} = \frac{\text{apparent mass}}{\text{volume}}$$

or

$$\rho' = \frac{G_s - 1 - e(1 - s)}{1 + e} \rho_w \tag{3.13}$$

If all voids are filled with water, $s = 1$ and this equation becomes

$$\rho'_{sat} = \frac{G_s - 1}{1 + e} \rho_w \tag{3.14}$$

SUMMARY OF EQUATIONS

If we use SI units, the density of water, ρ_w, is 1 g/cm^3 for all practical purposes, and the relationships derived above become

$$\rho_D = \frac{G_s}{1 + e}$$

$$\rho_{sat} = \frac{G_s + e}{1 + e}$$

$$\rho = \frac{G_s + se}{1 + e}$$

$$\rho' = \frac{G_s - 1 - e(1 - s)}{1 + e}$$

$$\rho'_{sat} = \frac{G_s - 1}{1 + e}$$

However, if sea-water is present, its density (about $1.04\,\text{Mg/m}^3$) must be inserted for ρ_w in Equations (3:4)–(3.14).

The following equations are worth remembering:

$$S = \frac{G_s w\%}{e} \tag{3.11}$$

$$\rho_D = \frac{100}{100 + w\%}\rho \tag{3.12}$$

3.3.3   Principle of Archimedes

This principle is explained in textbooks on elementary physics (for example, Abbott, 1969). If a solid body is wholly or partially immersed in a liquid, the upthrust force (or buoyancy force) acting on the body is equal to the weight force of the liquid displaced by the body, and acts vertically upwards through the centre of gravity of the displaced liquid. The apparent mass of a solid body immersed in water is therefore equal to the mass of the body less the mass of water displaced. This is the basis of the water immersion method for measuring density (Section 3.5.5).

The same principle applies to a body immersed in a gas, including the atmosphere, but this effect may be ignored for most practical purposes.

When a body floats in water, the mass of water displaced is equal to the mass of the body, whether it is totally or partially immersed. If it floats wholly immersed, its mean density must be the same as the density of water.

3.3.4   Specific Gravity

A soil consists of an accumulation of particles which may be of a single mineral type, such as clean quartz sand, or more usually a mixture of a number of mineral types, each with a different specific gravity. For a single mineral type the specific gravity of the solids comprising a mass of the soil is that of the mineral itself (for example, 2.65 for quartz). But for a soil consisting of a variety of minerals we are concerned only with the mean specific gravity of the mass as a whole, and this is the sense in which the term is used here. Specific gravities of a few common mineral types are listed in Table 3.1.

The specific gravity of most soils generally lies between 2.60 and 2.80, and for sands which consist wholly or mainly of quartz it is often sufficient to assume that the specific gravity is 2.65. The presence of particles consisting of other minerals will result in a different value. Clays consist of various minerals, most of which are heavier than quartz, and have higher specific gravities, typically 2.68–2.72 for many British soils. The presence of heavy metals in the form of oxides or other compounds can give even higher values. On the other hand, soils which contain appreciable quantities of peat or organic material may have specific gravities considerably less

Table 3.1. ABSOLUTE SPECIFIC GRAVITIES OF SOME COMMON MINERALS

Mineral	Composition	Absolute specific gravity
Anhydrite	$CaSO_4$	2.9
Barytes	$BaSO_4$	4.5
Calcite, chalk	$CaCO_3$	2.71
Feldspar	$KAlSi_3O_8$, etc.	2.6–2.7
Gypsum	$CaSO_4 \cdot 2H_2O$	2.3
Haematite	Fe_2O_3	5.2
Kaolinite	$Al_4Si_4O_{10}(OH)_8$	2.6
Magnetite	Fe_3O_4	5.2
Quartz (silica)	SiO_2	2.65
Peat	Organic	1.0 or less
Diatomaceous earth	'Skeletal' remains of microscopic plants	2.00

than 2.65, and sometimes below 2.0. Soils consisting of particles which contain small cavities (bubbles of gas), such as pumice, also show a low 'apparent' specific gravity, although the 'absolute' specific gravity is higher. Tropical soils may have specific gravities which are unexpectedly high or low.

Soils containing substantial proportions of heavy or light particles can give erratic values of specific gravity. A number of repeat tests may be needed to obtain a reliable average value for this type of soil.

Specific gravity is related to the density of water at 4 °C, but most laboratory tests are carried out at an ambient temperature of about 20 °C. However, the difference in the density of water between 4 °C and 20 °C is less than 0.003 g/cm^3 (i.e. within 0.3%), so for practical purposes this discrepancy can be neglected.

There are several different kinds of specific gravity in use, especially in American practice, and they are defined in Section 3.2. The 'apparent specific gravity' is generally used in Britain for soils and is referred to throughout this book as 'specific gravity' or (SG), and is denoted by G_s.

The 'absolute' SG gives the highest specific gravity value which can be obtained for a given soil. The 'apparent' SG may be equal to, but is usually slightly smaller than, the 'absolute' SG. The other specific gravity terms are listed in order of magnitude of their values, the 'bulk' SG (oven-dry) giving the lowest value, because it includes the volume of all voids associated with the particles. The 'absolute' SG can be used as a guide for the identification of rock minerals, and is the parameter used in Table 3.1. The three 'bulk' specific gravities apply mainly to concrete aggregates, and will not be referred to again.

3.3.5 Limiting Densities

STATES OF PACKING

The limiting densities, as applied to sands, are the dry densities corresponding to the two extreme states of packing (loosest and densest) at which the particles can be placed in nature. The maximum density represents the densest packing of particles without crushing the grains, and corresponds to minimum porosity. The minimum density represents the loosest packing, and corresponds to maximum porosity.

If we consider a collection of equal spheres, their extreme states of packing can be represented diagrammatically in two dimensions as shown in Fig. 3.5. They can be densely packed as at (a), or loosely packed as at (b). In each case the spheres are in face-to-face contact. The densities achieved depend upon the density of the spheres (i.e. SG of the particles), but

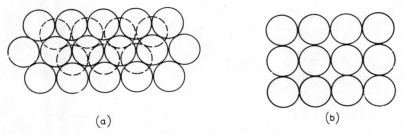

Fig. 3.5 *States of packing of equal spheres*

Table 3.2. THEORETICAL LIMITING VOIDS RATIOS AND POROSITIES FOR EQUAL SPHERES

	Densest state	*Loosest state*
Voids ratio e	0.35	0.91
Porosity n	26%	48%
Symbols used	e_{min}, n_{min}	e_{max}, n_{max}
	ρ_{max}	ρ_{min}

voids ratios and porosities do not, and can be calculated theoretically. These values are given in Table 3.2.

Real soils consist of particles of many sizes, and at their densest packing the voids between large particles contain smaller particles, and the voids between these contain yet smaller particles, and so on (Fig. 3.6a). The Fuller grading curve, referred to in Section 4.4.2, is based on the idealised limit of this concept (Fig. 3.6c). It might be expected that a real soil with a variety of particle sizes would show smaller values of e_{min} and n_{min} than can be obtained with equal spheres, but this is partially offset by the more irregular shape of real sand grains.

In the loosest state it is possible for groups of real soil particles to form 'arch' structures, as represented in Fig. 3.6(b), which can be sustained if left undisturbed. The possible higher values of e_{max} and n_{max} compared with equal spheres is again partially offset by the effect of irregularity of grain shape. For these reasons the voids ratio and porosity limits of many natural sands do not differ greatly from the theoretical values for equal spheres shown in Table 3.2.

In Fig. 3.6, sample (a) is dense, and is stable. Sample (b) is unstable, and the grain structure is likely to collapse under the influence of a sudden shock, vibration or inundation.

The minimum density referred to here must not be confused with the 'bulking' phenomenon of damp sand. It is possible to deposit damp sand at a porosity greater than that achieved by the standard procedures for determining the maximum porosity. This effect is due to a film of water on the particles preventing direct contact between them, and so increasing the porosity. This is a condition not found in nature, because natural and deposits have usually been either laid down under water, or wind-blown in dry conditions.

RELATIVE DENSITY

The voids ratio of a granular soil can be used with reference to the limiting voids ratios (e_{min}, e_{max}) as an index, in much the same way as the moisture content of clays is referred to the Atterberg limits (Chapter 2; Section 2.3.3). The term used here, corresponding to the liquidity index for clays, is relative density (RD) and is the ratio derived as follows. If the voids ratio of the soil is denoted by e, then

$$RD = \frac{e_{max} - e}{e_{max} - e_{min}}$$

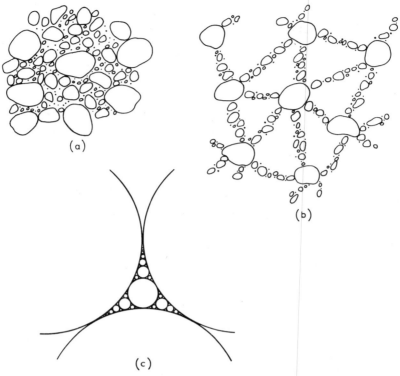

Fig. 3.6 *States of packing of soil particles: (a) densely packed, (b) loosely packed and (c) idealised 'Fuller' packing*

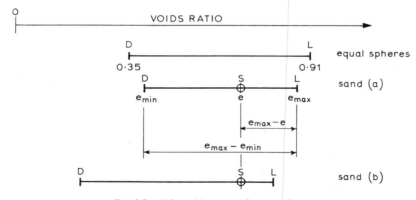

Fig. 3.7 *Relative densities of two sands*

Relative density is illustrated diagrammatically in Fig. 3.7, in which values of voids ratio are plotted from left to right. Zero voids ratio relates to a density equal to the density (SG) of the individual particles. The range of voids ratios for equal spheres is shown on the top line.

Considering sand (a), the voids ratio in its natural state is e, and is denoted by the point S. Its relative density (RD), as defined by the above equation, is near the middle of the range — that is, sand (a) has a 'medium' relative density.

Another material, sand (b), with different values of limiting voids ratios, but at the same natural voids ratio, would have a 'low' relative density — that is, it would be at a less dense state of packing than sand (a). It follows that voids ratio (or density), without reference to the

Table 3.3. TYPICAL DENSITIES AND OTHER PROPERTIES OF SOILS

Soil type	Moisture content w (%)	Bulk density, ρ (Mg/m³)	Dry density, ρ_D (Mg/m³)	Voids ratio, e	Porosity, n (%)	Degree of saturation, S (%)
Dry uniform sand, loose	0	1.36	1.36	0.95	49	0
Well-graded sand	7.5	1.95	1.81	0.46	32	43
Soft clay	56	1.67	1.07	1.52	60	99.5
Firm clay	22	1.96	1.61	0.68	41	87
Stiff glacial till	9.5	2.32	2.12	0.27	21	95
Peat	220*	0.98	0.31	3.67	78	85

* Can be up to 10 times greater in bog peat.

limiting voids ratio (or densities), is not sufficient to define the state of compaction of a sand.

To calculate the relative density from dry densities instead of voids ratios, the following expression is used:

$$\mathrm{RD} = \frac{\rho_D - \rho_{D,\min}}{\rho_{D,\max} - \rho_{D,\min}} \times \frac{\rho_{D,\max}}{\rho_D}$$

3.4 APPLICATIONS

3.4.1 Density

The *in situ* density of soil is an important property having many applications in soil engineering. In practical problems associated with earthworks and foundations, the weight of the soil itself exerts forces which have to be taken into account in the analysis. It is therefore necessary to know the bulk density of the soil, from which these forces can be calculated. In the analysis of the stability of a slope, such as an embankment or the sides of a cutting, the weight of soil provides the main force, but it is also significant in calculating the bearing capacity and settlement of foundations for other structures. The bulk density of natural soil is usually determined from laboratory measurements on undisturbed samples.

When soil is used as a construction material, such as in embankments or road sub-bases, measurement of dry density provides an important means of quality control. Direct measurement of compacted density *in situ* is beyond the scope of this book. The methods most often used are the sand-replacement method, and the core-cutter method, which are given in BS 1377:1975 under Tests 15 (A), (B), (C), (D). Undisturbed samples of compacted fill can be taken for laboratory measurements of density, moisture content and other properties.

In general, all undisturbed soil samples (whether from natural ground or from compacted soil) used for laboratory tests are measured for the determination of bulk density and moisture content, from which the dry density can also be calculated (Section 3.3.2).

Some typical values of density, voids ratio and porosity for a few soil types are given in Table 3.3.

3.4.2 Voids Ratio and Porosity

The main application of voids ratio is in the analysis of data from the oedometer consolidation tests, which is given in Volume 2.

Voids ratio, and saturation, are important factors relating to the compaction of soils, which is dealt with in Chapter 6.

Voids ratio is used as an index for assessment of the relative density of sands (see Section 3.3.5).

Porosity can be useful in some soils as an indication of permeability characteristics.

3.4.3 Specific Gravity

It is rarely possible to use specific gravity as an index for soils classification. But knowledge of the SG is essential in relation to some other soil tests, especially for calculating porosity and voids ratio, and is particularly important when compaction and consolidation properties are considered. The SG must also be known for the computation of particle size analysis from a sedimentation procedure (Chapter 4; Section 4.8).

3.4.4 Limiting Densities

The application of limiting densities relates only to granular soils, and to sands in particular. Limiting densities of sands are not so widely used as are the Atterberg limits of clays. Nevertheless the assessment of relative density can give some indication of the possible effects of loading or of disturbance on sand. This factor is particularly important where a sand deposit is likely to be affected by vibration, such as from machinery, or by earthquake shock. In loose sand or silt deposits below water table, vibration or shock could result in liquefaction and collapse of the grain structure. It is therefore desirable to have some means of assessing how loose a 'loose' deposit is, so that appropriate remedial measures (such as compaction by vibration) can be taken if necessary.

Classification of granular soils, based on relative density values, as suggested by Lambe and Whitman (1969) is indicated in Table 3.4.

Table 3.4.   RELATIVE DENSITY DESCRIPTION

Description	Relative density range, RD (%)
Very loose	0–15
Loose	15–35
Medium	35–65
Dense	65–85
Very dense	85–100

3.5   DENSITY TESTS

3.5.1 Scope

Measurement of the density of soils is often overlooked, but it can be just as important as the measurement of moisture content. It is good practice to measure the density of all undisturbed samples tested in the laboratory, and sometimes it may be the only test, apart from moisture content.

The tests to be described fall into three categories, which are based on three different principles for the measurement of volume. These are

(1) Linear measurement.
(2) Water displacement.
(3) Water immersion (weighing in water).

In many instances the specimen prepared for some other test, such as compressive strength or consolidation, will be used for the measurement of density. Alternatively, an undisturbed sample may be specially prepared for a density test. Sometimes a relatively undisturbed sample will be too friable to handle and prepare for other tests, and a density measurement in the sampling tube or container itself is the only possibility. A density measurement on an irregular lump sample will require a method different from that used for a regular block or cylinder.

3.5.2   Linear Measurement

APPARATUS

(1) Cutting and trimming tools (knife, wire saw).
(2) Steel straight-edge and square.
(3) Steel rule.
(4) Vernier calipers.
(5) Balance, readable to 0.05% of the mass of the sample.
(6) Oven and moisture content apparatus (Section 2.5.2).

PROCEDURAL STAGES

(1) Trim specimen.
(2) Weigh.
(3) Measure.
(4) Calculate.
(5) Report results.
(6) Determine dry density if needed.

TEST PROCEDURE

(1) *Preparation*

A specimen of regular shape is prepared from the undisturbed sample, the shape depending upon the type of sample. From a tube sample (e.g. a U100 tube) a cylindrical specimen is prepared. From a block sample, unless a 38 mm diameter specimen can be extracted without disturbance, it is preferable to prepare a cube or rectangular block.

Check all flat surfaces against the light with the straight-edge and trim until they are plane (no 'daylight'; see Fig. 3.8). Corners and angles must be checked with a try-square. The end faces of a cylindrical specimen must be at right angles to the axis. Use a sharp knife for trimming, *never* the straight-edge or blade of the square.

(2) *Weighing*

Use a balance of a capacity appropriate to the size of specimen, and weigh the specimen as accurately as possible (see Section 1.2.3).

(3) *Measurements*

Several linear measurements should be made on each face, and the readings averaged. For example, a cylindrical specimen should be measured as follows. Take three measurements of length, using vernier calipers, equally spaced around the cylindrical surface (L_1 to L_3, Fig.

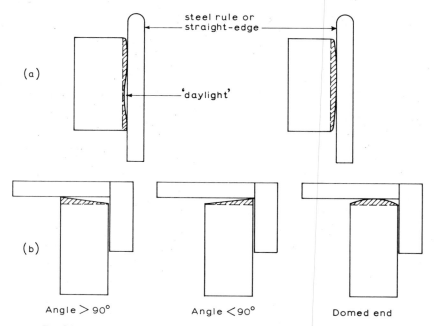

Fig. 3.8   *Trimming specimens: (a) checks for flatness and (b) checks for squareness*

3.9). Take two measurements of diameter, using the calipers, at each end and at mid-length (D_1 to D_6). Read the vernier to 0.1 mm and record each measurement.

When using vernier calipers, first check the zero reading and record it. Close the jaws carefully on to the specimen so that they make contact without bedding in. The calipers must be perpendicular to the surfaces being measured, not on the skew. Read the vernier, record it, and subtract the zero reading to obtain the length being measured.

The specimen should not be exposed to the air for any longer than necessary, and should not be unduly handled, as otherwise it will lose moisture and may shrink. Wear thin plastics gloves to minimise moisture loss due to the warmth of the hands.

(4)   *Calculation*

Calculate the mean value (mm) of each dimension from the several measurements, to the nearest 0.1 mm. For a rectangular block, if L = mean length (mm), B = mean breadth (mm) and H = mean height (mm), the volume V is given by

$$V = LBH \quad \text{mm}^3$$

$$= \frac{LBH}{1000} \quad \text{cm}^3$$

For a cylinder of diameter mean D and length L,

$$V = \frac{\pi D^2 L}{4000} \quad \text{cm}^3$$

If the mass of the specimen is m grams, the bulk density ρ is given by the equation $\rho = m/V$ Mg/m^3.

An example calculation for a cylindrical specimen is given in Fig. 3.9.

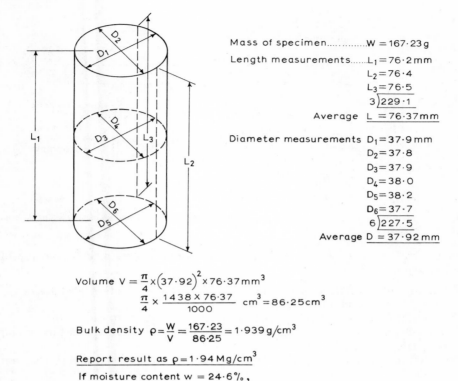

Mass of specimen.............W $= 167 \cdot 23\,g$

Length measurements......$L_1 = 76 \cdot 2\,mm$

$L_2 = 76 \cdot 4$

$L_3 = 76 \cdot 5$

3)229·1

Average L $= 76 \cdot 37\,mm$

Diameter measurements $D_1 = 37 \cdot 9\,mm$

$D_2 = 37 \cdot 8$

$D_3 = 37 \cdot 9$

$D_4 = 38 \cdot 0$

$D_5 = 38 \cdot 2$

$D_6 = 37 \cdot 7$

6)227·5

Average D $= 37 \cdot 92\,mm$

Volume V $= \frac{\pi}{4} \times \left(37 \cdot 92\right)^2 \times 76 \cdot 37\,mm^3$

$\frac{\pi}{4} \times \frac{1438 \times 76 \cdot 37}{1000}\ cm^3 = 86 \cdot 25\,cm^3$

Bulk density $\rho = \frac{W}{V} = \frac{167 \cdot 23}{86 \cdot 25} = 1 \cdot 939\,g/cm^3$

Report result as $\rho = 1 \cdot 94\,Mg/cm^3$

If moisture content w $= 24 \cdot 6\%$,

Dry density $\rho_D = \rho \times \frac{100}{100+w} = 1 \cdot 939 \times \frac{100}{124 \cdot 6}$

$= 1 \cdot 556\,Mg/m^3$

Report result as $\rho_D = 1 \cdot 56\,Mg/m^3$

Fig. 3.9   *Measurements of a cylindrical soil specimen and calculation of density*

(5)   *Result*

The result is reported to the nearest 0.01 Mg/m^3.

(6)   *Dry density*

If the dry density is also required, and no other test is to be done, the whole specimen can be broken up and placed in the oven to dry, in order to measure the moisture content. Make sure that no fragments of soil are lost.

The moisture content may be determined instead by using small representative portions of the specimen, or by using trimmings placed directly into a moisture content container when the specimen is being prepared. The trimmings should be taken from immediately adjacent to the test specimen.

If the measured moisture content is $w\%$, the dry density ρ_D is calculated from the equation

$$\rho_D = \frac{100}{100+w}\rho \quad Mg/m^3$$

3.5.3   Sample in Tube

If the density of the sample has to be measured in the sampling tube, the following outline procedure is used. The details refer to a U100 tube sample, but the same principle can be used

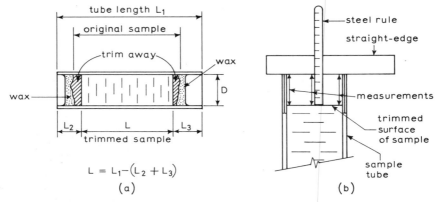

Fig. 3.10   *Measurements of a sample in a tube: (a) trimming ends and (b) measurements for sample length*

for tube samples of other sizes if weighings and measurements are made to the appropriate degree of accuracy.

(1)   Clean the outside of the tube, but do not remove the identification label. Record sample number, and tube number if marked.
(2)   Remove end caps and wax or other protective material from each end.
(3)   Trim each end of the sample to give flat surface normal to the axis of the tube. Remove all loose material (see Fig. 3.10a).
(4)   Measure the overall length of the sampling tube (L_1) to the nearest mm, using a metre-stick or steel tape.
(5)   Using a steel rule and straight-edge, measure from the ends of the tube to the trimmed surface of the sample to 0.5 mm, taking three or four readings at each end (Fig. 3.10b). Calculate the mean of each set to the nearest 0.5 mm (L_2 and L_3).
(6)   Using the internal-measuring jaws of the vernier calipers, measure the internal diameter of the tube on two perpendicular diameters at each end, to 0.1 mm. Calculate the average to 0.1 mm (D). Make sure that the calipers are opened to the maximum extent possible, to give a diametrical reading.
(7)   Weigh the tube and sample, to an accuracy of 5 g (m_1). A 20 kg capacity balance may be required if the sample is contained in a steel tube. The sample is weighed in the tube as a precautionary measure in case the sample cannot be kept together on extrusion.
(8)   Extrude the sample and collect all material together.
(9)   Weigh the extruded material (m_2).
(10) Clean out the tube, and weigh it (m_3).
(11) Check that $m_1 - m_3 = m_2$.
(12) Calculate sample length L:

$$L = L_1 - (L_2 + L_3) \text{ mm};$$

sample volume V

$$V = \frac{\pi D^2 L}{4000} \text{ cm}^3$$

bulk density

$$\rho = \frac{m_2}{V} \text{ or } \frac{m_1 - m_3}{V} \text{ Mg/m}^3$$

Fig. 3.11   *Water displacement apparatus* (*siphon can*)

Report the result to the nearest 0.01 Mg/m³, and that the density was measured in a U100 tube.

(13)   Measure the moisture content $w\%$ of the extruded sample, and calculate dry density ρ_D as in Section 3.5.2, stage (6), if it is required.

3.5.4   Water Displacement (BS 1377:1975, Test 15 (F))

When it is not possible to trim a specimen to a regular cylindrical or prismoidal shape, or when the only available sample is an irregular lump, determination of the volume by simple linear measurement is not practicable. The volume is then measured by one of the immersion methods. Water displacement is the simplest method, though less accurate than weighing in water (Section 3.5.5).

APPARATUS

(1)   Water-displacement apparatus, as shown in Fig. 3.11. A cross-section of the apparatus showing the siphon tube is given in Fig. 3.12.
(2)   A watertight container to act as a receiver for water siphoned from the above apparatus. A convenient size is 250 mm diameter and 250 mm deep. A large beaker may be used if appropriate.
(3)   Balance reading to 1 g.
(4)   Paraffin wax and thermostatically controlled bath for melting wax. Alternatively, a double pot such as a glue-pot will serve to prevent the wax being overheated.
(5)   Oven and moisture-content apparatus.

PROCEDURAL STAGES

(1)   Prepare specimen.
(2)   Weigh.
(3)   Fill voids.

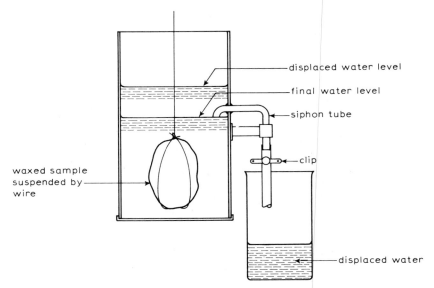

Fig. 3.12   *Principle of water displacement apparatus*

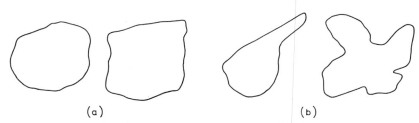

Fig. 3.13   *Trimmed lump specimens: (a) correct and (b) not satisfactory*

(4)   Coat with wax.
(5)   Weigh.
(6)   Prepare displacement apparatus.
(7)   Immerse in water.
(8)   Measure displaced water.
(9)   Measure moisture content.
(10) Calculate.
(11) Report result.

TEST PROCEDURE

(1) *Preparation of specimen*

Trim the specimen to a convenient size and shape, preferably about 100 mm in each direction.
Avoid making a long, narrow section, and avoid the formation of re-entrant angles (see Fig.
3.13). Use the largest specimen possible, compatible with the capacity of the apparatus.

(2) *Weighing*

Weigh the specimen to the nearest 1 g (m_s).

(3) Filling the voids

Fill all the surface air voids with Plasticine or putty, and trim off level with the surface of the specimen. Only air voids, and not cavities from which stones have been removed, should be filled. Paraffin wax could be used for this purpose if it is applied carefully (see stage 4), and if the voids are not large. Weigh the specimen after filling, to the nearest 1 g (m_f).

The reason for this operation is to ensure that naturally occurring voids are included as part of the sample volume, and are not penetrated by the protective coating of wax applied in stage (4).

(4) Coating with wax

Apply a first coat of paraffin wax, by brushing molten wax on to the surface. This must be done carefully, avoiding trapping air bubbles in deep cavities. The wax must be heated to just above its melting point; it must not be overheated, as otherwise shrinkage and cracking will result. When the wax has dried, immerse the whole sample in the wax-bath, remove and allow the wax to set, and repeat the immersion several times.

(5) Weighing

When cool, weigh the waxed sample to the nearest 1 g (m_w). The mass of wax (m) is given by

$$m = m_w - m_f$$

(6) Preparation of apparatus

Place the water-displacement apparatus on a level base, close the clip on the outlet tube, and pour water to a level about 50 mm above the siphon tube. Release the clip on the outlet tube and allow excess water to run to waste. Retighten the clip without moving the can.

Weigh the receiver container to the nearest 1 g (m_1).

(7) Immersion of sample

Place the receiving container below the siphon outlet. Lower the waxed specimen carefully into the apparatus until it is completely immersed, taking care to ensure that no air bubbles are trapped underneath the specimen. This operation is easier if the specimen is suspended by a length of fine wire looped around it.

Release the clip on the outlet tube, allowing displaced water to siphon into the receiver. Wait until all excess water has been siphoned off, then retighten the clip.

(8) Measurement of displaced water

Weigh the receiver and water to the nearest 1 g (m_2).

(9) Measurement of moisture content

Remove the specimen from the can, dry the surface and break it open. Take representative samples, free from filler and wax, for the determination of the moisture content (w).

(10) Calculations

The volume of the immersed waxed specimen (V_b cm^3) is equal to the volume of water displaced (cm^3), which is equal to its mass in grams:

$$V_b = m_2 - m_1$$

The volume of paraffin wax is equal to m/ρ_P, where ρ_P is the density of the wax (typically 0.91

Mg/m^3, but the actual value should be verified). Therefore the volume of specimen, V_s, is given by

$$V_s = V_b - \frac{m}{\rho_P}$$

The bulk density, ρ, of the soil specimen is equal to its mass divided by its volume — i.e.

$$\rho = \frac{m_s}{V_s} \quad Mg/m^3$$

The dry density, ρ_D, is calculated as in Section 3.5.2, stage (6).

(11)   *Results*

The bulk density and dry density are reported to the nearest $0.01 \ Mg/m^3$, and the moisture content to two significant figures. The method is reported as the BS water-displacement method.

ALTERNATIVE PROCEDURE

A simpler method can be used if the sample is small.

(1)   Take the smallest glass graduated measuring cylinder into which the waxed specimen will conveniently fit.
(2)   Add water up to a known mark, and record the reading.
(3)   Prepare, wax and weigh the specimen as for stages (1)–(5) of the siphon can method.
(4)   Immerse the specimen in the water in the cylinder, avoiding trapping any air.
(5)   Read and record the new water level in the cylinder. The difference between the two readings is the volume ($V_b \ cm^3$) of the waxed specimen.
(6)   Measure moisture content, and calculate bulk density and dry density, as for stages (9)–(11) above. Report as the measuring cylinder water-displacement method.

3.5.5   Weighing in Water (BS 1377:1975, Test 15 (E))

This method requires relatively simple weighing apparatus, and can be used for determining the density of quite large lump samples or compacted samples. It makes use of the principle of Archimedes (Section 3.3.3) for the measurement of volume.

APPARATUS

(1)   Two-pan balance, such as a semi-self-indicating type, 15 or 20 kg capacity, reading to 1 g. For convenience it is preferable for it to be fitted with a flat pan instead of a scoop pan.
(2)   Suitable cradle and supporting frame for suspending the sample below the balance. An apparatus developed from the arrangement outlined in Figs. 31 and 32 of BS 1377:1975 is shown in Fig. 3.14.
(3)   Paraffin wax and wax-bath.
(4)   Watertight container (metal or plastics) such as a dust-bin.
(5)   Oven and moisture-content apparatus.

PROCEDURAL STAGES

(1)   Prepare apparatus.
(2)   Prepare specimen.

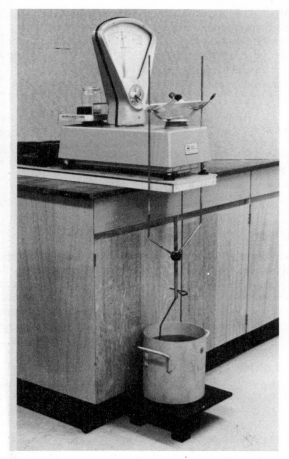

Fig. 3.14   Apparatus for weighing in water

(3)   Weigh.
(4)   Fill voids.
(5)   Coat with wax.
(6)   Weigh.
(7)   Weigh immersed in water.
(8)   Measure moisture content.
(9)   Calculate.
(10) Report result.

TEST PROCEDURE

(1)   *Preparation of apparatus*

(a)   Place the balance on a board or battens projecting over the edge of a firm bench, so that
      the cradle can be suspended over the middle of the water container standing on the floor
      (see Fig. 3.14). Place a counterbalance weight on the board or battens to ensure that it
      will not tip over when carrying the specimen.
(b)   Adjust the height of the container and support it so that the largest specimen to be tested,
      when placed in the cradle, will be completely immersed, and will not touch the sides or
      bottom of the container.

(c)   Fill the container with water to about 80 mm below the top.
(d)   With the cradle completely immersed, add a counterweight to the scale pan to bring the
      balance reading to zero. It is preferable to use a jar or tin containing sand, water or lead-
      shot for this purpose, so that it will not be confused with the weights added later. If
      balance weights are used, they should be clearly identified as counterweights — for
      example, by standing them on a sheet of thin coloured paper.

Stages (2)–(6)

Similar to stages (1)–(5) in Section 3.5.4, using the balance in the normal manner.

   mass of prepared sample $= m_s$ g
   mass of sample and filler $= m_f$
   mass of wax-coated sample $= m_w$
   mass of wax $= m = m_w - m_f$

(7)   *Weighing in Water*

Place the specimen in the cradle and suspend the cradle from the supporting frame so that the
sample is completely immersed in the water in the container. Check that no air bubbles have
been trapped underneath the specimen. Add more water or adjust the supporting frame, if
necessary, to ensure complete immersion of the specimen. Stage (1) should then be repeated
with the sample removed, followed by a check-weighing of the immersed specimen.
   Measure the apparent mass of the specimen in water to the nearest 1 g (m_g).

(8)   *Measurement of moisture content*

Remove the sample from the cradle, dry the surface and break it open. Take representative
samples, free from filler and wax, for the determination of moisture content (w).

(9)   *Calculations*

From Archimedes' principle, the volume of the immersed wax specimen, V_g (cm³), is given by

$$V_g = m_w - m_g \text{ (masses in grams)}$$

The volume of the wax coating is equal to its mass divided by its density — i.e. m/ρ_P.
The volume of the specimen itself (V_s) is therefore given by

$$V_s = V_g - \frac{m}{\rho_P} = (m_w - m_g) - \frac{m}{\rho_P}$$

The bulk density of the specimen is given by

$$\rho = \frac{m_s}{V_s} \text{ g/cm}^3 \ (= \text{Mg/m}^3)$$

The dry density is calculated as in Section 3.5.2, Stage (6).

(10)   *Results*

The bulk density and dry density are reported to the nearest 0.01 Mg/m³, and the moisture
content to two significant figures. The method is reported as the BS immersion in water (or
weighing in water) method.

3.6 SPECIFIC GRAVITY

3.6.1 Scope

Three tests for the measurement of specific gravity of soil particles are described. These have different applications, which may be summarised as follows:

(1) Density bottle method — for fine-grained soils only.
(2) Gas jar method — for all soils.
(3) Pycnometer method — for use in a site laboratory on medium- and coarse-grained soils.

The first two methods are given in the 1975 British Standard. The thrid was given in British Standards up to and including 1967, but has now been superseded by the gas jar method. However, it is useful as a fairly simple on-site procedure where full laboratory facilities are not available.

3.6.2 Density Bottle Method (BS 1377:1975, Test 6 (B))

This is the traditional method for the accurate measurement of specific gravity of particles heavier than water, or of the density of liquids.

Distilled water is normally used as the density bottle fluid, but if the soil contains soluble salts, an alternative liquid should be used. The usual liquid is kerosene (paraffin) or, alternatively, white spirit. The SG of the liquid must be measured separately (Section 3.6.3). The test procedure which follows is the same whether water or another fluid is used, but if the latter, the calculation must allow for the SG of the liquid.

APPARATUS

(1) Density bottles (50 ml) with stoppers, numbered and calibrated (three for each sample). (See Fig. 3.15.)
(2) Constant-temperature water-bath, with shelf for holding density bottles maintained at 25 °C (see Fig. 3.16).
(3) Vacuum desiccator (Figs. 1.16 and 3.15).
(4) Oven and moisture-content apparatus.
(5) Analytical balance reading to 0.001 g.
(6) Small riffle-box.
(7) Source of vacuum, and vacuum tubing.
(8) Chattaway spatula 150 × 3 mm.
(9) Wash bottle.
(10) Rubber-coated tongs, or rubber gloves.

MATERIALS

De-aired distilled water or de-aired kerosene (or white spirit), the specific gravity of which must be known or measured.

Although dissolved air does not greatly affect the accuracy of the result, there is the danger of air coming out of solution in the form of small bubbles when the temperature is raised to 25 °C.

PROCEDURAL STAGES

(1) Prepare density bottles.
(2) Prepare test specimen.
(3) Weigh and dry specimen in bottle (three bottles).
(4) Add liquid and apply vacuum.
(5) Stir and repeat until air is removed.

Fig. 3.15   Density bottles in vacuum desiccator

Fig. 3.16   Density bottles in constant-temperature bath

(6)   Top up, transfer to temperature bath.
(7)   Weigh bottle + soil + liquid.
(8)   Weigh bottle + liquid.
(9)   Calculate SG values.
(10)   Report results.

TEST PROCEDURE

(1)   *Preparation of density bottles*

Wash each density bottle with stopper, and dry in the oven at 105–110 °C. Cool in the desiccator and weigh to the nearest 0.001 g (m_1). Repeat to ensure that the bottle has been dried to constant weight.

If density bottles conforming to BS 733:1975 are used, do not dry in an over, otherwise distortion may occur. Rinse with acetone or an alcohol–ether mixture and dry by blowing warm air through them.

Each density bottle should carry a unique identification number. If this is not engraved by the manufacturer, a laboratory number should be lightly scratched on the neck with a diamond-tipped pencil.

(2) *Preparation of test specimen*

Obtain a sample of 50–100 g by quartering down the original sample. A representative proportion of gravel-size particles, if present, must be included, but these must be ground with pestle and mortar to pass a 2 mm BS sieve. Riffle the sample to obtain a test specimen of about 30 g, oven dry at 105–110 °C, and cool in the desiccator.

If at this temperature the SG is likely to be changed owing to loss of water of hydration, a drying temperature of 80 °C should be used, and the drying time extended if necessary. This fact must be reported.

(3) *Placing in bottles*

Divide the dried specimen into three approximately equal parts, and place each into a density bottle using tongs or wearing rubber gloves. This should be done directly from the desiccator. Weigh each bottle with soil to 0.001 g (m_2).

(4) *Adding liquid and applying vacuum*

Add the de-aired liquid (distilled water or kerosene) carefully to each bottle so that the soil is covered and the bottles are about half-full.

Pour the liquid down the side of the bottle to avoid disturbing the soil and thereby entrapping the air. Place the bottles (without stoppers) in the vacuum desiccator. Reduce the pressure gradually, and leave under vacuum overnight. Air trapped in the soil will bubble out, but this must not occur too violently, as otherwise small drops of suspension may be lost through the mouth of the bottle. The vacuum should be maintained until no further loss of air can be seen.

(5) *Removal of air*

Release the vacuum, remove the desiccator lid and carefully stir the soil in the bottle with the Chattaway spatula. Wash any soil particles adhering to the blade back into the bottle with a little of the de-aired liquid. Replace the desiccator lid and apply the vacuum again until no further loss of air can be seen.

This process is repeated until no more air is evolved from the soil. Complete removal of air is essential for obtaining reliable results from this test. If in doubt, leave the bottles for a further period under vacuum.

(6) *Transfer to constant-temperature bath*

Remove each bottle from the desiccator and add further air-free liquid until full. Insert the stopper, and place the stoppered bottle in the constant-temperature bath so that it is immersed up to the neck. Leave for 1 h at least, or longer if necessary for it to attain the temperature of the bath.

Surplus liquid will exude through the capillary tube in the stopper. If the liquid is other than water, it should be carefully absorbed on a filter paper to avoid contamination of the water-bath.

If there is a decrease in volume of the liquid, remove the stopper and top up. Repeat if necessary after a further 1 h in the water-bath.

(7)   *Weighing*

Remove the bottle from the bath and wipe it dry. Avoid prolonged contact with the hands, which may raise the temperature further. Weigh (bottle + stopper + soil + liquid) to 0.001 g (m_3).

(8)   *Weighing bottle with liquid*

Clean out each bottle, and fill completely with de-aired liquid. Insert the stopper and immerse in the constant-temperature bath as before. Repeat stages (6) and (7) as necessary, and weigh (bottle + stopper + liquid) to 0.001 g (m_4).

(9)   *Calculation*

Calculate the specific gravity G_s of the soil in each bottle from the following equation:

$$G_s = \frac{G_L(m_2 - m_1)}{(m_4 - m_1) - (m_3 - m_2)}$$

where G_L = specific gravity of the liquid used, at the constant temperature (if distilled water is used, G_L may be assumed to be equal to 1.000 for most purposes); m_1 = mass of density bottle (g); m_2 = mass of bottle + dry soil (g); m_3 = mass of bottle + soil + liquid (g); m_4 = mass of bottle + liquid only (g).

The average of the three values obtained is calculated. If any one value differs from the average value by more than 0.03, the test should be repeated.

(10)   *Results*

The average value derived above is reported to the nearest 0.01 as the specific gravity (SG) of the soil. Since it is a ratio, it is a dimensionless number.

The liquid used in the test should be reported with the result.

3.6.3   Calibrations

CALIBRATION OF DENSITY BOTTLE

Each density bottle is dried to constant weight, cooled and weighed to 0.001 g. The bottle is then filled with air-free distilled water. Insert the stopper and place in the constant-temperature bath until it has attained a temperature of 25 °C, as for stage (6) in Section 3.6.2. Take the bottle out of the bath, wipe it dry (care being taken not to raise the temperature of the bottle by prolonged contact with the hands) and weigh to 0.001 g. The test is repeated twice more.

The volume of each density bottle is calculated as follows:

m_1 = weight of density bottle with stopper (g)
m_4 = weight of density bottle with stopper full of distilled water at 25 °C

Specific gravity of water at 25 °C = 0.997 08. Then

$$V_d = \frac{m_4 - m_1}{0.997\,08}$$

where V_d = volume of density bottle (ml).

Fig. 3.17   Gas jars and mechanical shaker

SPECIFIC GRAVITY OF LIQUID

Calibrate three density bottles, using distilled water as described above. Empty and dry them, and allow them to cool.

Fill with liquid, insert stopper, place in temperature-bath, top up as for stage (6) in Section 3.6.2.

Remove, dry and weigh (m_5).

The specific gravity, G_L, of the liquid is calculated from the equation

$$G_L = \frac{m_5 - m_1}{m_4 - m_1}$$

where m_1 = mass of density bottle; m_4 = mass of bottle filled with distilled water; m_5 = mass of bottle filled with liquid.

Calculations should be done using 5-figure logarithmic tables or a calculator of 5-figure accuracy. A slide rule is not sufficiently accurate. The result should be rounded off to the third decimal place.

The value of G_L is used in the equation given in step (9) of Section 3.6.2 for calculating the SG of soil particles, when a fluid other than distilled water is used.

3.6.4   Gas Jar Method (BS 1377:1975, Test 6 (A))

This test appeared for the first time in the 1975 British Standard. It can be used for soils containing particles up to 37.5 mm size. Stones larger than this should first be broken down so as to pass a 37.5 mm sieve.

APPARATUS

(1)   Gas jar, 1 litre capacity, with rubber bung and ground-glass cover-plate.
(2)   Mechanical shaker, giving end-over-end shaking motion (Fig. 3.17).

(3)   Balance, 5 kg capacity, accurate to 0.2 g.
(4)   Thermometer.

PROCEDURAL STAGES

(1)   Select sample.
(2)   Prepare gas jar.
(3)   Weigh jar and sample.
(4)   Add water and shake.
(5)   Top up and fit glass plate.
(6)   Weigh jar, sample, water, plate.
(7)   Weigh jar and water.
(8)   Calculate.
(9)   Report result.

TEST PROCEDURE

(1)   *Selection of sample*

A representative sample of about 800 g is obtained by quartering the original sample. Any particles larger than 37.5 mm are broken down. The sample is oven dried to constant mass and stored in an airtight container until used. For a fine-grained soil about 400 g is sufficient.

(2)   *Preparation of gas jar*

The gas jar and ground-glass plate are dried in the oven, cooled and weighed to the nearest 0.2 g (m_1).

(3)   *Weighing*

About half of the prepared dry sample is placed in the gas jar, and the jar, contents and ground-glass plate are weighed to 0.2 g (m_2).

(4)   *Adding water and shaking*

About 500 ml of water at room temperature is added to the soil in the jar. The rubber bung is inserted, and the jar is allowed to stand for 4 h. The gas jar is shaken by hand to put the particles in suspension, and is then placed in the shaking apparatus and shaken for 20–30 min.

The jar and bung must be clamped tightly enough to prevent loss of water, but not so tightly that the jar might split while being shaken. Adhesive plastics tape wrapped around the jar and supporting frame of the shaker will ensure that the jar is held in place.

(5)   *Topping up and fitting glass plate*

After the required shaking period the jar is taken out of the shaker. Remove the rubber bung carefully to avoid losing fine particles. Any soil particles adhering to the bung must be washed carefully back into the jar with a jet of distilled water. Any froth that has formed is dispersed with the water jet.

Water is added to the gas jar to within about 2 mm of the top, with the jar standing on a level surface. After standing for about 30 min to allow the soil to settle, the gas jar is filled to the brim with water. The ground-glass plate is slid carefully on to the top of the jar, with the jar slightly tilted to avoid trapping any air under the plate.

(6)   *Weighing jar with sample, water and plate*

The outside of the jar and plate are carefully dried with an absorbent cloth or blotting paper, without disturbing the plate. Jar, contents and plate are weighed to 0.2 g (m_3).

(7) *Weighing jar and water*

The gas jar is emptied and washed out thoroughly. It is filled to the brim with water at room temperature, and the ground-glass plate is slid on without entrapping any air. The jar is dried, and the whole is weighed to 0.2 g (m_4).

Stages (3)–(6) are repeated on the other half of the sample of the same soil, so that two values of the SG are obtained.

(8) *Calculations*

The specific gravity, G_s, of the soil particles for each determination is calculated from the equation

$$G_s = \frac{m_2 - m_1}{(m_4 - m_1) - (m_3 - m_2)}$$

where G_s = specific gravity of soil particles; m_1 = mass of jar and plate; m_2 = mass of jar, plate and dry soil; m_3 = mass of jar, plate, soil and water; m_4 = mass of jar filled with water, and plate.

(9) *Results*

If the two values of specific gravity differ by 0.03 or less, the average of the two values is reported to the nearest 0.01 as the SG of the soil. If the difference is greater than 0.03, the tests should be repeated.

The method is reported as the British Standard gas jar method.

3.6.5 Pycnometer Method

This test was described in BS 1377:1967 as Test 5 (B), and the notation used below is the same as was used there. This method is satisfactory for use with non-cohesive soils in a site laboratory with limited facilities. It is not recommended for general laboratory use, because it is less accurate than the gas jar method (Section 3.6.4). It is not suitable for clay soils.

APPARATUS

(1) Pycnometer, as illustrated in Fig. 3.18. This consists of a screw-top glass jar of about 1 kg capacity with a brass conical cap, screw ring and rubber sealing ring.
(2) Balance reading to 0.5 g.
(3) Oven and moisture-content apparatus.
(4) Glass stirring rod, thermometer.

PROCEDURAL STAGES

(1) Prepare pycnometer.
(2) Prepare sample.
(3) Place in pycnometer and weigh.
(4) Add water and stir.
(5) Remove air and top up.
(6) Weigh.
(7) Fill pycnometer with water.
(8) Weigh.
(9) Repeat test.
(10) Calculate.
(11) Report results.

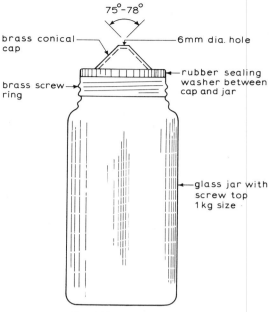

Fig. 3.18   Pycnometer

TEST PROCEDURE

(1)   *Preparation of Pycnometer*

Clean and wash the pycnometer, including the cap assembly, dry and weigh to the nearest
0.5 g (W_1).

Check that the rubber sealing ring is in good condition. If it has gone hard, replace it. Screw
the securing ring so as to make the seal watertight. Make locating marks adjacent to one
another on the conical cap, screw ring and glass jar. When the pycnometer is used, the ring
should be tightened to this position every time so that the pycnometer volume remains
constant.

If a watertight seal is difficult to make, grind the rim of the pycnometer jar by rubbing it on a
sheet of fine carborundum paper laid on a perfectly flat surface, such as the glass plate used for
liquid limits. Leakage can give rise to serious errors.

(2)   *Preparation of sample*

About 1 kg of oven-dried sample, cooled in the desiccator, is required. Stones larger than
37.5 mm should first be broken down to below this size.

A rather smaller sample will be adequate for sandy or silty soils without stones.

(3)   *Placing in Pycnometer*

Remove the screw top from the pycnometer and insert about 400–500 g of the sample direct
from the desiccator. Weigh the jar and sample together with the cap assembly to 0.5 g (W_L).

(4)   *Adding water*

Add water at about room temperature to about half-fill the jar. Stir thoroughly with the glass
rod to remove entrapped air. Replace the cap assembly and fill with water. Check that the
locating marks coincide.

(5) *Removing air and topping up*

Shake the pycnometer, holding one finger over the hole in the conical cap, and allow any further air to escape and froth to disperse. Allow to stand for 24 h if this appears to be necessary. Top up with water to the hole. Entrapped air is a major source of error in this test.

By holding a finger over the hole in the cap, the pycnometer can be rolled on the bench to provide additional agitation.

(6) *Weighing*

Carefully dry the pycnometer on the outside, and weigh it with the water and soil to 0.5 g (W_3).

(7) *Filling with water*

Empty the pycnometer, wash it out thoroughly, and fill with water at room temperature up to the hole in the cap. Check that the screw cap locating marks coincide when refitting the cap.

(8) *Weighing*

Dry the pycnometer on the outside and weigh to the nearest 0.5 g (W_4).

(9) *Repeat test*

Repeat stages (3)–(6) using the other portion of the sample.

(10) *Calculation*

Calculate the specific gravity, G_s, of the soil particles from the equation

$$G_s = \frac{W_2 - W_1}{(W_4 - W_1) - (W_3 - W_2)}$$

where W_1 = mass of pycnometer; W_2 = mass of pycnometer and soil; W_3 = mass of pyncometer, soil and water; W_4 = mass of pycnometer full of water only.

Calculate the average of the two results obtained. If the two results differ by more than 0.05, repeat the test.

(11) *Results*

Report the specific gravity to the nearest 0.05. The method is reported as the pycnometer jar method.

3.7   LIMITING DENSITIES

3.7.1   Scope

Tests to determine the limiting densities of granular soils have not yet been specified in the British Standard. The procedures generally accepted for practical purposes have been based on those proposed by Kolbuszewski (1948a). Other investigators since then, especially in the USA (for example, Selig, 1963), have used more elaborate equipment, such as vibrating tables for measuring the maximum density. The tests described here require only a vibrating hammer and standard soil-testing apparatus, and they give results which are reliable enough for most engineering applications.

Kolbuszewski's tests were developed with clean sands — that is, sand particle sizes from 0.06 mm to 2 mm containing no silt or clay. Silty sands and silts cause difficulties in the maximum density test due to segregation and 'boiling out' of the finest particles. This results in a density lower than the true maximum density being measured; in some cases measured 'maximum' densities less than the dry densities corresponding to optimum moisture contents in a standard compaction test have been reported. The procedure suggested here (Section 3.7.3) for silty soils may not, perhaps, give the highest density which it is possible to achieve, but it can generally be relied upon to give a repeatable result at a high density, and to avoid the above anomaly.

For the determination of the minimum density, slow pouring of dry sand in air through a funnel is often recommended. However, slow pouring in air does not result in the lowest possible density, as this allows each grain time to drop into voids between other grains. Minimum density is achieved when well-dispersed grains of dry sand fall with a high intensity through a small drop so that they immediately 'lock' into position in an open structure caused by the upward movement of air they displace. This is the basis of the dry shaking test (Section 3.7.4).

When pouring in water, the terminal velocities of falling particles are much less than in air and a slower rate of pouring is required. Too high a rate of deposition would cause excessive turbulence, resulting in possible disturbance of particles already deposited. The deposition-in-water procedure (Section 3.7.5) enables a steady rate of deposition to be maintained.

3.7.2 Maximum Density (Sands)

This test is suitable only for sands containing very little or no silt, and consisting of particles which are not susceptible to crushing. It makes use of some of the apparatus listed in BS 1377:1975, Test 14, the vibrating-hammer method for the determination of the dry density/moisture content relationship (see Chapter 6). The maximum density is determined by compacting the soil, submerged under water, into a standard compaction mould with a vibrating hammer. The procedure is similar to that recommended by Kolbuszewski.

APPARATUS

(1)   Electric vibrating hammer, such as a Kango hammer of the type referred to in Section 6.5.4 (Fig. 3.19).
(2)   Steel tamper for attachment to the vibrating hammer, with a circular plate 100 mm diameter (see Fig. 3.20).
(3)   Cylindrical metal mould (compaction mould) 105 mm internal diameter and 115.5 mm high, with detachable base-plate and extension collar (see Fig. 6.7).
(4)   Balance 10 kg capacity, reading to 1 g.
(5)   Watertight container large enough to take the compaction mould, standing on a firm base. If an ordinary galvanised bucket is used, place it on a block so that it is supported by the bottom and not just on the flange.
(6)   Metal tray, say 600 mm square with sides 80 mm deep.
(7)   Stop clock.
(8)   Straight-edge and small tools.
(9)   Oven and moisture-content apparatus.

PROCEDURAL STAGES

(1)   Prepare mould.
(2)   Assemble mould.
(3)   Place in water.
(4)   Place layer of soil in mould.
(5)   Compact by vibration under water.
(6)   Repeat stages (4) and (5) on second layer of soil.

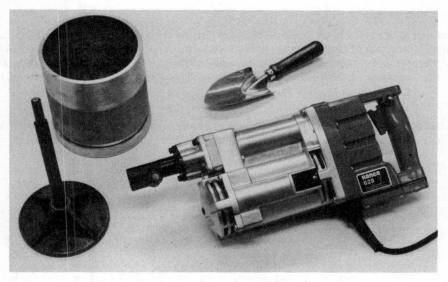

Fig. 3.19   Electric vibrating hammer (Kango)

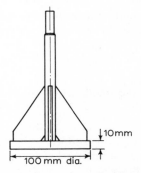

Fig. 3.20   Tamper for maximum density test on sands

(7)   Repeat stages (4) and (5) on third layer of soil.
(8)   Remove and allow free water to drain from sample.
(9)   Trim sample flush with mould.
(10) Weigh.
(11) Extrude sample.
(12) Measure moisture content.
(13) Calculate.
(14) Report results.

PROCEDURE

(1)   *Preparation of sample*

About 4 kg of soil of sand size is required. If the sample contains a few pieces of fine gravel up to 6.3 mm, it does not matter, but any larger pieces should be broken to pass a 6.3 mm sieve, or removed. If the soil is dry, it should be thoroughly mixed with enough water to saturate the grains, and stirred to remove entrapped air.

(2)   *Assembly of mould*

Clean and dry the mould body, base-plate and extension collar. Check and record the diameter and length of the mould body, using calipers, to 0.1 mm. Apply a thin coat of oil to the internal faces of all three components. Attach the base-plate firmly to the mould, and weigh to 1 g (m_1). Attach the extension collar to the mould body.

(3)   *Placing in water*

Lower the mould assembly into the water container and add water to about the level of the top of the mould body, both inside and outside the mould. See that the container is on a firm solid base.

(4)   *Placing soil in mould*

Add soil to nearly half-fill the mould, and spread it level. The amount of soil should be such that when compacted it will fill the mould to about one-third.

(5)   *Compaction of first layer*

Place the circular tamper on the sample and compact with the vibrating hammer for 5 min, timed with the stop-clock, keeping the stem vertical. Throughout this time apply a steady downward pressure to give a force of about 350 N on the sample. The pressure required is greater than that necessary to prevent the hammer bounding on the soil. It may be checked by applying the vibrating hammer (without vibration) on to a platform scale and pressing down until a reading of about 35 kg is registered. With practice one can judge the feel of the correct pressure.

(6)   *Second layer*

Lift out the vibrating hammer, add a second layer of soil equal to the first and repeat stage (5).

(7)   *Third layer*

Add a third layer of soil and repeat stage (5). Add water to the container, if necessary, to ensure that the soil being compacted remains submerged. The final level of the soil should be about 5 mm above the top of the mould body (see Fig. 3.21).

(8)   *Removal of mould*

Remove the mould containing the soil, wipe off loose soil from the outside and allow free water to drain from the sample.

(9)   *Trimming sample*

Remove the extension collar carefully, and trim off the compacted soil level with the top of the mould.

(10)   *Weighing*

Weigh soil with mould and base-plate to 1 g (m_2).

(11)   *Extrusion of sample*

Take off the base-plate, and remove the soil from the mould either with an extruder or by breaking it up. Place the sample in a tray.

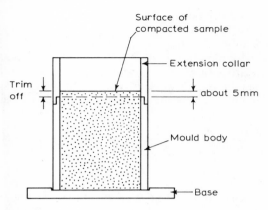

Fig. 3.21   Soil in mould after compaction

(12)   *Measurement of moisture content*

Weigh the whole sample, oven dry it and weigh it dry, to determine the moisture content. For a sandy soil this procedure is preferable to taking small moisture-content specimens.

(13)   *Calculations*

The volume of the standard compaction mould is exactly $1000\,\text{cm}^3$ if the dimensions are exactly as shown in Fig. 6.7. The wet density of the soil in the mould is equal to

$$\frac{m_2 - m_1}{1000} \quad \text{g/cm}^3 \;(=\text{Mg/m}^3)$$

If the measured moisture content is $w\%$, the maximum dry density is given by

$$\rho_{D_{\max}} = \left(\frac{m_2 - m_1}{1000}\right)\left(\frac{100}{100 + w}\right) \quad \text{Mg/m}^3$$

If the dimensions are not standard (for instance, owing to wear of the mould), then its volume ($V\,\text{cm}^3$) is given by

$$V = \frac{\pi D^2 L}{4000} \quad \text{cm}^3$$

where D = mean diameter of mould and L = mean length of mould.
The maximum dry density, $\rho_{D,\max}$, is given by

$$\rho_{D,\max} = \left(\frac{m_2 - m_1}{V}\right)\left(\frac{100}{100 + w}\right) \quad \text{Mg/m}^3$$

(14)   *Results*

The maximum dry density is reported to the nearest $0.01\,\text{Mg/m}^3$, and the fact that the vibrating-hammer procedure was used is recorded.

3.7.3   Maximum Density — Silty Soils

For the reasons given in Section 3.7.1, the above method has been found to be unsatisfactory for soils containing appreciable amounts of silt. It is suggested that a dynamic compaction test, based on the British Standard 'heavy' (4.5 kg rammer) compaction test, be used instead to establish a density approaching the 'maximum' density. As with the vibrating-hammer test (Section 3.7.2), this method is not suitable for soils containing particles susceptible to crushing.

APPARATUS

As for Test 13 of BS 1377:1975 (see Chapter 6, Section 6.5.3).

PROCEDURE

(1)   Carry out the test described in Section 6.5.3 to determine the optimum moisture content and corresponding dry density for this degree of compaction. (Compaction in five layers, 27 blows per layer, using 4.5 kg rammer with 450 mm drop.)
(2)   Take about 5 kg of the soil and adjust the moisture content to about 2% less than the optimum determined in stage (1).
(3)   Compact the soil into the mould as for the 'heavy' compaction test, but applying 80 blows instead of 27 on each of the five layers.
(4)   Measure the resulting density and moisture content.
(5)   Calculate the dry density, and report to the nearest 0.02 Mg/m^3 as an approximation to the maximum dry density.

The application of 80 blows per layer is suggested as a practical expedient, because with some soils there has been found to be very little increase in density when the number of blows is increased to 100 or 150 per layer. However, some soils may require a greater compactive effort to achieve the highest possible density, and this can be assessed only on the basis of experience. If appreciably greater effort is required, the moisture content may need to be more than 2% below the optimum determined from the 'heavy' compaction test. The moisture content used should be close to the 'optimum' for the degree of compaction applied, and optimum moisture content decreases with increasing compactive effort. (See Fig. 6.3.)

ALTERNATIVE PROCEDURE

A more satisfactory method for determining the maximum density of sands, including those containing fines, may be to use the apparatus referred to in Section 6.6.6, but this equipment is not likely to be commercially available until the proposed new British Standard has been published. The same apparatus would also be suitable for the maximum-density test on gravelly soils described in Section 3.7.6.

3.7.4   Minimum Density—Dry

This test, like the one which follows it, was devised by Kolbuszewski. It is very simple, and gives a density which is almost as low as the minimum density obtained by pouring in water without entrapped air. It is widely accepted as a 'standard' test for sands.

APPARATUS

(1)   Glass graduated cylinder, 2000 cm^3, approximately 75 mm diameter.
(2)   Rubber bung to fit the cylinder.
(3)   Balance reading to 0.1 g.
(4)   Elastic band to fit around the cylinder.

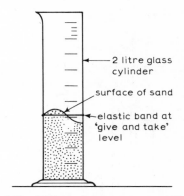

2 litre glass
cylinder

surface of sand

elastic band at
'give and take'
level

Fig. 3.22   Minimum density test on sand (dry)

PROCEDURE

(1)   Weigh out 1000 g of oven-dried sand, and place in the cylinder. Fit the bung.
(2)   Shake the cylinder by inverting it a few times, to thoroughly loosen the sand.
(3)   Turn it upside down, pause until all the sand is at the top end, then very quickly turn it back again and stand it on the bench without jarring it.
(4)   If the resulting surface of the sand is level, record the volume of sand on the graduated scale to the nearest 10 ml. If the surface is not level, adjust the elastic band around the cylinder to the best give-and-take level, and record the volume at that level. Do not shake or jolt the cylinder while this is being done. (See Fig. 3.22.)
(5)   Repeat about six times.
(6)   Take the highest volume reading V cm^3 and calculate the minimum dry density from the equation

$$\rho_{D,\min} = \frac{1000}{V} \quad \text{Mg/cm}^3$$

Report the result to the nearest 0.02 Mg/m^3, and the procedure as the dry shaking method.

3.7.5 Minimum Density — In Water

APPARATUS

The apparatus for this test consists essentially of a 2000 cm^3 measuring cylinder (as in Section 3.7.4) and a funnel, arranged as shown in Fig. 3.23. The top of the bend in the outlet tube should be only a little below the surface of water in the funnel at the start. The end of the outlet tube should not be below the underside of the bung in the cylinder, as otherwise it will act as a siphon.

PROCEDURE

(1)   Set up the apparatus shown in Fig. 3.23, ensuring that the bung is a watertight fit.
(2)   Weigh out 100 g of oven-dry sand, and place it in water in the funnel above the stopper.
(3)   Thoroughly stir the sand and water to displace air bubbles.
(4)   When the sand has settled down, lift the cork with the wire handle to allow the sand to trickle steadily into the cylinder. At the same time open the outlet tap to allow displaced water to escape.

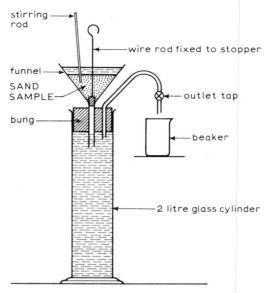

Fig. 3.23   Apparatus for minimum density test on sand in water

(5)   When the sand has settled in the cylinder, observe the volume reading (V cm^3) at its surface. If the sand contains silt, it will take 30 min or more to settle so that the surface can be seen. The level of the top of the silt may be taken as the volume reading, because the effect of the silt in unsegregated loose soil is to hold the sand grains apart as well as to occupy the voids between them. Measure the thickness, or estimate the volume, of the silt.

(6)   Calculate the minimum dry density $\rho_{D,\min}$ from the equation

$$\rho_{D,\min} = \frac{1000}{V} \quad \mathrm{Mg/m^3}$$

Report the result to the nearest 0.02 Mg/m^3, and the procedure as the pouring-in-water method. Any loss of fines through the outlet should be reported, as well as the approximate volume of segregated silt.

3.7.6   Gravelly Soils

An indication of the possible range of densities of a gravelly soil may sometimes be required. The principles used for the tests described in Sections 3.7.2 and 3.7.4 may be used, with the modifications described below to allow for the larger grain size and greater quantity of material necessary. The tests are suitable for material passing a 20 mm sieve. If larger stones are present, they should be broken down to about that size.

MAXIMUM DENSITY

The apparatus required is that used for Test 14 of BS 1377:1975 (see Section 6.5.4). The main items are:

Cylindrical metal mould (CBR mould), 152 mm diameter and 127 mm high, with base-plate and extension collar.
Tamper with 145 mm diameter plate for the Kango hammer.

Watertight container large enough to take the CBR mould.
Other items are as used for the maximum density tests on sands (Section 3.7.2).

The procedure is similar to that used for sands, except that the soil is placed in the mould in five approximately equal layers instead of three. Each layer should be compacted for 5 min.

When the soil is being levelled off to the top of the mould body, any voids resulting from the removal of stones should be filled with fine material pressed well in.

Calculations are as for the test on sands. If the mould is of the standard dimensions, its volume is 2305 cm^3 and the compacted wet density is equal to

$$\frac{m_2 - m_1}{2305} \quad \text{Mg/m}^3$$

If the dimensions differ from the standard, the volume must be calculated.

Before weighing, leave the sample to drain long enough to allow the 'free' water to escape. The whole sample should be dried after weighing, to determine the moisture content. Calculate the dry density, $\rho_{D,\text{max}}$, as for sands.

Report the result to the nearest 0.01 Mg/m^3, stating that the vibrating hammer with a CBR mould was used. The percentage of particles larger than 20 mm should also be reported, and whether they were removed, or broken down and put back with the sample.

MINIMUM DENSITY

A glass measuring cylinder is impracticable for gravelly soils. The CBR mould referred to above, or a rigid metal mould of about the same size, is used instead.

(1) Measure the mould, and weigh the mould and base-plate to the nearest 5 g (m_1).
(2) Thoroughly mix the oven-dried soil, to ensure an even distribution of the larger particles

(3) Place it loosely so as to fill a bucket or similar container at least 50% larger than the mould being used.
(4) Place the mould on a large tray on the floor.
(5) Empty the bucketful of soil rapidly into the mould from a height of about 0.5 m, so that the mould is filled almost instantaneously. The surplus soil will overflow on to the tray.
(6) Level the soil to the surface of the mould by removing it carefully so as not to disturb the soil in the mould. Check the top surface with the straight-edge.
(7) Weigh soil, mould and base-plate to the nearest 5 g (m_2).
(8) Repeat stages (2)–(7) about six times.
(9) Use the lowest value of m_2 to calculate the minimum dry density $\rho_{D,\text{min}}$ from the equation

$$\rho_{D,\text{min}} = \frac{m_2 - m_1}{V} \quad \text{Mg/m}^3$$

where V is the volume of the mould (cm^3).
    If the mould is the standard size, $V = 2503$ cm^3.
(10) Report the result to the nearest 0.02 Mg/m^3, and the procedure as rapid dry pouring. State the percentage of particles larger than 20 mm in the original sample, and whether they were removed or broken down and put back.

BIBLIOGRAPHY

Abbott, A. F. (1969). *Ordinary Level Physics*, 2nd edition (SI units), Chapter 12, Heinemann, London
American Society for Testing and Materials (1972). Designation D2049-69. 'Standard method of test for relative density of cohesionless soil'. ASTM, Philadelphia

BS 733:1965, 'Density bottles, British Standards Institution, London

Kaye, G. W. C. and Laby, T. H. (1973). *Tables of Physical and Chemical Constants*, 14th edition. Longmans, London

Kolbuszewski, J. (1948a). 'An experimental study of the maximum and minimum porosities of sands'. *Proc. 2nd Ind. Conf. Soil Mech. and Found. Eng., Rotterdam*, Vol. 1.

Kolbuszewski, J. (1948b). 'General investigation of fundamental factors controlling loose packing of sands'. *Proc. 2nd Int. Conf. Soil Mech. and Found. Eng., Rotterdam*, Vol. 7.

Lambe, T. W. and Whitman, R. V. (1969). *Soil Mechanics*. Wiley, New York.

Loos, W. (1936). 'Comparative studies of the effectiveness of different methods for compacting cohesionless soil'. *Proc. 1st Int. Conf. Soil Mech. and Found. Eng., Cambridge, Mass., USA*, Vol. III

Pauls, J. T. and Goode, J. F. (1970). 'Suggested method of test for maximum density of noncohesive soils and aggregates'. *ASTM STP 479*.

Selig, E. T. (1963). 'Effect of vibration on density of sand'. *Proc. 2nd Panamerican Conf. on Soil Mech. and Found. Eng., Brazil*, Vol. 1

US Department of the Interior, Bureau of Reclamation (1974). *Earth Manual*, 2nd edition. Test designation E-12. US Government Printing Office, Washington, DC

Yemington, E. G. (1970). 'Suggested method of test for minimum density of noncohesive soils and aggregates'. *ASTM STP 479*

Chapter 4

Particle size

4.1 INTRODUCTION

4.1.1 Scope

A soil consists of an assemblage of discrete particles of various shapes and sizes. The object of a particle size analysis is to group these particles into separate ranges of sizes, and so determine the relative proportions, by dry weight, of each size range.

4.1.2 Test Procedures

Particle size analyses are also referred to as particle size distribution (PSD) tests, sizing tests or mechanical analysis (MA) tests. Two separate and quite different procedures are used, in order to span the very wide range of particle sizes which are encountered. These are the sieving and the sedimentation procedures. Sieving is used for gravel and sand size ('coarse') particles, which can be separated into different size ranges with a series of sieves of standard aperture openings (Section 4.6). Sieving cannot be used for the very much smaller silt and clay size ('fine') particles, so a sedimentation procedure is used instead. Measurements are made either by sampling a suspension with a special pipette (the pipette method, Section 4.8.2), or by determining its density with a special hydrometer (the hydrometer method, Section 4.8.3).

For soils containing both coarse and fine particles, composite tests using both sieving and sedimentation methods have to be carried out if a full particle size analysis is required. Particle size testing can range from a simple sieving test on a 'clean' sand and gravel, to elaborate composite tests on clay–silt–sand–gravel mixtures.

Test procedures for different types of materials are similar in principle but vary in detail, and are described separately. The most difficult type of material to deal with is that referred to as 'boulder clay', which is treated as a special case.

4.1.3 Presentation of Data

Results of a particle size analysis may be presented as a table showing the percentages of particles finer than certain standard sizes. Usually, however, the results are presented graphically to show the percentages finer than any given size, plotted against the particle size on a logarithmic scale. This graphical presentation is known as the particle size distribution (PSD) curve, or the grading curve, for the material, and is described in Section 4.3.4.

There are other ways of presenting particle size distribution data graphically, such as by means of a grain size–frequency curve in which the percentages by mass between certain sizes are plotted against the logarithm of the grain size. These methods are used in other industries, such as powder technology (see, for example, Allen, 1974), but the semi-logarithmic plot to be described is the only procedure generally used for soils.

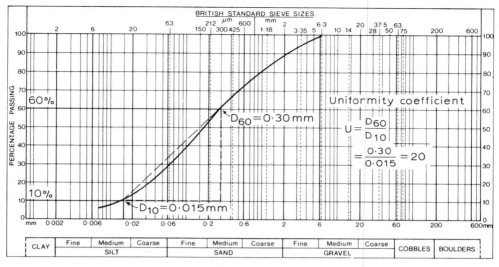

Fig. 4.1   *Particle size distribution chart*

4.2   DEFINITIONS

PARTICLE SIZE ANALYSIS expresses quantitatively the proportions by weight of the various sizes of particles present in the soil.

PARTICLE SIZE is usually given in terms of the equivalent particle diameter:

GRAVEL: particles from 60 mm to 2 mm.
SAND: particles from 2 mm to 0.06 mm.
SILT: particles from 0.06 to 0.002 mm.
CLAY: particles (clay minerals) smaller than 0.002 mm (2 μm).
FINES are particles which pass a 63 μm sieve.
CLAY FRACTION is the percentage of particles smaller than 2 μm, as determined by a standard sedimentation procedure.

QUARTERING   The process of obtaining a small but representative portion from a large sample of cohesionless material. The sample is first divided into quarters by a standard procedure, two of which are retained, which reduces the size of sample by half. The process is repeated until a sample of the required size is obtained.

RIFFLING   A similar process to quartering but using a RIFFLE BOX of appropriate size. However, this term is often used to indicate subdivision of material with or without a riffle box.

SIEVE SIZE   Refers to the aperture size of the sieve mesh — that is, the length of the side of the square openings through which particles pass.

SIEVE DIAMETER   Refers to the diameter of the body of the sieve (the frame) which holds the mesh.

EFFECTIVE SIZE (D_{10})   The particle size for which 10% of particles are finer, and 90% are coarser. It corresponds to $P = 10\%$ on the particle size curve (see Fig. 4.1).

UNIFORMITY COEFFICIENT (U)   The ratio of the 60% particle size to the 10% particle size:

$$U = \frac{D_{60}}{D_{10}}$$

It is a measure of the slope of the line joining these two points (see Fig. 4.1).

FINENESS MODULUS   The sum of the cumulative percentages *retained* on each of a standard set of sieves, divided by 100. It is used as a numerical indication of grading characteristics, and is applied only to concrete aggregates. The relevant sieves are those used for testing concrete aggregates — namely 150, 300, 600 μm; 1.2, 2.4, 4.76 mm aperture.

SPECIFIC SURFACE   The total surface area of particles contained in a unit mass.

FLOCCULATION   The coagulation of particles together in a suspension, giving the effect of larger particles.

DISPERSION   The separation of discrete particles so that they will remain separated and not flocculate, when in suspension.

4.3   THEORY

4.3.1   Limitations

A particle size distribution analysis is a necessary index test for soils, especially coarse soils, in that it presents the relative proportions of different sizes of particles. From this it is possible to tell whether the soil consists of predominantly gravel, sand, silt or clay sizes, and to a limited extent which of these size ranges is likely to control the engineering properties. Particle size curves are of greater value if supplemented by descriptive details such as colour and particle shape, together with grain packing and fabric when observed in the undisturbed state. But engineering behaviour also depends on factors other than size of particles, such as mineral type, structure and geological history, which have a significant effect on engineering properties and cannot be assessed from particle size tests alone.

For the coarse soil fraction the proportions of the various sizes are determined from the sizes of square openings in sieves, whereas for fine soil particles they are based upon diameters of equivalent spheres obtained from a sedimentation analysis, notwithstanding that particles in the fine silt to clay size range are far from spherical. Indeex, many clay minerals consist of thin flat plate-like or elongated particles. The measurement of the true size of particles therefore becomes less accurate towards the fine end of the scale. However, the absolute magnitude of particle size is of less importance than the distribution of sizes as determined by a recognised standard procedure, and this is the sense in which particle size is applied to soils. In particular, the 'clay fraction' (the proportion of material consisting of particles smaller than 0.002 mm) is often used as an index for correlating with other engineering properties.

With some soils it is difficult to define what is meant by an 'individual particle', such as those in which the size of particles depends upon the degree of disaggregation achieved prior to testing. Examples are residual soils (see Chapter 7) and soils containing fragments of weakly cemented sandstone or similar material. With such soils it is necessary to control the extent to which particles are broken down in the preparation stage. For soil types such as peat and weathered chalk the concept of individual particles does not exist in the same sense as for sands, silts and clays.

4.3.2   Grouping by Particle Size

Soils may be divided on the basis of their dominating particle size into six arbitrary categories which are called boulders, cobbles, gravel, sand silt, clay. The size ranges of particles comprising each category, as used in Britain, were originally proposed by Glossop and Skempton (1945). These size ranges, as defined in the British Standard and in Code of Practice CP 2001, are given in Table 4.1, which also shows the subdivision of gravel, sand and silt into coarse, medium and fine sizes. Classification by particle size in other countries may differ from British practice, and a comparison of American (ASTM) systems with the British system is shown diagrammatically in Fig. 4.2.

Table 4.1. CLASSIFICATION BY PARTICLE SIZE (Based on BS 1377:1975

Particle size (mm)		Designation	Test procedure
> 200 200 —		BOULDERS	Measurement of separate pieces
60 —		COBBLES	
20 —	Coarse	GRAVEL	Sieve analysis
6 —	Medium		
2 —	Fine		
0.6 —	Coarse	SAND	
0.2 —	Medium		
0.06 —	Fine		
0.02 —	Coarse	SILT	Sedimentation analysis
0.006 —	Medium		
0.002 —	Fine		
and less		CLAY	

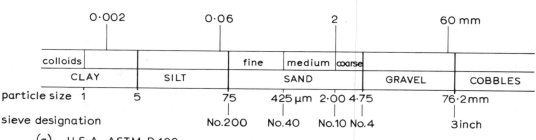

(a) U.S.A. ASTM D422

(b) Great Britain BS 1377 : 1975

Fig. 4.2 *Comparison of systems for classifying particle size ranges of soils*

Usually the boulder and cobble sizes, if present, are removed and measured separately before test, although in some instances it may be necessary to include cobbles in the analysis (see Section 4.6.3). It is the gravel, sand, silt and clay sizes which are generally recognised and tested as 'soils'.

Table 4.2. PARTICLE SIZE, MASS AND SURFACE AREA OF EQUAL SPHERES

Equivalent soil category	Particle size (mm)	Approximate mass of particle (g)	Approximate number of particles per gram	Approximate surface area	
				(mm²/g)	(m²/g)
Small cobble (largest 'soil' particle)	75	590	(1.7/kg)	30	
Coarse sand	1	0.0014	720	2300	
Fine sand	0.1	1.4×10^{-6}	7.2×10^5	23 000	0.023
Medium silt	0.01	1.4×10^{-9}	7.2×10^8	23×10^5	0.23
Clay (smallest size measured)	0.001	1.4×10^{-12}	7.2×10^{11}	2.3×10^6	2.3

4.3.3 Influence of Particle Size

Particles encountered in most soils testing range from a maximum size of 75 mm down to about 0.001 mm, which is the normal limit of measurement in the sedimentation test. The ratio between these extreme sizes is 75 000 to 1.

As the particle size decreases, the number of particles contained in each gram of material increases, in inverse proportion to the cube of the particle size. The mass of each particle decreases in the same ratio. The total surface area of particles per unit mass (known as 'specific surface', and expressed as mm²/g or m²/g) increases in inverse proportion to the particle size. These characteristics are shown in Table 4.2 for the two extreme sizes referred to above, together with three intermediate sizes, assuming spherical particles of specific gravity 2.65. The ratio of the extreme sizes of spherical particles in terms of mass is enormous — about 4.2×10^{14} to 1. There can be few test procedures which embrace such a wide range of values.

Because natural soil particles are rarely spherical, their specific surface is higher than those shown in Table 4.2. Thus fine sand particles may have a specific surface of about 0.03 m²/g. Individual particles of clay are flat and plate-like, which gives them specific surfaces many times greater than for spheres, depending on the type of clay mineral. Typical values for three of the commonest types are approximately as follows (Lambe and Whitman, 1969):

Kaolinite	10 m²/g
Illite	100 m²/g
Montmorillonite	1000 m²/g

The extremely high specific surface of clay particles is one of the factors responsible for the cohesive properties of clay.

4.3.4 Particle Size Distribution Curve

The type of form recommended in the British Standard for presenting particle size distribution is shown in Fig. 4.1. The advantage of plotting particle size data on a standard chart of this kind is that it enables engineers to recognise instantly the grading characteristics of a soil far more easily than from tabulated figures. Moreover the position of a curve on the chart indicates the fineness or coarseness of the grains; the higher and further to the left the curve lies, the finer the grains, and vice versa. The steepness, flatness and general shape indicate the distribution of grain sizes within the soil. Examples are discussed in Sections 4.4.2 and 4.4.3.

While this type of chart is used in the UK and many European countries, the practice in the USA until recently has been to use the chart the other way round, with cobbles and gravel to the left and clay to the right. Particle size curves so drawn are the mirror image of those plotted according to British practice.

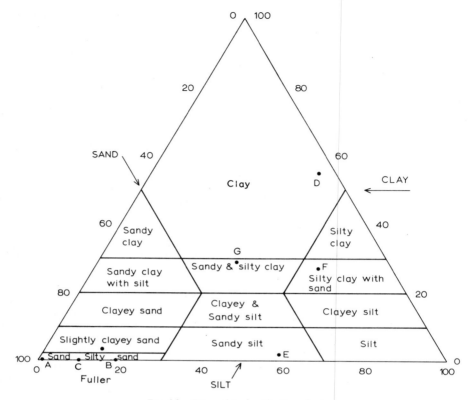

Fig. 4.3   *Triangular classification chart*

In Fig. 4.1 is also shown the derivation of the 'effective size' (D_{60}) and the 'uniformity coefficient' (U) from a typical grading curve.

4.3.5   Triangular Classification Chart

A triangular classification chart is not often used but it can be convenient for comparing clay–silt–sand mixtures on the basis of the proportions of each constituent (see Fig. 4.3). Each side of the triangle is divided into 100 parts, representing the percentages of clay, silt and sand. A point within the triangle indicates the percentages of these three constituents, all of which total 100%

The triangular chart originally introduced by the US Bureau of Reclamation used the term 'loam', which is not recognised by the British Standard classification system. The version shown in Fig. 4.3 uses the British Standard terminology.

This type of chart can also be used to show (clay + silt):sand:gravel, or any other three main constituents of soil, on the three sides, if this is an aid to classification. In any case, the three percentages shown must always add up to 100.

4.4   APPLICATIONS

4.4.1   Engineering Classification

The analysis of soils by particle size provides a useful engineering classification system for soils, provided that the limitations referred to in Section 4.3.1 are taken into account.

Fig. 4.4   *Samples of granular soils: (a) uniformly graded, (b) well graded, (c) poorly graded (gap graded)*

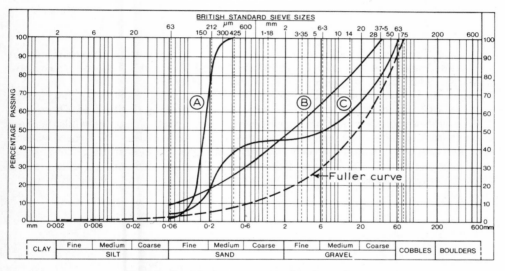

Fig. 4.5   *Particle size curves — sands and gravels*

For engineering purposes in Britain, soils are divided on the basis of particle size into the six categories shown in Table 4.1. Each heavy horizontal dividing line in the 'Designation' column separates materials of significantly differing engineering properties.

In practice, most natural soils do not fall entirely within one of these main size ranges, but consist of a mixture of two or more of these categories. Nevertheless grading curves provide a means by which soils can be classified and their engineering properties broadly assessed.

4.4.2   Classification of Sands and Gravels

Grading curves enable sands and gravels to be identified as being of three main types, based on the distribution of particle sizes, as follows. Typical examples are shown in Fig. 4.5.

(1)   *Uniform soils*, in which the majority of the grains are very nearly the same size (Fig. 4.4a). The grading curve is very steep, as shown by Curve A in Fig. 4.5, which represents a uniform sand. The uniformity coefficient is not much more than 1.0 (the lowest theoretically possible). A synonymous description is 'narrowly graded'.

Table 4.3. EFFECTIVE SIZES AND UNIFORMITY COEFFICIENTS

Grading curve	Effective size, D_{10} (mm)	D_{60} (mm)	Uniformity coefficient, D_{60}/D_{10}	Description
A	0.12	0.18	1.5	Uniform fine SAND
B	0.070	4.5	64	Well-graded silty SAND and GRAVEL
C	0.14	15.	107	Poorly graded fine to medium SAND and GRAVEL
Fuller curve	0.66	24	36	'Ideal' grading
D	–	0.0025	–	CLAY
E	0.0051	0.060	12	Sandy SILT
F	–	0.019	–	Sandy and silty CLAY
G	0.0013 (estimated)	2.0	1500 approx.	Gravelly sandy silty CLAY (boulder clay)

(2)   *Well-graded soils*, containing a wide and even distribution of particle sizes (Fig. 4.4b). A well-graded silty sand and gravel is shown by curve B in Fig. 4.5. The smooth concave upward grading curve is typical of well-graded material.

The dashed curve in Fig. 4.5 represents a theoretical grading of a material in which the particles fit together in the densest possible state of packing (Fuller and Thompson, 1907). In this idealised material the largest particles just touch each other, while there are enough intermediate-size particles to occupy the voids between the largest without holding them apart; smaller particles occupy voids between intermediate sizes; and so on. The arrangement is illustrated diagrammatically in Fig. 3.6(c). The Fuller grading curve has the characteristic smooth shape referred to above, and is derived from the equation

$$P = \sqrt{\left(\frac{D}{D_{max}} \times 100 \right)}$$

where P is the percentage (by mass) of particles smaller than diameter D and D_{max} is the maximum particle size (75 mm in the example shown).

(3)   *Poorly graded soils*, deficient in certain sizes (Fig. 4.4c). The grading curve has two distinct sections separated by a near-horizontal portion, as shown by curve C in Fig. 4.5. This is described as a gap-graded material, and in natural soils the deficiency usually occurs in the coarse sand–fine gravel range. The term 'poorly graded' can also be applied to any soil (including uniformly graded soil) which does not comply with the description 'well graded'.

Points A, B, C corresponding to the above grading curves are plotted on the triangular chart, Fig. 4.3. For B and C the gravel fraction is ignored and percentages of clay, silt and sand are enhanced to express them as percentages of material smaller than 2 mm. Effective sizes, and uniformity coefficients, are summarised in the upper part of Table 4.3.

4.4.3   Classification of Clays and Silts

Soils consisting entirely of clay, or entirely of silt-size particles, are very rarely encountered. Most clays contain silt-size particles, and most material described as silt includes some clay or some sandy material, or both. A clay soil containing quite a high percentage of silt particles could have the properties of clay, and if so, would be described simply as 'clay' (see Chapter 7).

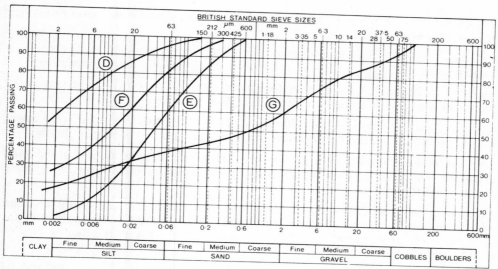

Fig. 4.6   Particle size curves — silts and clays

This is illustrated by the large area designated 'clay' on the triangular chart, Fig. 4.3. Some typical grading curves are shown in Fig. 4.6.

Curve D is described as 'clay', although it consists of 56% clay and 44% silt size particles. However, if the clay sizes (smaller than 2 mm) consist of ground-down silt particles instead of clay minerals, this material would exhibit the properties of silt and would be described as such.

Curve E shows a well-graded soil consisting mainly of silt, with a clay fraction of less than 2%, and is described as 'sandy silt (with a trace of clay)'. Curve F has a mixture of clay, silt and sand. The effect of the clay would predominate, and the soil is described as 'silty clay with sand'. Curve G represents a well-graded soil containing particles of all sizes from cobbles down to clay. This type of soil is found, for example, in glacial till, and is often loosely called 'boulder clay', which in geological terminology is considered to be inaccurate. However, 'boulder clay' is used in this book to mean a soil, usually of glacial origin, which contains enough clay to give it cohesion and is well-graded from clay through to gravel or cobble-size particles. 'Boulder clay' in this sense does not necessarily contain boulders, but it does have the engineering characteristics of a clay. The material represented by curve G could be described as 'gravelly sandy silty clay'.

Points D, E, F representing these three grading curves are plotted in the triangular chart, Fig. 4.3. Point G relates only to the material smaller than 2 mm in the boulder clay. Effective sizes and uniformity coefficients for gradings E and G are summarised in the lower part of Table 4.3. The '10% passing' sizes for gradings D and F are outside the range of the grading chart, so effective size and uniformity coefficient for these two examples cannot be determined.

As indicated in Section 4.3.1, the grading characteristics revealed above do not furnish a complete description of cohesive soils. The physical behaviour of a clay soil is more important than its particle size distribution, and for this the Atterberg limits provide information that is more significant (see Chapter 2).

4.4.4   Engineering Practice

Some applications of particle size analysis in geotechnology and construction are as follows.

Selection of fill materials   Soils used for the construction of embankments and earth dams, in addition to other requirements, are required to be within specified limits as defined by particle size distribution curves. The various zones of an earth dam, for instance, each require different grading characteristics.

Road sub-base materials   Each layer of a road or airfield runway sub-base must comply to a particular grading specification in order to provide a mechanically stable foundation.

Drainage filters   The grading specification for a filter layer must be related in a certain way to the grading of the adjacent ground, or of the next filter layer.

Groundwater drainage   The drainage characteristics of the ground are to a large extent dependent upon the proportion of fines (silt and clay size particles) present in the soil.

Grouting and chemical injection   The most suitable grouting process to be used in soils, and the extent to which the ground can be impregnated, depend mainly on the grading characteristics of the soil.

Concreting materials   Sands and gravels for use as concrete aggregates are divided into various types on the basis of zones on a particle size distribution chart. In the exploration for sand and gravel resources, particle size analysis is the main criterion for selection of sites for potential development.

Dynamic compaction   In some situations poor ground conditions can be improved by dynamic compaction, and a particle size analysis can give an indication of the feasibility of this process.

4.5   PRACTICAL ASPECTS

4.5.1   Choice of Procedure

The procedures to be used for determining the particle size distribution in soils depend upon:
(1)   Size of the largest particles present.
(2)   Range of particle sizes.
(3)   Soil characteristics.
(4)   Stability of soil grains.

Item (1) determines the size of sample required. Item (2) determines the method to be used, such as whether or not a sedimentation test is required. The complexity of the testing is governed by (3) and depends on whether the soil is granular and 'clean' (no fines); predominantly granular with fines; or decidedly cohesive. The initial preparation of the soil for testing is dependent on (4), because special care is needed if particles are easily broken down (see Section 1.5.4).

The procedures to be described are as follows:

Simple dry sieving.
Composite dry sieving.
Wet sieving.
Special procedures for 'boulder clay' types of soil.
Sedimentation by hydrometer measurements.
Sedimentation by pipette analysis.
Composite sieving and sedimentation.

For clean sands and gravels (i.e. those containing no silt or clay) dry sieving can be used. If the material consists of sand sizes only, this can be a straightforward operation on a sample of convenient size. If the sample contains particles of medium to large gravel size, a larger sample will be required initially, but this can be reduced by quartering at some stage of the sieving process. This is referred to as composite sieving. If larger particles (cobbles) are present, a very large sample will be needed, which may be quartered two or three times at suitable stages. The quantity of material required is discussed in Section 4.5.2.

Soils containing silt or clay must first be washed to remove the fine particles which pass a 63 μm sieve. This is known as wet sieving. The retained material can then be dry sieved. Since the fine material is usually run to waste with the washing water, the total quantity of fines can be determined only from the difference between the unwashed and washed masses after drying. Composite sieving is nearly always necessary unless sand is the coarsest material present.

Predominantly clayey and silty soils containing sand are first pretreated for a sedimentation test, and then washed through a 63 μm sieve. Material passing the 63 μm sieve is collected and used for a sedimentation test. The material retained is dry sieved, and the sieving curve added to the sedimentation curve to give the grading curve for the whole sample.

Cohesive soils containing gravel or larger particles require special treatment, which is described separately.

A sieving analysis should not be started without first considering which of the described procedures will be the most appropriate for the type of soil to be tested. When the procedure has been selected, sketch out the sequence of operations as a flow diagram, in the manner shown in Figs. 4.10, 4.13, 4.16, 4.18, 4.21 or 4.24. If the procedure is subsequently modified, amend the flow diagram accordingly. Calculations should be set out in a manner similar to the flow diagram, as shown in Figs. 4.11, 4.14, 4.19, 4.22 or 4.25. The sieve analysis sheet used for recording the masses retained on the sieves (Fig. 4.9) may not be suitable for this purpose except for the simplest type of calculations.

4.5.2 Quantity of Material for Test

For clays, silts and sands the size of particles are such that a sample of about 100 g is sufficient for particle size tests to be representative. However, with gravel-size particles much larger quantities are required for representative results to be obtained. If the size of the sample is too small, the inclusion or exclusion of only a few large particles can influence the whole particle size analysis considerably.

BS1377:1975, in Section 1.5.4.2 (5), gives a guide to the minimum quantity of material to be taken for sieve analysis. These recommendations are given in Table 4.4. This relates to the maximum size of particle present 'in substantial proportion' (more than 10% of the total

Table 4.4. MINIMUM QUANTITIES FOR PARTICLE SIZE TESTS

Maximum size of material present in substantial proportion retained on BS sieve (mm)	Minimum mass of sample to be taken for sieving	Source
Pass 2 mm or smaller	100 g	Based on
6.3	200 g	BS 1377:1975,
10	500 g	Section 1.5.4.2 (5)
14	1 kg	
20	2 kg	
28	6 kg*	
37.5	15 kg	
50	35 kg	
63	50 kg	
75	70 kg	
100	150 kg	Author's suggestion
150	500 kg	
200	1000 kg	

*   The British Standard states 5 kg, but 6 kg is consistent with the linear graph in Fig. 4.7.

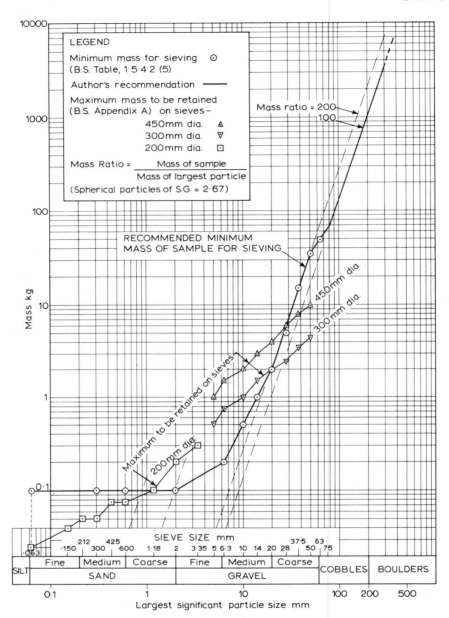

Fig. 4.7   *Minimum quantity of material required for sieving*

sample). These minimum quantities are plotted against the size of particles, to a log scale, in Fig. 4.7. The British Standard table only goes as far as 63 mm — that is, the upper end of the gravel range.

When larger material (cobbles) is present, a useful rule is to start with a sample whose dry mass is not less than 100 times the mass of the largest particle. This relationship is indicated by one of the lines on the graph in Fig. 4.7. It can be seen that the upper ends of the British Standard recommendations follow a parallel line, which corresponds to a sample mass of 200 times the mass of the largest particle. These lines are based on the assumption that particles are spherical, with a specific gravity of 2.67.

A general rule recommended here is indicated graphically by the heavy line in Fig. 4.7. This complies with the requirements of the BS below 63 mm, and above that size represents a mass ratio of about 100.

It is evident that the mass of sample required increases rapidly with increasing particle size in the cobbles range. For particles up to 200 mm across, about 1000 kg (1 tonne) of material is required. For 300 mm particles this increases to 3 tonne. A method of dealing with very large samples of this kind of material is described in Section 4.6.3.

4.5.3   Dispersing Agents

A dispersing agent is used with a soil suspension in water in order to ensure separation, or dispersion, of discrete particles of soil, especially in the silt to clay range. A small quantity of a soluble chemical is added before the sedimentation test, usually in the form of a measured quantity of a prepared stock solution.

Numerous substances have been tried for use as dispersing agents, many of them very successfully for most soils. They include

Sodium polyphosphate	Sodium hydroxide
Sodium tripolyphosphate	Sodium silicate
Sodium hexametaphosphate	Tannic acid
Sodium tetraphosphate	Starch
Sodium oxalate	Trisodium phosphate
Sodium carbonate	Tetrasodium phosphate
Sosium bicarbonate	

For most purposes it has been found that sodium hexametaphosphate (known commercially as Calgon) is one of the most suitable and convenient dispersants. The stock solution recommended by the British Standard is made up of

35 g sodium hexametaphosphate
7 g sodium carbonate
Distilled water to make 1 litre of solution

This stock solution is referred to as the *standard dispersant*, and is added to the water used for sieving or sedimentation in the proportions specified. A fresh stock solution should be prepared each month, because it is unstable, and will not keep for any length of time. The date of preparation should be marked clearly on the container, which should have a tight-fitting stopper.

This general-purpose dispersant is not suitable for lateritic soils (defined in Chapter 7). For these soils trisodium phosphate or tetrasodium phosphate is suggested. The concentration of dispersant necessary is best determined by trial with each particular soil. These two dispersants are not effective with other types of soil.

Incomplete dispersion results in the formation of relatively large crumbs or 'flocs' of soil particles (aggregations of smaller particles) which in a sedimentation test fall relatively rapidly through the water, leaving a clear layer above the suspension. If this occurs, a higher concentration (two or three times the standard) of dispersing agent may need to be used. If this is not effective, a different dispersant may be necessary.

Use of a dispersant is also necessary before wet sieving of soils containing clay or silt. For this purpose the British Standard specifies covering the soil with water containing 2 g per litre of sodium hexametaphosphate, or 50 ml of 'standard dispersant' solution per litre (see Sections 4.6.6 and 4.6.7).

Table 4.5.   METRIC SIEVES

Construction	Aperture size: Full Set (A)	'Standard' set (B)	'Short' set (C)	Suitable sieve diameters 450 mm	300 mm	200 mm
Perforated	75 mm	+		+		
steel place	63	+	+	+		
(square hole)	50			+		
	37.5	+		+	+	
	28			+	+	
	20	+	+	+	+	
	14				+	
	10	+			+	
	6.3	+	+		+	
	5				+	
	3.35	+			+	+
	2	+	+	(+)	+	+
	1.18	+				+
	600 μm	+	+	(+)	+	+
	425					+
Woven wire	300	+				+
	212		+			+
	150	+				+
	63	+	+	(+)	+	+
Lid and receiver	+	+	+	+	+	+
	19 sieves	13 sieves	7 sieves			

Many other test sieves, up to 125 mm and down to 38 μm aperture size, are available. They are manufactured to BS 410:1969'
(+) indicates sieves useful for wet sieving of large samples

4.5.4   Sample Selection and Preparation

The general procedure for the selection and preparation of a small representative sample for test from a large sample is outlined in Section 1.5. Detailed procedures for sample preparation are given in the section describing each type of test, as appropriate.

4.5.5   Selection of Sieves

The complete range of sieves specified by the British Standard is given in Table 4.5. It is not necessary to use every sieve for every test, but the sieves used should adequately cover the range of aperture sizes for each particular soil.

Three selections of sieves are suggested in Table 4.5.

(A)   Full set — i.e. every sieve listed in the BS (19 sieves).
(B)   'Standard' set, which cover most requirements (13 sieves).
(C)   'Short' set — i.e. those sizes which subdivide soil types. These are the sizes indicated by the heavy lines on the particle size sheet (Fig. 4.1), and comprise 63 μm, 20 mm, and their submultiples (7 sieves).

For most purposes the 'standard' set is sufficient to provide a reasonable grading curve, but if the soil is uniformly graded, whether as a whole or over part of the range (giving a steeply inclined grading curve) the use of intermediate sieves will give a better-defined curve. An extra sieve can easily be used after completion of the test if the material retained on each sieve is kept separate.

The largest aperture sieve necessary is the one through which all the material just passes. The mass retained in this sieve is zero, and is the size corresponding to 100% passing. The next smallest aperture sieve is the first one to retain material in the test.

Fig. 4.8   *Test sieves*

Test sieves are made in three standard diameters — namely 450 mm, 300 mm, 200 mm — and typical sieves are shown in Fig. 4.8. The diameter selected should be appropriate to the quantity of material to be sieved. It is usual to riffle the material immediately before changing to sieves of smaller diameter.

4.6 SIEVING PROCEDURES

4.6.1 Simple Dry Sieving (BS1377:1975, Test 7 (B))

SCOPE

Dry sieving is the simplest of all methods of particle size analysis. The apparatus used, the test procedure and the method of calculation are described here in some detail because they are relevant to all other sieving methods.

According to the British Standard, dry sieving may be carried out only on materials for which this procedure gives the same results as the wet-sieving procedure. This means that it is applicable only to clean granular materials, which usually implies clean sandy or gravelly soils — that is, soils containing negligible amounts of particles of silt or clay size. If in doubt about the validity of the dry-sieving method, the wet-sieving procedure (Section 4.6.4) should be followed instead.

If particles of medium gravel size or larger are present in significant amounts, the initial size of the sample required may be such that riffling is necessary at some stage to reduce the sample to a manageable size for fine sieving. The procedure is then referred to as 'composite sieving', which is described in Section 4.6.2.

APPARATUS

(1)   Test sieves to BS 410. The aperture sizes specified in BS 1377:1975 are listed in Table 4.5, which also indicates the diameters in which these sieves are normally kept available for use. A set of each diameter requires a lid and receiver. It is advantageous to keep one set of sieves of each diameter for dry sieving and another set for wet sieving. Comments on selection of sieves are given in Section 4.5.5.
(2)   Mechanical sieve shaker (optional), preferably with timing device.
(3)   Balances appropriate to the quantities of material to be used.
(4)   Riffle-box.
(5)   Standard drying oven, 105–110 °C.

SIEVE ANALYSIS OF SOIL (~~WET*~~/DRY* SIEVING)

Job: *2567*   Operator: *A.B.Smith*
Sample No.: *3/12*   Date: *6.12.78*
Site: *Elmbridge*   Depth *3.75m*   Description: *Light brown*
Total mass of dry sample: *500* g   *fine to medium*
                                                   sand

BS test sieve size	Mass retained	Mass retained	Total Mass retained	Per cent retained	Total passing	Remarks
	g	g		%	%	
75 mm						
63 mm						
50 mm						
37.5 mm						
28 mm						
20 mm						
Passing 20 mm						

Riffled Sample Passing 20 mm

	g	g		%	%	
14 mm						
10 mm						
6.3 mm						
Passing 6.3 mm						

Riffled Sample Passing 6.3 mm *500g*

	g			%	%	
5 mm						
3.35 mm						
2 mm						
1.18 mm	0			0	100	
600 micron	20			4	96	
425 micron						
300 micron	170			34	62	
212 micron						
150 micron	235			47	15	
63 micron	71			14.2	0.8	
Passing 63 micron	3.5			0.7		
Total	499.5			99.9		

* *Delete the inappropriate word*

Fig. 4.9   Sieve analysis data sheet

(6)   Sieve brushes, double-ended, brass or nylon bristles.
(7)   Metal trays.
(8)   Rubber pestle and mortar.
(9)   Scoop, and miscellaneous small tools.
(10) Sieve analysis work sheet (Fig. 4.9).

PROCEDURAL STAGES

(1)   Select and prepare test specimen.
(2)   Oven dry, cool, weigh.

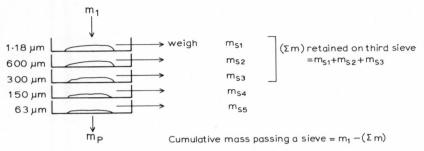

Fig. 4.10 *Sequence diagram, simple dry sieving*

(3) Select sieves.
(4) Pass through sieves.
(5) Weigh each size fraction.
(6) Calculate cumulative percentages passing each size.
(7) Plot grading curve.
(8) Report results.

The sieving process is illustrated diagrammatically in Fig. 4.10.

PROCEDURE

(1) *Selection*

The specimen to be used for the test is obtained from the original sample by riffling, or by sub-division using the cone-and-quarter method (Section 1.5.5). The appropriate minimum quantity of material depends upon the maximum size of particles present, and is indicated in Table 4.4. See also Section 4.5.2.

(2) *Preparation*

The specimen is placed on a tray and is allowed to dry, preferably overnight, in an oven maintained at 105–110 °C. After drying to constant weight, the whole specimen is allowed to cool, and is weighed to an accuracy within 0.1% or less of its total mass (m_1).

(3) *Selection and assembly of sieves*

The sieves to be used are selected to suit the size of sample and type of material, as discussed in Section 4.5.5. In this example five sieves from the 'standard' set (Table 4.5) have been used.

Sieve frames must not be out of true, and should fit snugly one inside the other, to prevent escape of dust. Sieves are nested together with the largest aperture sieve at the top, and a receiving pan under the smallest aperture sieve at the bottom.

(4) *Sieving*

The dried soil sample is placed in the topmost sieve and is shaken for long enough for all particles smaller than each aperture size to pass through. This can be achieved most conveniently by using a mechanical sieve shaker, as described in paragraph (A) below. If a shaker is not available, sieving can be done by hand, using each sieve separately, as described in paragraph (B). General notes on sieving procedures which apply to either method are given in paragraph (C).

(A) *Use of mechanical shaker* The whole nest of sieves with receiving pan is placed in the shaker; the dried soil is placed in the top sieve, which is then fitted with the lid; and the sieves are securely fastened down in the machine (see Figs. 1.26 and 1.27). Agitation in the shaker

Table 4.6.   MAXIMUM MASS TO BE RETAINED ON EACH SIEVE

	Maximum mass		
Sieve aperture	*450 mm diameter sieves (kg)*	*300 mm diameter sieves (kg)*	*200 mm diameter sieves (g)*
50 mm	10	4.5	
37.5	8	3.5	
28	6	2.5	
20	4	2.0	
14	3	1.5	
10	2	1.0	
6.3	1.5	0.75	
5	1.0	0.5	
3.35			300
2			200
1.18			100
600 μm			75
425			75
300			50
212			50
150			40
63			25

should be for a minimum period of 10 min. Some shakers have a built-in timing device which can be preset to switch off the motor automatically after the desired period.

(B)   *Hand sieving*   The largest aperture sieve to be used is fitted with a receiver, and the dried sample is placed on the sieve. The lid should be fitted to prevent escape of dust. The sieve must be agitated by shaking so that the particles roll in an irregular motion, and until no more particles pass through the openings.

The material in the receiver is transferred to a tray and the receiver is fitted to the next sieve in the series. The process is repeated, using the material from the tray, and so on through all the sieves to be used, down to and including the 63 μm sieve.

The foregoing procedure is the ideal, but in practice several sieves can often be nested together for shaking.

(C)   *Notes on procedure*   Whether shaking is carried out in a machine or by hand, it is essential to ensure at each stage that sieving is complete. On the other hand, an unduly long period of sieving must be avoided, because this would give particles additional opportunity to pass through any openings which may be slightly oversize. The British Standard specifies a minimum sieving time of 10 min, but a maximum time of 15 or 20 min should be standard practice in order to obtain consistent results.

The material retained on each sieve should be examined to make sure that only individual particles are retained. Any agglomerations of particles not naturally cemented together should be broken down with a rubber pestle in a mortar and resieved. On the larger aperture sieves individual particles may be assisted through by hand placing, but they must not be pushed through.

Sieves must not be overloaded. The mass retained on each sieve must not exceed the masses given in Table 4.6. To prevent overloading, a large test sample should be split into two or more parts before sieving. The material on an individual overloaded sieve should be split into parts each not exceeding the mass referred to in Table 4.6 and resieved separately.

With continual use, abrasion causes wear of the sieve material (whether metal plate or woven wire fabric), resulting in the apertures increasing in size. A periodic check should be made by sieving an appropriate soil through the sieves and comparing the results with those obtained using new or little-used sieves. If the cumulative percentage retained on any sieve differs by more than, say, 0.5% as calculated, the worn sieve should be discarded and replaced by a new one.

(5) *Weighing*

The material retained on each sieve is transferred in turn to the pan of a suitable balance, or to a weighed container. Any particles lodged in the apertures of the sieve should be carefully removed with a sieve brush, the sieve being first placed upside-down on a tray or a clean sheet of paper. These particles are added to those retained on the sieve.

Weighing of each size fraction should be to an accuracy of at least 0.1% of the total initial test sample mass. The masses retained (m_{s1}, m_{s2}, etc.) are recorded against the sieve aperture size on the particle size test work-sheet (Fig. 4.9). The mass (m_F) passing the 63 μm sieve is also measured and recorded.

(6) *Calculations*

In order to draw up a particle size distribution curve, or to tabulate the data, it is necessary to calculate the cumulative percentage (by mass) of particles finer than each sieve aperture size — that is, passing each sieve. This can be done in two ways.

(A) *British Standard Method*   The mass retained on each sieve is first calculated as a percentage of the total initial mass m_1, from which the cumulative percentages passing each sieve are derived.

Let m_{s1} denote the mass retained on the first sieve used; m_{s2} that on the second sieve used; and so on. The percentage retained on the first sieve, denoted by R_1, is given by

$$R_1 = \frac{m_{s1}}{m_1} \times 100$$

The percentage passing this sieve, P_1, is given by

$$P_1 = 100 - R_1 = 100 - \left\{ \frac{m_{s1}}{m_1} \times 100 \right\}$$

The percentage retained on the second sieve is

$$R_2 = \frac{m_{s2}}{m_1} \times 100$$

and the percentage passing is

$$P_2 = P_1 - R_2 = P_1 - \left\{ \frac{m_{s2}}{m_1} \times 100 \right\}$$

Similarly,

$$P_3 = P_2 - R_3$$

and so on.

Note that each percentage retained, R, must be subtracted from the previous percentage passing, P, and not from 100, because we are calculating cumulative percentages.

(B) *Alternative Method*   The cumulative mass passing each sieve is calculated first, from which the percentages passing are derived.

The mass retained on the first sieve is denoted by m_{s1}. The mass *passing* the first sieve $= m_1 - m_{s1}$. The percentage passing the first sieve is given by

$$P_1 = \frac{(m_1 - m_{s1})}{m_1} \times 100$$

The mass passing the second sieve $= m_1 - m_{s1} - m_{s2}$. The percentage passing the second sieve is given by

$$P_2 = \frac{m_1 - (m_{s1} + m_{s2})}{m_1} \times 100$$

and so on.

The percentage passing any subsequent sieve can be written as

$$P = \frac{m_1 - \Sigma m}{m_1} \times 100$$

where Σm denotes the sum of the masses retained on all sieves down to and including the one in question: $\Sigma m = m_{s1} + m_{s2} + \dots$, etc.

The calculated mass passing the last sieve should be equal, or very nearly equal, to the mass collected in the receiving pan. If this is denoted by m_p, the percentage of fines, P_F, passing the last sieve is

$$P_F = \frac{m_p}{m_1} \times 100$$

Advantages of method (B) are:

(1)   Comparison of the calculated and measured masses passing the last sieve provides a check on both the weighings and the arithmetic.
(2)   Observed readings, instead of calculated percentages, are successively subtracted. There is a tendency to round off the latter, and this could lead to cumulative errors.

   Both calculation procedures are shown by the examples given in Fig. 4.11, in which the arithmetic has been made simple. The discrepancy of 0.5 g between the calculated and weighed mass passing the 63 μm sieve is insignificant. If the difference is appreciable, the calculations, and if necessary the weighings, should be rechecked.

(7)   *Plotting*

The special graph sheet referred to in Section 4.3.4 is used for plotting the particle size distribution curve. On this sheet the standard sieve sizes are marked by vertical dashed lines. The percentage smaller than any given size (i.e. the percentage passing each sieve) is plotted as the ordinate, to a linear scale, against 'sieve aperture size', and the points are connected with a smooth curve or by straight lines. The curve from the example given above is shown in Fig. 4.12, and the calculated percentages passing are shown against each plotted point. Note that the first point plotted represents 100% passing.

   Several particle size curves may be drawn on one sheet, but different symbols should be used for both the plotted points and the connecting lines. Each curve should be clearly labelled, as in the examples shown in Figs. 4.5 and 4.6. Up to four particle size curves may be conveniently plotted on one sheet.

Initial mass used: 500 grams = m_1

Method A

$$P\% = 100 - \frac{\Sigma m}{m_1} \times 100$$

Sieve size mm	Mass retained g	Percentage retained	Percentage passing
1.18	0	0	100%
0.600	20	$\frac{20}{500} \times 100 = 4$	$100 - 4 = 96\%$
0.300	170	$\frac{170}{500} \times 100 = 34$	$96 - 34 = 62\%$
0.150	235	$\frac{235}{500} \times 100 = 47$	$62 - 47 = 15\%$
0.063	71	$\frac{71}{500} \times 100 = 14.2$	$15 - 14.2 = 0.8\%$
Pass (m_p)	3.5	(Check:	$\frac{3.5}{500} \times 100 = 0.7\%$)

Method B

$$P\% = \frac{m_1 - \Sigma m}{m_1} \times 100$$

Sieve size mm	Mass retained g	Cumulative mass passing g	Percentage passing
1.18	0	$500 - 0 = 500$	100%
0.600	20	$500 - 20 = 480$	$\frac{480}{500} \times 100 = 96\%$
0.300	170	$480 - 170 = 310$	$\frac{310}{500} \times 100 = 62\%$
0.150	235	$310 - 235 = 75$	$\frac{75}{500} \times 100 = 15\%$
0.063	71	$75 - 71 = 4$	$\frac{4}{500} \times 100 = 0.8\%$
Pass (m_p)	3.5	(Discrepancy $4 - 3.5 = 0.5$)	

Fig. 4.11   *Calculation procedures, simple dry sieving*

(8)   *Reporting results*

In addition to the particle size curve and the usual sample identification data, the sheet should include the visual description of the sample. This should be the description of the sample before testing, and modified as necessary as a result of the additional information revealed by the test result. Any material removed before sieving, such as vegetation or an isolated cobble, should be reported.

Tabulated data showing the percentages passing each sieve are sometimes required instead of, or in addition to, the grading curve. An example is given in Fig. 4.12, together with the description of the sample.

The method of test is reported as dry sieving in accordance with BS 1377:1975, Test 7(B).

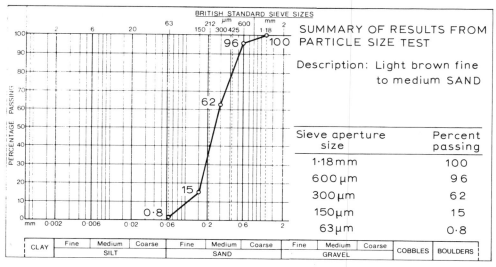

Fig. 4.12   *Results from particle size test and grading curve*

4.6.2   Composite Sieving

SCOPE

The simple procedure described above should be used only with clean sands, or with the washed sand-size fraction of other soils.

When gravel is present in appreciable quantity, it is necessary to start with a large sample and then at some stage subdivide it to give a smaller sample which is more manageable when the finer sieves are used. This process is referred to here as composite sieving. The mass of sample required initially depends upon the size of the largest particles present, as indicated in Table 4.4 and Fig. 4.7. It may be necessary with some materials to subdivide twice, or even three times. Each subdivision must be made by proper riffling or cone-and-quartering, using the procedures described in Section 1.5.5 and avoiding segregation of larger particles. The term 'riffle' is used below to indicate subdivision of the sample by any appropriate method.

Riffling of the test sample to a smaller quantity can conveniently be done after passing the 20 mm and 2 mm sieves, although the British Standard makes reference to subdivision at 20 mm and 6.3 mm. The point at which riffling is done is immaterial, and should be judged to suit the sample under test. In any case, it is convenient to riffle before changing to sieves of a smaller diameter. The process of composite sieving is shown diagrammatically in Fig. 4.13.

APPARATUS

The apparatus required is the same as listed in Section 4.6.1 for simple dry sieving. In addition, two large metal trays are required on which to mix and quarter the material.

PROCEDURE

The initial dry-sieving procedure, using the 'coarse gravel' size sieves down to 20 mm aperture, is similar to that for simple sieving (Section 4.6.1). Overloading of individual sieves must be avoided (see Table 4.6). Either hand shaking or the shaking machine may be used. The material passing through the 20 mm sieve is collected together and weighed. It is then riffled to a convenient size for the remaining sieves, and is weighed carefully before proceeding. Sieving through the smaller sieves using the riffled portion is carried out as for simple sieving. A second

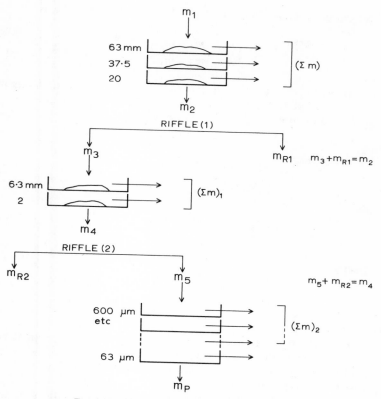

Fig. 4.13   *Sequence diagram, composite sieving*

riffling may be desirable after passing, say, the 2 mm sieve.

CALCULATIONS

The percentage passing each sieve before riffling is calculated as for simple sieving (Section 4.6.1). After riffling the calculation is modified as follows. Let m_1 = mass of original sample; m_2 = mass passing the 20 mm sieve; m_3 = mass of riffled fraction of m_2 used for subsequent sieves. The percentage passing each sieve in terms of the mass at the start of that set of sieves (m_3) is calculated, exactly as for simple sieving. These percentages are then multiplied by the correction factor m_2/m_1 to express them as a percentage of the whole original sample. This factor, muliplied by 100, is equal to the percentage passing the 20 mm sieve. The percentages so calculated can be plotted directly on to the grading curve.

If it is necessary to subdivide the sample a second time, after passing, say, the 2 mm sieve, a further correction must be applied to the weighings on subsequent sieves. Let m_4 = mass passing the 2 mm sieve; m_5 = mass of riffled fraction of m_4 used for subsequent sieves. The percentage, P, passing each sieve after subdividing is first calculated in terms of m_5. These percentages must then be multiplied by m_4/m_5 to express them as a percentage of the sample after the first subdivision, and then by m_2/m_3 to correct them back to the initial mass. The factor to apply after the second subdivision is therefore

$$\left(\frac{m_4}{m_3} \times \frac{m_2}{m_1} \right)$$

The percentages can be plotted directly on to the grading curve.

If a third subdivision is necessary (as in Fig. 4.16), the correction factor is:

$$\left(\frac{m_6}{m_5} \times \frac{m_4}{m_3} \times \frac{m_2}{m_1}\right)$$

and is applied to percentages calculated in terms of m_7, where $m_6 = $ mass passing the sieve before riffling; $m_7 = $ mass of riffled fraction of m_6 used for subsequent sieves. Percentage calculations can be represented by the following equations:

Before riffling:

$$P = \left[\frac{m_1 - (\Sigma m)}{m_1}\right] \times 100\%$$

After first riffle:

$$P_1 = \left[\frac{m_3 - (\Sigma m)_1}{m_3}\right]\left(\frac{m_2}{m_1}\right) \times 100\%$$

After second riffle:

$$P_2 = \left[\frac{m_5 - (\Sigma m)_2}{m_5}\right]\left(\frac{m_4}{m_3}\right)\left(\frac{m_2}{m_1}\right) \times 100\%$$

and so on.

In the above equations P represents the percentage retained on a particular sieve (before riffle); (Σm) denotes the sum of the masses retained on all sieves down to and including the one in question; P_1 and $(\Sigma m)_1$ denote similar quantities for the sieves following the first riffle; P_2 and $(\Sigma m)_2$ similarly after the second riffle; and so on.

If the mass passing the 63 μm sieve (after the second riffle) is denoted by m_p, the percentage finer than 63 μm, P_F, is given by the equation

$$P_F = \frac{m_p}{m_1} \times \left(\frac{m_4}{m_3} \times \frac{m_2}{m_1}\right) \times 100\%$$

A worked example with two stages of riffling is shown in Fig. 4.14. Percentages are calculated using method (B) of Section 4.6.1, but method (A) could also be used. Percentages retained on each sieve are shown. After riffling, the correction factor, m_2/m_1, can be replaced by the percentage passing the preceding (20 mm) sieve, and it is not then necessary to multiply by 100. This is because the percentage passing the 20 mm sieve is equal to $(m_2/m_1) \times 100$. Similarly, the correction factor, $(m_4/m_3) \times (m_2/m_1)$, after the second riffle can be replaced by the percentage passing the 2 mm sieve.

The grading curve for this example is given in Fig. 4.15, and indicates where riffling took place. This material is sand and gravel with cobbles and virtually no silt. The presence of cobbles exceeding 63 mm size called for a large sample, but washing has not been included in this example.

4.6.3 Cobbles and Boulders

An extreme example of composite sieving is the particle size analysis of non-cohesive material containing large cobbles or boulders, such as river terrace deposits. For this type of material it may be necessary to start with a sample consisting of a tonne or several tonnes excavated from

Initial dry mass: 15 kg = 15 000 g = m_1

Sieve size	Mass retained	Cumulative mass passing	Percentage passing	Calculation
75 mm	0	$m_1 = 15\,000$	100%	
63	300	$15\,000 - 300 = 14\,700$	$\dfrac{14\,700}{15\,000} \times 100 = 98.0\%$	
37.5	900	$14\,700 - 900 = 13\,800$	$\dfrac{13\,800}{15\,000} \times 100 = 92.0\%$	(1)
20	1250	$13\,800 - 1250 = 12\,550 = m_2$	$\dfrac{12\,550}{15\,000} \times 100 = 83.7\%$	
		$\downarrow$ RIFFLE (1)		
		$m_3 = 2275 \quad (10\,275) = m_{R1}$		
6.3	550	$2275 - 550 = 1725$	$\dfrac{1725}{2275} \times \dfrac{12\,550}{15\,000} \times 100 = 63.4\%$	(2)
2	450	$1725 - 450 = 1275 = m_4$	$\dfrac{1275}{2275} \times 83.7 = 46.9\%$	
		$\downarrow$ RIFFLE (2)		
		$m_5 = 200 \quad (1075) = m_{R2}$		
600 μm	90	$200 - 90 = 110$	$\dfrac{110}{200} \times \dfrac{1275}{2275} \times \dfrac{12\,550}{15\,000} \times 100 = 25.8\%$	
212	67	$110 - 67 = 43$	$\dfrac{43}{200} \times 46.9 = 10.1\%$	(3)
63	39	$43 - 39 = 4$	$\dfrac{4}{200} \times 46.9 = 0.9\%$	
Pass 63	$m_p = 4$			

Calculations:

$$(1) \quad P = \frac{m_1 - (\Sigma m)}{m_1} \times 100$$

$$(2) \quad P = \frac{m_3 - (\Sigma m)_1}{m_3} \times \frac{m_2}{m_1} \times 100$$

$$(3) \quad P = \frac{m_5 - (\Sigma m)_2}{m_5} \times \frac{m_4}{m_3} \times \frac{m_2}{m_1} \times 100$$

Fig. 4.14   *Calculations, composite sieving*

a trial pit. The quantity required can be assessed from Fig. 4.7. The procedure is illustrated in Fig. 4.16.

As the material is excavated, it is placed on a platform or tarpaulin, and pieces larger than 75 mm are set aside. If necessary, they are brushed clean of adhering fines, which must be returned to the main sample.

The large cobbles are gauged for size with a series of square wood or metal frames. Suitable aperture sizes are 100, 150, 200, 300, 400 mm, and even larger, if necessary. Each size range is weighed on site with platform scales of, say, 250 kg capacity. It is essential to mount the platform scales on a rigid level base, and to shield the scales from hot sun, wind or rain.

The main sample is weighed in batches on the platform scales. The whole weighing

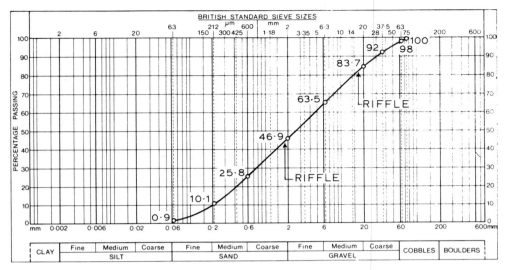

Fig. 4.15   *Grading curve, composite sieving*

operation should be repeated, if practicable, to ensure that there has been no mistake in obtaining the total mass of the whole sample. The percentage of each of the cobbles and boulders sizes can be calculated immediately if the material is, for all practical purposes, dry. If not, a correction for moisture content must be made later to obtain the total dry mass.

The main sample, consisting of minus 75 mm material, can then be riffled, with a large riffle-box, or by cone-and-quartering. A check should be made by visual inspection that the largest particles have been divided up in representative proportions. A riffled sample of at least 50 kg is required. For convenience it should be placed in two or more lined sacks or bags for transfer to the main laboratory. Each bag should not be too heavy for one man to handle. Leakage or loss of fines during transit must be avoided; therefore unlined hessian or similar sacks are not suitable.

The rest of the particle size analysis can take place in the laboratory, the riffled sample being used. Further subdivision will be necessary at subsequent stages. A worked example is not shown, but the procedure to be followed, and the calculations, are similar to those described in Section 4.6.2, allowance being made in the calculations for every riffling stage. If a sedimentation test is carried out on the fines fraction, a complete particle size curve ranging from cobbles or boulders down to the clay size can be constructed.

4.6.4   Wet Sieving — Fine Soils (BS 1377:1975, Test 7 (A))

SCOPE

If a soil contains silt or clay, or both, even in small quantities, it is necessary to carry out a wet sieving procedure in order to measure the proportion of fine material present. Even when dry, fine particles of silt and clay can adhere to sand-size particles and cannot be separated by dry sieving, even if prolonged. Washing is the only practicable means of ensuring complete separation of fines for a reliable assessment of their percentage. If clay is present, or if there is evidence of particles sticking together, the material should be immersed in a dispersant solution before washing, as described in Section 4.6.6 under soil category (2).

The procedure is described in detail below for non-cohesive soils containing little or no gravel.

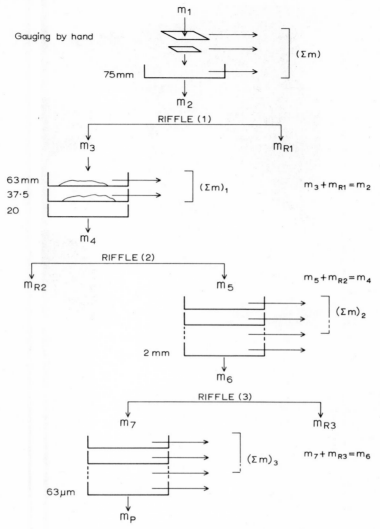

Fig. 4.16   *Sequence diagram, sieving with cobbles and boulders*

APPARATUS

As for simple sieving (Section 4.6.1), with the addition of the following:

(9)   Evaporating dishes.
(10) Rubber tubing connected to water tap, fitted at the other end with a small spray such as
      the rose from a small watering can (see Fig. 4.17).

PROCEDURAL STAGES (Non-cohesive soil)

(1)   Select and prepare test specimen.
(2)   Oven dry, cool, weigh.
(3)   Wash through 2 mm and 63 μm sieves.
(4)   Dry retained material.
(5)   Weigh.

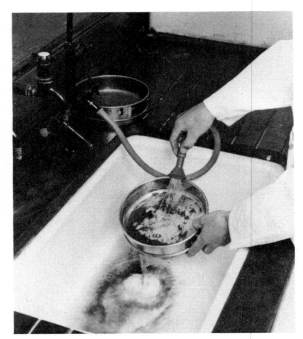

Fig. 4.17   *Washing soil on* 63 μm *sieve*

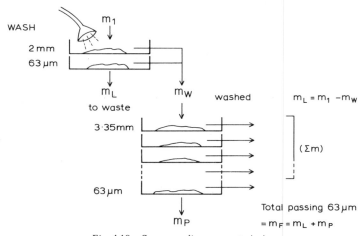

Fig. 4.18   *Sequence diagram, wet sieving*

(6)   Pass through range of sieves.
(7)   Weigh each size fraction.
(8)   Calculate percentages passing each sieve and percentage of fines.
(9)   Plot grading curve.
(10) Report result.

The procedure is illustrated in Fig. 4.18.

PROCEDURE

(1) *Selection and Preparation*

As for dry sieving (Section 4.6.1).

(2) *Oven drying, cooling, weighing*

As for dry sieving. The total initial dry mass (m_1) must be carefully measured and verified, because the fines will be washed away and their mass determined by difference. There will be no overall check by addition, as there is in dry sieving.

(3) *Wash*

The 2 mm sieve is nested in the 63 μm sieve, but the lid and receiver are not used. An additional intermediate sieve may be included to protect the 63 μm sieve from overloading if the soil contains a high proportion of coarse or medium sand.

The soil is placed a little at a time on the 2 mm sieve, and washed over a sink with a jet or spray of clean water, as shown in Fig. 4.17. The silt and clay passing the 63 μm sieve is allowed to run to waste. When the material on the 2 mm sieve has been washed free of fines, washing on the 63 μm sieve is continued until the waste water is seen to run clear.

During this operation the sieve must not be allowed to become overloaded with soil or to overflow with water. The mass of soil retained on the 63 μm should not exceed 150 g at any one time. Table 4.6 gives the recommended maximum quantities which may be retained on each sieve. If this is likely to be exceeded, the material should be sieved in two or more portions.

The sink used for this operation should be fitted with a silt trap, as referred to in Section 1.6.6.

(4) *Drying*

The whole of the material retained on each sieve is allowed to drain, and is carefully transferred to trays or evaporating dishes. These are placed in an oven to dry at 105–110 °C, preferably overnight.

(5) *Weighing*

After cooling, the whole of the dried material is put together and weighed to an accuracy of 0.1% (m_w).

(6) *Sieving*

The dry soil is passed through a nest of the complete range of sieves to cover the sizes of particles present, down to the 63 μm sieve. This operation may be carried out by hand, or preferably on a sieve shaker, exactly as in the dry sieving procedure (Section 4.6.1).

If the fraction passing a particular sieve is large i.e. substantially greater than 150 g, it should be accurately weighed and then subdivided to give a sample of 100–150 g, as described for composite sieving (Section 4.6.2).

(7) *Weighing*

The portion retained on each sieve is weighed, each to an accuracy of 0.1%.

(8) *Calculations*

The cumulative percentage passing each sieve is calculated in the same way as for simple sieving (Section 4.6.1, stage 6), either method (A) or method (B) being used. Note that percentages must be expressed as a percentage of the *total* initial dry mass (m_1), not the mass after washing (m_w).

The process of washing removes the clay and the finer silt particles from the material, but some particles only slightly smaller than 63 μm sieve may be retained on that sieve owing to the effect of surface tension of water. In the subsequent dry sieving these particles pass through the 63 μm sieve, and the presence of this fine material does not necessarily imply that washing was inadequate.

Therefore the total amount of fines (i.e. material passing the 63 μm sieve), m_F, in the original sample is made up of two parts: (1) that which is lost in the washing process (m_L); (2) that which passes the 63 μm sieve when dry sieved (m_p). The mass of fines lost by washing is calculated by difference:

$$m_L = m_1 - m_w$$

The mass of dry-sieve fines, m_p, is determined by weighing. Hence,

$$m_F = m_L + m_p$$

This provides a check on the sieving calculations, because m_F should be equal to the calculated cumulative mass passing the 63 μm sieve.

The percentage fines can be calculated directly, and is equal to

$$\frac{m_F}{m_1} \times 100\%$$

A calculation for a silty fine to coarse sand, which did not require riffling, is given in Fig. 4.19, and the grading curve is shown in Fig. 4.20.

4.6.5   Wet Sieving — Gravelly Soils (non-cohesive)

OUTLINE OF PROCEDURE

If the sample contains particles of coarse gravel and cobble size, a large sample (15 kg or more) is necessary. After drying and weighing (total mass m_1), the material is sieved on a large-diameter 20 mm sieve, with a portion being taken at a time, so as not to overload the sieve (see Table 4.6). Particles retained are brushed to remove finer material which may be adhering to them, but individual particles must not be broken down. The material retained on the 20 mm sieve, after drying, if necessary, is then sieved on appropriate larger aperture sieves and the amount retained on each is weighed.

The fraction passing the 20 mm sieve, including 'brushings' from larger particles, is then weighed (m_2), and is subdivided to give a convenient mass (say 2 kg) for the remainder of the sieving operation (m_3). Washing on a 63 μm sieve, drying and dry sieving are carried out as described in Section 4.6.2. The procedure is illustrated in Fig. 4.21.

CALCULATIONS

Even when a sample is large and is weighed initially in kilograms, calculations are easiest if all masses are expressed in grams. The percentages retained on the larger sieve sizes are calculated as described previously.

After riffling and washing on the 63 μm sieve, percentages must be calculated on the basis of the dry mass (m_3) of material before washing (immediately after riffling), *not* the dry mass after washing (m_w).

As in the previous example (Section 4.6.4), the total amount of fines is made up of two parts, denoted by m_L and m_p. In this case, however, the mass m_p relates only to the riffled portion m_5, so it must be increased by the ratio m_4/m_5 to obtain the equivalent mass of dry-sieved fines m_E in the total washed sample m_w.

Initial dry mass:	$500\ g = m_1$
Dry mass after washing on 63 μm sieve:	$370\ g = m_W$
Mass lost by washing:	$\overline{130\ g} = m_L$

Sieve size	Mass retained g	Percent retained	Cumulative mass passing	Percentage passing
3.35 mm	0	0	500	100%
2	20	4	$\dfrac{-20}{480}$	$\dfrac{480}{500} \times 100 = 96\%$
1.18	35	7	$\dfrac{-35}{445}$	$\dfrac{445}{500} \times 100 = 89\%$
600 μm	60	12	$\dfrac{-60}{385}$	$\dfrac{385}{500} \times 100 = 77\%$
212	145	29	$\dfrac{-145}{240}$	$\dfrac{240}{500} \times 100 = 48\%$
63	100	20	$\dfrac{-100}{140}$	$\dfrac{140}{500} \times 100 = 28\%$
Pass 63 μm	$m_p = 10$		$m_L = 130$ $m_P = 10$ $\overline{}$ $m_F = 140$	28%

Calculation: $P = \dfrac{m_1 - (\Sigma\, m)}{m_1} \times 100\%$

Fig. 4.19   Calculations, simple wet sieving

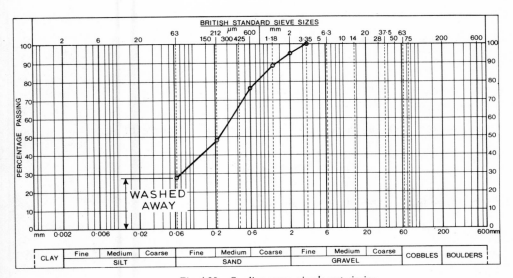

Fig. 4.20   Grading curve, simple wet sieving

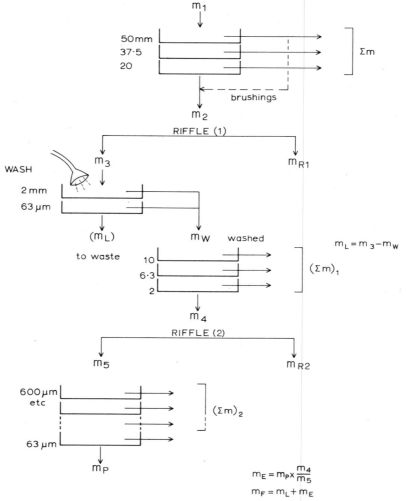

Fig. 4.21   *Sequence diagram, wet sieving of non-cohesive soil with gravel*

$$m_E = m_p \times \frac{m_4}{m_5}$$

The equivalent total fines, m_F, contained in m_3 is therefore given by the equation

$$m_F = m_L + m_E = m_L + \left(m_p + \frac{m_4}{m_5} \right)$$

The percentage fines in the whole sample is equal to

$$\frac{m_F}{m_3} \times \frac{m_2}{m_1} \times 100\%$$

which provides a check on the sieving calculations, as before.

Initial mass = 20 kg = 20 000 g = m_1

Sieve size	Mass retained	Mass passing	Total mass passing	Percentage passing
75 mm	0	m_1 = 20 000		100%
50	1000	$\dfrac{-1000}{19\ 000}$		$\dfrac{19\ 000}{20\ 000}$ x 100 = 95.0%
37.5	800	$\dfrac{-800}{18\ 200}$		$\dfrac{18\ 200}{20\ 000}$ x 100 = 91.0%
20	2200	$\dfrac{-2200}{m_2 = 16\ 000}$		$\dfrac{16\ 000}{20\ 000}$ x 100 = 80.0%

↓

RIFFLE

m_3 = 2400   (13 600) = m_{R1}

↓

─WASH ON 63 μm

───→ (m_L = 400 to waste)

		m_W = 2000	2400	
10	410	$\dfrac{-410}{1590}$	$\dfrac{-410}{1990}$	$\dfrac{1990}{2400}$ x $\dfrac{16\ 000}{20\ 000}$ x 100 = 66.3%
6.3	400	$\dfrac{-400}{1190}$	$\dfrac{-400}{1590}$	$\dfrac{1590}{2400}$ x 80 = 53.0%
2	420	$\dfrac{-420}{m_4 =\ \ 770}$	$m_4 + m_L = \dfrac{-420}{1170}$	$\dfrac{1170}{2400}$ x 80 = 39.0%

↓

RIFFLE

m_5 = 160   (610) = m_{R2}

Adjusted
mass retained

1170

600 μm	73	73 x $\dfrac{770}{160}$ = 351	$\dfrac{-351}{819}$	$\dfrac{819}{2400}$ x 80 = 27.3%
212	51	51 x $\dfrac{770}{160}$ = 245	$\dfrac{-245}{574}$	$\dfrac{574}{2400}$ x 80 = 19.1%
63	29.5	29.5 x $\dfrac{770}{160}$ = 142	$\dfrac{-142}{432}$	$\dfrac{432}{2400}$ x 80 = 14.4%
Pass 63 μm	m_p = 6.4	6.4 x $\dfrac{770}{160}$ = 30.8 = m_E		

Check: Total passing 63 μm = m_L + m_E = 400 + 30.8 = 430.8 = m_F

Percentage passing 63 μm = $\dfrac{430.8}{2400}$ x 80 = 14.4%

Fig. 4.22   Calculations, wet sieving of non-cohesive soil with gravel

This example is similar to that given in Fig. 4.14, except that the presence of silt meant that washing was necessary. Washing of the whole sample without first riffling would be an unnecessarily lengthy process by the usual means. However, a large sieve-shaker can be fitted with an attachment for washing while shaking, as described by West and Dumbleton (1972). This procedure is relatively quick, and is suitable for medium gravel and sand containing small proportions of silt and clay.

The calculations related to the procedure described above (washing after riffling) are set out in Fig. 4.22, and the grading curve is shown in Fig. 4.23. The procedure for washing before riffling is given in Section 4.6.7.

4.6.6   Wet Sieving – Cohesive Soils

SCOPE

The procedures for carrying out particle size tests on cohesive soils depend upon the range of particle sizes present, and for this purpose the soils can be divided into three categories:

(1)   Cohesive soils containing particles no larger than 2 mm (coarse sand size); for example, sandy and silty clays.
(2)   Cohesive soils containing particles up to 20 mm (medium gravel size); for example, gravelly sandy and silty clays.
(3)   Cohesive soils containing larger particles, such as boulder clay.

The first two categories are dealt with separately below, and the third is described in Section 4.6.7.

COHESIVE SANDY SOILS

Cohesive soils containing particles no larger than sands present little difficulty, because a test can be carried out on a relatively small sample (100–150 g). The material is first pretreated for a sedimentation test (Section 4.8.1) and then washed on a 63 μm sieve. The retained portion is dried, then sieved as described in Section 4.6.1. The material passing the 63 μm sieve is collected for a sedimentation analysis (Section 4.8.2 or 4.8.3).

COHESIVE SOILS WITH FINE GRAVEL

Cohesive soils with up to medium-size gravel require a sample of about 2 kg. A representative sample is dried and carefully weighed, and a separate sample of the 'matrix' material (i.e. with most gravel particles removed) is set aside for the sedimentation pretreatment procedure which is described in Section 4.8.1.

The dried representative sample is spread out on a tray and covered with water containing 2 g/litre of sodium hexametaphosphate. The soil is allowed to stand for at least an hour, and is stirred frequently. This disperses the clay fraction, so that clay and silt will not adhere to larger particles.

The material is then washed a little at a time through the 2 mm sieve nested in a 63 μm sieve as described for wet sieving (Section 4.6.4), all the fines being washed to waste. When the water passing through the 63 μm sieve is virtually clear, the material retained in the sieve is dried and shaken dry through the complete range of sieves from 20 mm downwards. The material retained on each sieve is weighed to an accuracy of 0.1%. Calculations are similar to those given in Section 4.6.4. However, if riffling is carried out at some stage after washing, the calculations in Section 4.6.5 are applicable.

The grading curve from the sedimentation test, if required, is added to the sieving curve after the appropriate proportionate adjustment referred to in Section 4.8.5 has been made. The grading sheet should include a note indicating that the soil was pretreated by immersion in the dispersant solution.

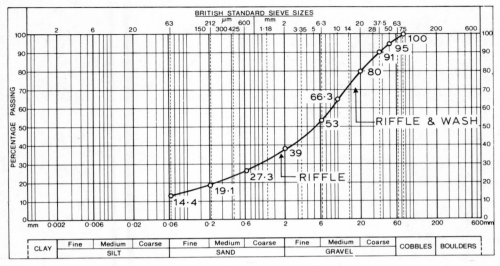

Fig. 4.23   *Grading curve, silty sand and gravel*

4.6.7   'Boulder Clay'

SCOPE

In this context 'boulder clay' refers to soils of the type defined in Section 4.4.3. The clay content, though perhaps only about 15% of the whole, is sufficient to give the material *en masse* sufficient cohesion for it to behave as a clay. This description also applies to colliery spoil materials and the like.

A soil of boulder clay type is the most difficult on which to carry out a particle size analysis, for the following reasons:

(1)   The presence of gravel or cobbles necessitates a large initial sample, probably exceeding 20 kg.
(2)   Because it is cohesive, it must be pretreated with dispersant and the fines fraction washed away. This can be very time-consuming.
(3)   It is of no use to dry the whole sample initially, because the clay when dried becomes almost brick-hard. It is then very difficult to break down and remove from larger particles.
(4)   Because of the complexity of the test procedure, the calculations are less straightforward than those previously described.

The following procedure is the author's adaptation of the general principles outlined in BS 1377:1975, Test 7(A). In this method the material is washed before riffling. The procedure is shown diagrammatically in Fig. 4.24.

SELECTION OF SAMPLE

The mass of the sample taken initially for testing should not be less than that indicated in Table 4.4 or Fig. 4.7 for the maximum significant particle size. If cobble sizes are present in significant quantity, a sample of 50 kg or more is required, and this is indeed a large sample for cohesive material. The quantity indicated relates to dry mass. Since this type of material is not dried at the outset, allowance must be made for the amount of water present in the soil, which can only be estimated at this stage.

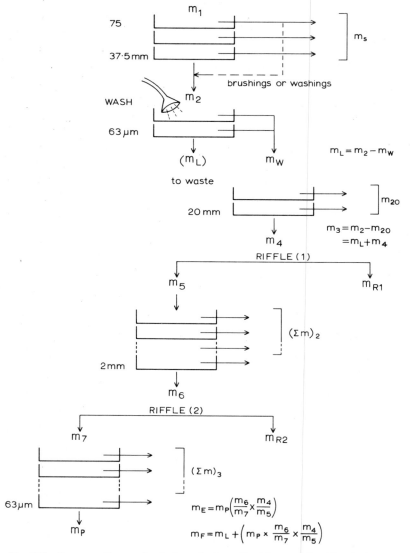

Fig. 4.24   *Sequence diagram for sieving cohesive soils with gravel ('boulder clay' type)*

PROCEDURE

(1)   Before weighing the sample to be tested, take several small specimens (each of about 50 g)
      representative of the fine fraction from different points of the sample. Break or chop them
      up, mix together, and from this material set aside about 100 g of soil representative of the
      fraction finer than 2 mm for sedimentation analysis. Return the remainder to the main
      sample.

(2)   Weigh the whole sample (W_1) in its natural undried state, and spread out on a tray or
      trays.

(3)   Take three representative 'lumps' of material, each of about 300 g, for moisture content
      measurements. Remove any particles larger than 20 mm and return them to the main
      sample.

(4)   Weigh each sample, dry overnight, weigh, and calculate the moisture contents. The average of the three ($w\%$) is the moisture content of the minus 20 mm fraction. Retain the dried material.

(5)   Add dispersant solution (water containing 2 g/litre of sodium hexametaphosphate) to the sample on the tray. Stir frequently and break down any clay lumps.

(6)   Allow to stand overnight, and add the dried material from the moisture content determination (stage 4), which should be agitated until completely broken down.

(7)   Remove all particles larger than 37.5 mm by hand, using a 37.5 mm sieve as a gauge. Brush them, if necessary, to ensure that these particles (referred to here as 'stones') have been washed free of adhering fine material, which is to remain with the main sample.

(8)   Dry the stones in the oven, and weigh when cool (m_s).

(9)   Separate the stones into size fractions on the 50 mm, 63 mm and 75 mm sieves, using, if necessary, wood frames for gauging larger sizes, such as 100 mm, 150 mm, 200 mm as required.

(10) Weigh each size fraction.

(11) Take a little at a time of the sample from stage (6), and wash on a 20 mm sieve nested on a 2 mm sieve nested on a 63 μm sieve. Use 450 mm diameter sieves if available. Allow material finer than 63 μm to run to waste. Collect about a litre of washings periodically, and continue washing until this water is seen to be clear.

(12) Oven dry the material retained on these sieves.

(13) Weigh the whole of the retained washed material (m_w).

(14) Sieve through 28 mm and 20 mm sieves (450 mm diameter).

(15) Weigh material retained on each sieve (the total denoted by m_{20}), and material passing the 20 mm sieve (m_4).

(16) Riffle m_4 to obtain a suitable size sample (m_5) for the next set of sieves (300 mm diameter).

(17) Dry sieve on appropriate sieves down to 2 mm.

(18) Weigh material retained on each sieve, and material passing the 2 mm sieve (m_6).

(19) Riffle (m_6) to obtain a suitable sample (m_7) for the fine sieves (200 mm diameter).

(20) Dry sieve on selected sieves down to 63 μm size.

(21) Weigh material retained on each sieve, including any passing the 63 μm sieve (m_p).

CALCULATIONS

The notation used in the following explanation is summarised in Table 4.7. Undried masses are denoted by W; dry masses by m; moisture content by w. Symbols in parentheses represent quantities which are calculated; the others are measured.

The complete calculations are set out in Fig. 4.25.

It is assumed that particles larger than 20 mm contain no absorbed water, or that the quantity held is negligible. The undried mass of material passing 20 mm (W_3) is therefore equal to the undried mass of the original sample with the particles larger than 20 mm removed. These consist of the stones retained on 37.5 mm, and the washed portion passing 37.5 mm and retained on 20 mm (m_{20}); therefore

$$W_3 = W_1 - (m_s + m_{20})$$

If the particles larger than 20 mm contain appreciable absorbed water in their natural state, this can be measured separately and allowed for by adjusting m_s and m_{20}.

Since the moisture content of the minus 20 mm fraction is known ($w\%$), the equivalent dry mass (m_3) of this fraction is given by

$$m_3 = \frac{100}{100+w} \times W_3$$

Table 4.7. NOTATION FOR BOULDER CLAY SIEVE ANALYSIS

Size fraction	Dry mass	Undried mass	Remarks
Initial total mass of sample tested	(m_1)	W_1	
'Stones' larger than 37.5 mm	m_s		
Total passing 37.5 mm	(m_2)		
Passing 37.5 mm after washing	m_w		
Washed away through 63 μm	(m_L)		$m_L = m_2 - m_W$
Retained on 20 mm after passing 37.5 mm	m_{20}		$m_{20} = m_W - m_4$
Total material finer than 20 mm	(m_3)	(W_3)	$m_3 = m_4 + m_L$
Initial moisture content of W_3	$w\%$		
Passing 20 mm after washing	m_4		
Riffled portion of m_4	m_5		
Washed material passing 2 mm	m_6		after dry sieving
Riffled portion of m_6	m_7		
Passing 63 μm	m_p		
Total finer than 63 μm	(m_F)		$m_F = m_L + m_p\left(\dfrac{m_6}{m_7} \times \dfrac{m_4}{m_5}\right)$

By adding back the dry mass of particles larger than 20 mm, the initial dry mass (m_1) of the whole sample is obtained:

$$m_1 = m_3 + m_s + m_{20}$$

The mass is used as the starting point for the calculations set out in Fig. 4.25, which shows only the dry masses.

The mass of material passing the 37.5 mm sieve (m_2), and used in the washing process, is equal to the initial mass less the mass of stones — that is,

$$m_2 = m_1 - m_s$$

$$= m_3 + m_{20}$$

The calculation of percentages passing the sieves down to 20 mm is as described in Section 4.6.5. Washing on the 63 μm sieve has no effect on the masses retained on these sieves. The mass lost in the washing (m_L) is obtained by difference — that is,

$$m_L = m_2 - m_W$$

The calculation procedure after riffling is similar to that given in Section 4.6.5, and corrected percentages are related to the dry mass before washing, as shown in Fig. 4.25. After the second riffle, a second multiplying factor is introduced, as explained in Section 4.6.2.

The total amount of fine material (m_F) passing the 63 μm sieve is made up of two parts, as in Section 4.6.5, comprising m_L and m_p. The latter must be increased by the ratio $(m_6/m_7) \times (m_4/m_5)$ to give the equivalent mass to add to m_L. The total mass finer than 63 μm in the whole sample is given by

$$m_F = m_L + m_p\left(\frac{m_6}{m_7} \times \frac{m_4}{m_5}\right)$$

and the percentage of fines is equal to

$$\frac{m_F}{m_1} \times 100\%$$

Initial undried mass: $\quad$ 24 970 g $= W_1$

Estimated total initial dry mass: 23 160 g $= m_1$

Sieve size	Mass retained	Cumulative mass	Mass passing		Percentage passing	Calculation

100 mm — 0 — 0

$m_1 = 23\ 160$

75 — 714 — 714 $\quad = \dfrac{-714}{22\ 446}$

W_1	$= 24\ 970$	100%

50 — 1275 — 1989 $\quad = \dfrac{-1275}{21\ 171}$

$m_s + m_{20} = \dfrac{3894}{21\ 076}$, W_3

$\dfrac{22\ 446}{23\ 160} \times 100 = 96.9\%$

37.5 — 762 — 2751 $= m_s$ $\quad \dfrac{-762}{m_2 = 20\ 409}$

$w = 9.4\%$

$\dfrac{21\ 171}{23\ 160} \times 100 = 91.4\%$

$m_3 = \dfrac{100}{109.4} \times 21\ 076$

$= 19\ 265$

$\dfrac{20\ 409}{23\ 160} \times 100 = 88.1\%$ $\quad$ (1)

WASH ON 63 μm

$m_s + m_{20} = \dfrac{3894}{}$
$\therefore m_1 = 23\ 159$

retained $m_W = 11\ 958$ $\qquad$ lost $(8451) = m_L$

Total mass passing $\quad \dfrac{8451}{23\ 160} \times 100 = 36.5\%$ lost

11 958 $\qquad\qquad m_2 = 20\ 409$

28 — 387 — 387 $\quad \dfrac{-387}{11\ 571}$ $\qquad \dfrac{-387}{20\ 022}$ $\quad \dfrac{20\ 022}{23\ 160} \times 100 = 86.5\%$

20 — 756 — 1143 $= m_{20}$ $\quad \dfrac{-756}{10\ 815} = m_4$ $\qquad m_3 = \dfrac{-756}{19\ 266}$ $\quad \dfrac{19\ 266}{23\ 160} \times 100 = 83.2\%$ $\quad$ (2)

RIFFLE (1)

3894 $= m_s + m_{20}$ $\qquad m_5 = 2504 \qquad (8311) = m_{R1}$

Corrected mass retained $\qquad m_3 = 19\ 266$

14 — 117.4 — 117.4 $\quad 117.4 \times \dfrac{10\ 815}{2504} = 507.1$ $\qquad \dfrac{-507.1}{18\ 758.9}$ $\quad \dfrac{18\ 759}{23\ 160} \times 100 = 81.0\%$

6.3 — 331.2 — 448.6 $\quad 331.2 \times \dfrac{10\ 815}{2504} = 1430.5$ $\qquad \dfrac{-1430.5}{17\ 328.4}$ $\quad \dfrac{17\ 328}{23\ 160} \times 100 = 74.8\%$

2 — 477.2 — 925.8 $\quad 477.2 \times \dfrac{10\ 815}{2504} = 2061.1$ $\qquad m_8 = \dfrac{-2061.1}{15\ 267.3}$ $\quad \dfrac{15\ 267}{23\ 160} \times 100 = 65.9\%$ $\quad$ (3)

$m_5 = 2504$
$ \dfrac{925.8}{1578.2}$ $\qquad m_6 = 1578.2$

RIFFLE (2)

$m_7 = 105.3 \qquad (1472.9) = m_{R2}$

15 267.3

600 μm — 38.2 — 38.2 $\quad 38.2 \times \dfrac{1578.2}{105.3} \times \dfrac{10\ 815}{2504} = 2472.5$ $\quad \dfrac{-2472.5}{12\ 794.8}$ $\quad \dfrac{12\ 795}{23\ 160} \times 100 = 55.2\%$

212 — 41.3 — 79.5 $\quad 79.5 \times \dfrac{1578.2}{105.3} \times \dfrac{10\ 815}{2504} = 2673.1$ $\quad \dfrac{-2673.1}{10\ 121.7}$ $\quad \dfrac{10\ 122}{23\ 160} \times 100 = 43.7\%$

63 — 22.5 — 102.0 $\quad 102.0 \times \dfrac{1578.2}{105.3} \times \dfrac{10\ 815}{2504} = 1456.3$ $\quad \dfrac{-1456.3}{8665.4}$ $\quad \dfrac{8665}{23\ 160} \times 100 = 37.4\%$ $\quad$ (4)

Passing 63 — 3.3 $= m_p$ — 105.3 $= m_7$ $\quad 3.3 \times \dfrac{1578.2}{105.3} \times \dfrac{10\ 815}{2504} = 214$

$m_L = 8451$
Check: $m_F = \dfrac{}{8665}$

Fig. 4.25 Calculations for 'boulder clay' type soil

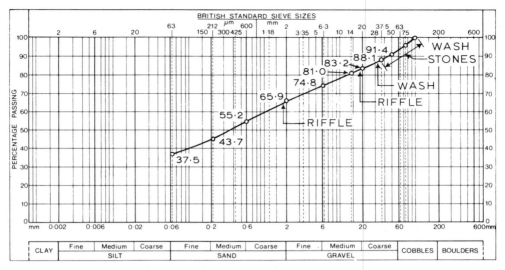

Fig. 4.26   Grading curve from sieving test on 'boulder clay'

References to the mathematical equations for the percentage calculations at each stage are indicated in the right-hand column of Fig. 4.25. They are as follows:

$$P = \frac{m_1 - (\Sigma m)}{m_1} \times 100 \tag{4.1}$$

$$P = \frac{m_2 - (\Sigma m)_1}{m_1} \times 100 \tag{4.2}$$

$$P = \frac{m_3 - (\Sigma m)_2 \times (m_4/m_5)}{m_1} \times 100 \tag{4.3}$$

$$P = \frac{m_8 - (\Sigma m)_3 \times (m_6/m_7) \times (m_4/m_5)}{m_1} \times 100 \tag{4.4}$$

Equations (4.3) and (4.4) are more complicated than the step-by-step arithmetical procedure shown in Fig. 4.25, which is easier to follow when calculating by hand. However, the equations may be useful when a computer program for these and similar sieving calculations is being written.

The grading curve from the sieving test on this material is shown in Fig. 4.26. The complete grading curve, including a sedimentation analysis, is given in Fig. 4.49, and shows a well-graded mixture of all particle sizes from cobbles down to and including clay.

4.7   SEDIMENTATION THEORY

4.7.1   Introduction

The theory of sedimentation is based on the fact that large particles in suspension in a liquid settle more quickly than small particles, assuming that all particles have similar densities and shapes. The velocity which a falling particle eventually reaches is known as its terminal velocity. If the particles are approximately spherical, the relationship between terminal velocity, V, and particle diameter, D, is given by *Stokes' Law*, named after Sir George Stokes (1891). This states that the terminal velocity is proportional to the square of the particle diameter, or

$$V \propto D^2$$

Although clay particles are far from spherical (see Section 4.3.1), the application of Stokes' Law based on diameters of equivalent spheres provides a basis for comparison of the particle size distribution in fine soils which is sufficiently realistic for most practical purposes.

4.7.2 General Principles

In a sedimentation test a suspension of a known mass of fine soil particles of various sizes is made up in a known volume of water. The particles are allowed to settle under gravity, and this is the process known as sedimentation. From certain measurements made at known intervals of time, the distribution of particle sizes can be assessed.

A model which illustrates the sedimentation process is represented diagrammatically in Fig. 4.27. Only four different sizes of particles are represented, their terminal velocities and approximate equivalent diameters being shown in Table 4.8.

A real soil contains particles of many different sizes, but the principles of sedimentation can be understood by considering what happens to each of these four sizes after various intervals of time as they are allowed to settle out of a suspension in water in a container just over 300 mm deep.

Immediately after the suspension has been shaken up (time $t = 0$), all particles are uniformly distributed throughout the depth of suspension, as in Fig. 4.27(a). If we assume that each particle reaches its terminal velocity within a very short time, after 10 s the coarse silt particles have each fallen 10 mm, the fine silt particles 1 mm and the particles finer than that hardly at all (Fig. 4.27b). After 100 s the situation is as in Fig. 4.27(c) and the coarse silt particles starting at the 200 mm mark have settled to the bottom. After 1000 s (about 17 min) all the coarse silt particles have reached the bottom (Fig. 4.27d). All the solid particles now remaining in suspension are smaller than 35 μm, so a sample taken from anywhere in the suspension would contain only those particles smaller than this size. At the same instant a sample taken from just above the 100 mm mark would only contain particles smaller than 12 μm (medium silt).

After 10 000 s (about $2\frac{3}{4}$ h) from the start all medium silt particles have reached the bottom and all fine silt particles are below the 100 mm mark (Fig. 4.27e). The upper 100 mm, therefore, contains only clay particles in suspension. After 30 000 s (about 8 h) silt particles of all sizes have reached the bottom, leaving only the clay sizes in suspension (Fig. 4.27f). The smallest size we are considering, 1.2 μm, will require 300 000 s (about $3\frac{1}{2}$ days) to reach the bottom, but smaller particles will take much longer. A suspension containing an appreciable amount of clay remains cloudy indefinitely.

By applying Stokes' Law to the sedimentation model described above, we can calculate the maximum diameter of particles remaining above a particular depth after a certain interval of time from the start. The mass of solid particles present can be determined either by sampling from a specified depth (as in the pipette test), or by measuring the density of the suspension with a hydrometer (the hydrometer test).

The sampling depth of 100 mm used in the standard pipette test is indicated in Fig. 4.27(a). The shaded area marked 'H_R zone' in Fig. 4.27(b) indicates the range of effective depths at which the density of suspension is measured during the period of a typical hydrometer test.

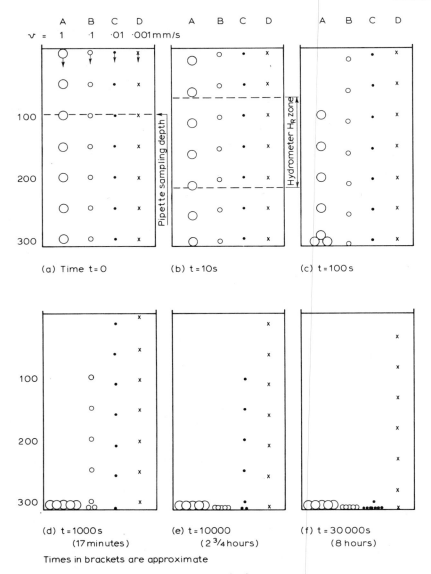

Fig. 4.27   *Representation of sedimentation process*

4.7.3   Stokes' Law

According to Stokes' Law, the terminal velocity v of a spherical particle falling freely in a fluid is given by

$$v = \frac{D^2 g(\rho_s - \rho_L)}{18\,\eta} \tag{4.5}$$

where D = diameter of particle; ρ_s = mass density of the solid particle; ρ_L = mass density of the fluid (liquid); η = dynamic viscosity of the fluid; g = acceleration due to gravity.

The application of Stokes' Law to the process of sedimentation is based on the following simplifying assumptions.

Table 4.8.   TERMINAL VELOCITIES OF PARTICLES IN SUSPENSION

Particle	Terminal velocity (mm/s)	Approximate diameter (μm)
Coarse silt	1	35
Medium silt	0.1	12
Fine silt	0.01	3.5
Clay	0.001	1.2

Table 4.9.   SYMBOLS, UNITS AND CONVERSION FACTORS FOR SEDIMENTATION FORMULA

Quantity	Symbol	Coherent unit	Practical unit	Multiplying factor
Viscosity of water (see Note (1))	η	Ns/m^2	mNs/m^2 ($=mPa\,s$)	$\frac{1}{1000}$
Height of fall	H	m	mm	$\frac{1}{1000}$
Acceleration due to gravity	g	m/s^2	m/s^2	1
Time	T	s	min	60
Densities (see Note (2))	ρ_s, ρ_w	kg/m^3	Mg/m^3	1000
Particle diameter	D	m	mm	$\frac{1}{1000}$

(1) *Viscosity.*   The practical SI unit of dynamic viscosity in mPa s (millipascal second) or mN s m^{-2} (millinewton second per square metre), which is equal to the traditional metric unit cP (centipoise).
(2) *Densities.*   The density of water (ρ_w) may be taken here as 1.000 Mg/m³ exactly, at the usual test temperature, because th3 SG of particles (G_s) is related to the same temperature.

(1)   The condition of viscous flow in a still liquid is maintained.
(2)   There is no turbulence — that is, the concentration of particles is such that they do not interfere with one another.
(3)   The temperature of the liquid remains constant.
(4)   Particles are small spheres.
(5)   Their terminal velocity is small.
(6)   All particles have the same density.
(7)   A uniform distribution of particles of all sizes is formed within the liquid.

Equation (4.5) can be re-written

$$D= \left\{ \frac{18\,\eta}{g} \cdot \frac{v}{(\rho_s-\rho_l)} \right\}^{1/2} \tag{4.6}$$

If the particle falls a distance H in time T, its mean velocity $=H/T$.
If the liquid is water, we may write

$$\rho_L=\rho_w$$

Equation (4.6) then becomes

$$D= \left\{ \frac{18\,\eta}{g} \cdot \frac{H}{T(\rho_s-\rho_w)} \right\}^{1/2} \tag{4.7}$$

Using the practical units given in Table 4.9, and writing t (min) for T, Eq. (4.7) can be written as

$$\frac{D}{1000} = \left\{ \frac{18 \times \eta}{1000\,g} \frac{H}{1000 \times t \times 60(\rho_s-\rho_w) \times 1000} \right\}^{1/2}$$

Table 4.10. VISCOSITY AND DENSITY OF WATER

Temperature (°C)	Dynamic viscosity, η (mPa s)	Density, ρ_w (Mg/m³)
0	1.7865	0.999 84
5	1.5138	0.999 95
10	1.3037	0.999 70
15	1.1369	0.999 09
20	1.0019	0.998 20
25	0.8909	0.997 04
30	0.7982	0.995 65
40	0.6540	0.992 22

Based on Kaye and Laby (1973).

or

$$D = \left\{ \frac{3 \times 10^{-4} \times \eta}{g} \cdot \frac{H}{t(\rho_s - \rho_w)} \right\}^{1/2} \tag{4.8}$$

Putting $g = 9.807 \, \text{m/s}^2$, $\rho_w = 1.000 \, \text{Mg/m}^3$ and $\rho_s = G_s$ (specific gravity of particles), this equation becomes

$$D = \left\{ \frac{30.59 \times 10^{-6} \times \eta H}{t(G_s - 1)} \right\}^{1/2}$$

or

$$D = 0.005\,531 \left\{ \frac{\eta H}{t(G_s - 1)} \right\}^{1/2} \tag{4.9}$$

Equation (4.9) provides the basis for the derivation of sampling times for the pipette test (Table 4.13), and of the nomographic chart for the hydrometer method (Fig. 4.40). Values of η and ρ_w for water temperatures from 0–40°C are given in Table 4.10. Intermediate values may be obtained by interpolation, either arithmetically or graphically.

4.8 SEDIMENTATION PROCEDURES

4.8.1 Pretreatment

Before proceeding with either the pipette or hydrometer sedimentation test, the soil must be chemically treated to remove organic matter, and possible calcareous matter, and treated with dispersant and thoroughly agitated to ensure that discrete particles are separated. This process is known as pretreatment. The principle is the same whether it is for a pipette test or a hydrometer test, but there are a few differences in detail which are mainly due to the different sizes of sample required for the two tests.

The procedure described below applies to both the pipette and hydrometer sedimentation tests. Where the two types of test require different procedures, these are given under the appropriate heading.

APPARATUS

Most of the items listed below are shown in Fig. 4.28.

(1)   Mechanical shaker (end-over-end or vibratory) capable of keeping up to 75 g of soil in suspension in 150 ml of water. The shaker used for the gas-jar specific gravity test (Fig.

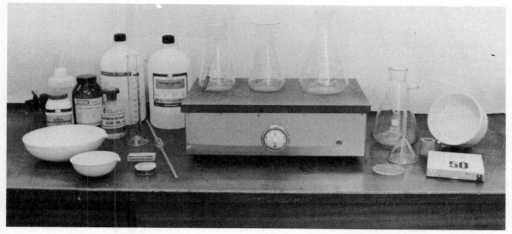

Fig. 4.28   Apparatus for pretreatment of sedimentation test sample

3.17) is suitable if fitted with an adaptor for holding a centrifuge bottle or conical flask.
(2)   Wide-mouth conical beaker, 1000 ml or 650 ml.
(3)   Measuring cylinder, 100 ml.
(4)   Wash bottle and distilled water.
(5)   Glass stirring rod with rubber 'policeman'.
(6*)   Centrifuge (if available) capable of holding bottles of 250 ml capacity.
(7*)   Polypropylene centrifuge bottles, 250 ml.
(8*)   Buchner funnel, 100 mm, and filtration flask.
(9*)   Filter pump or vacuum supply.
(10*) Whatman No. 50 filter paper.
(11*) Borosilicate glass evaporating dish.
(12)   Balance reading to 0.001 g.
(13)   Sieves 200 mm diameter; 63 μm, 212 μm, 600 μm, 2 mm apertures.
(14)   Riffle-box, and small tools for sample mixing and dividing.
(15)   Oven, 105–110 °C, and desiccator.

Required for pipette test only

(16)   Pipette, 25 ml.
(17)   Glass beaker, about 500 ml.
(18)   Glass filter funnel, 100 mm.

Required for hydrometer test only

(19)   Pipette, 100 ml.
(20)   Four porcelain evaporating dishes about 150 mm diameter.

REAGENTS

(1)   Hydrogen peroxide (20 volume solution).
(2)   'Standard' dispersant solution — that is, 33 g of sodium hexametaphosphate and 7 g of
       sodium carbonate in distilled water to make 1 litre (see Section 4.5.3).
(3)   Hydrochloric acid (N solution).

---

*   Items (8)–(11) are alternatives to items (6) and (7) if a centrifuge is not available.

PROCEDURAL STAGES

(1)   Select and prepare test specimen.
(2)   Treat for organic matter.
(3)   Treat for calcareous matter.
(4*)   Filter and dry.
           or
(5*)   Centrifuge.
(6)   Disperse.
(7)   Sieve.

TEST PROCEDURE

(1)   *Selection of sample*

The test specimen is obtained from the air-dried original sample by riffling, or by cone-and-quartering, using the fraction which passes a 2 mm sieve. The approximate size of specimen required is indicated in Table 4.11.

   It is not necessary to determine the exact mass at this stage unless the losses due to pretreatment are required. In that case the dry mass is measured either by oven drying and weighing or by weighing the air-dried sample and determining the moisture content from a separate subsample and calculating the dry mass.

Table 4.11. QUANTITY OF MATERIAL FOR
                SEDIMENTATION TESTS

Material	Pipette test	Hydrometer test
Sample after initial riffling	60 g	200 g
Test specimen:		
Sandy soils	30 g	100 g
Clayey soils	12 g	50 g

(2)   *Pretreatment for organic matter*

The air-dried test specimen is treated as follows.

Pipette method   Place the soil in a 650 ml conical beaker and add 50 ml of distilled water. Boil the suspension gently until the total volume is reduced to about 40 ml. Use a similar beaker alongside containing 40 ml of water as a comparison in order to judge the right amount. Allow to cool, and then add 75 ml of hydrogen peroxide.

Hydrometer method   Place the soil in a 1000 ml conical beaker, or a 500 ml beaker if it is known that a large amount of hydrogen peroxide will not be needed. Add 150 ml of hydrogen peroxide and stir gently for a few minutes with a glass rod.

   Cover with a cover-glass and allow to stand overnight. A tight-fitting cork must not be used, otherwise the pressure of evolved gas may burst the flask.
   Next morning heat the flask and contents gently, either on a low-heat hot-plate or on a low gas flame. Agitate frequently by stirring or by shaking with a rotary motion. Frothing over must be avoided. If necessary, add more hydrogen peroxide in increments of about 100 ml until the oxidation process is complete. Very organic soils may require several additions of hydrogen peroxide, and the oxidation process may take 2 or 3 days.
   As soon as frothing has subsided, the volume of liquid is reduced to about 50 ml by boiling,

---

*   See footnotes on pages 189 and 190.

which decomposes any excess hydrogen peroxide. The flask must not be boiled dry. The mixture is allowed to cool.

If appropriate, hydrochloric acid treatment is carried out next. (stage 3). If acid treatment is not used, the next stage is to extract the soil by filtration (stage 4) or by means of a centrifuge (stage 5).

(3)   *Pretreatment for calcareous matter*

This process is not called for in BS 1377:1975, but it was given in the 1967 Standard for use where appropriate. A check should always be made for reaction with hydrochloric acid by dropping a few spots of N HCl on to a small portion of the sample. If there is no effervescence, acid treatment is not required. A visible reaction indicates the presence of calcareous compounds which could act as a cementing agent, preventing separation of individual grains. If the effect of these compounds on the final result is to be assessed, parallel tests can be carried out on a pair of identical specimens with and without acid treatment.

When the soil and water mixture has cooled, add the required volume of hydrochloric acid (see Table 4.12). Stir with a glass rod for a few minutes and allow to stand for 1 h.

Check the pH of the suspension with a universal or wide-range pH paper, or with litmus paper. If the reaction has ceased, the suspension should show an acid reaction (pH less than 5, or blue litmus turns red). If not, add further increments of hydrochloric acid, as indicated in Table 4.12, followed by stirring and standing as above, until an acid reaction remains.

Table 4.12.   AMOUNT OF ACID FOR PRETREATMENT
(hydrochloric acid, N solution)

	Pipette test	Hydrometer test
Initial quantity	10 ml	100 ml
Subsequent increments	10 ml	25 ml

(4)   *Filtration and Drying

Set up a Buchner funnel fitted with a Whatman No. 50 filter paper on a vacuum filtration flask. It may be convenient to arrange simultaneous operation of, say, six filtration flasks by connecting them to a manifold outlet on the vacuum service line. The vacuum outlet should be fitted with a water trap to prevent water being inadvertently drawn into the main vacuum line and to the vacuum pump.

Transfer the contents of the conical beaker to the Buchner funnel, rinsing the beaker with a little distilled water to ensure that no soil is lost.

Wash thoroughly with distilled water. If acid treatment has been carried out, the washing must continue until all traces of acid are removed. When a pH paper indicator or blue litmus paper no longer gives an acid reaction, washing is complete.

Transfer the residue to a previously weighed borosilicate glass evaporating dish, using a fine jet of distilled water from a wash-bottle. Ensure that no soil is lost.

Dry in an oven, cool and weigh, and calculate the dry mass of prepared soil. This mass, denoted by m, is used in subsequent calculations, and must be measured accurately.

If the initial untreated dry mass, m_0, was measured, the pretreatment loss is equal to $(m_0 - m)$, and the percentage loss which would be reported is equal to $[(m_0 - m)/m_0] \times 100\%$.

---

*   This procedure may be used in place of stage (5) if a centrifuge is not available, and must be followed if acid treatment (stage 3) has been done. The apparatus is shown in Fig. 4.29.

Fig. 4.29   *Vacuum filtration apparatus*

(5)   *Centrifuging*

A centrifuge, if available, provides the most rapid and convenient means of recovering the soil after hydrogen peroxide pretreatment, and is the procedure recommended in the British Standard.

Weigh the dry 250 ml centrifuge bottle with stopper accurately to 0.001 g. Transfer the soil suspension from the conical beaker into the centrifuge bottle, using a fine jet of distilled water from a wash-bottle, taking care not to lose any soil particles. Adjust the volume of water in the bottle to about 200 ml, and fit the stopper. Place in the centrifuge. Usually four to six bottles can be fitted simultaneously, but opposite tubes must be balanced. Run the centrifuge for 15 min at about 2000 rev/min. Remove the bottle and decant the clear supernatant liquid (i.e. the layer of liquid containing no suspended solids). Place bottle and contents in the oven with the stopper removed and allow to dry overnight.

If a centrifuge which takes 250 ml bottles is not available, the soil suspension may be divided, if necessary, into two or more smaller bottles for use in a smaller centrifuge.

Next morning replace the stopper, cool in a desiccator, and weigh accurately. Calculate the dry mass of prepared soil (m).

Also calculate the pretreatment loss as in (stage 4) above, if required.

(6)   *Dispersion*

If the centrifuge (stage 5) has been used, dispersion by mechanical shaking can be carried out in the centrifuge bottle. If the soil has been filtered (stage 4), it must be transferred from the evaporating dish to a suitable container, such as a conical flask with stopper, as indicated below.

Pipette method   Add 100 ml of distilled water to the soil in the centrifuge bottle or evaporating dish. If in the evaporating dish, the suspension must then be transferred to the container without any loss of soil particles. Shake vigorously until all the soil is in suspension. Add 25 ml of the standard dispersing solution (Section 4.5.3), using a pipette.

Hydrometer method   Add 100 ml of the standard dispersing soluiton (Section 4.5.3), using a pipette, to the soil in the centrifuge bottle or evaporating dish. If in the evaporating dish, the

---

*   This procedure should not be used if acid treatment (stage 3) has been done, because it is then necessary to wash on a filter paper to remove all traces of acidity. The soil is retained on the filter paper as described in stage (4) above.

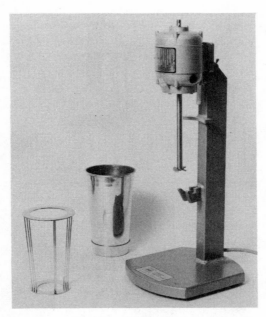

Fig. 4.30   *High-speed stirrer*

suspension must then be transferred to the container without any loss of soil particles. Shake vigorously until all the soil is in suspension.

Fit the centrifuge bottle or other container, tightly stoppered, on to the mechanical shaker. Shake for at least 4 h, or overnight if convenient. However, if excess shaking would cause breakdown of soil particles (e.g. with shales), the period of shaking should be less than 4 h. The British Standard is not specific about this, and a decision must be based on experience and knowledge of the soil.

If a mechanical shaker of this type is not available, the following alternative procedures could be used for dispersion.

(a)   The 1967 Standard specified dispersion in the cup of a mechanical stirrer (Horlicks mixer), fitted with a wire baffle (Fig. 4.30). The soil suspension is stirred at high speed for 15 min.

(b)   Alternatively, a Vibro-Stirrer has been found to be equally effective. This type of stirrer is fitted with a blade which oscillates through an angle of $10°$ at 5000 Hz (Fig. 4.31). One advantage of this device is that it can be used in the beaker or conical flask used for pretreatment.

(c)   A laborious hand method is described in the 1967 Standard. The soil suspension is placed in a mortar and rubbed vigorously with a rubber pestle. After settling for 2 min, the liquid suspension is decanted through a 75 μm sieve into the receiver to which it is fitted. More distilled water is added to the mortar and the pestling and decanting processes are repeated until the decanted liquid is clear. Care must be taken to prevent breakdown of individual particles.

(7)   *Wet Sieving*

Transfer the soil and suspension to a 63 μm sieve, nested on a receiver, without loss of soil.

Wash the soil with a jet of distilled water from a wash-bottle, until all fine material is washed through the sieve. The amount of water should not exceed 150 ml for the pipette test or 500 ml for the hydrometer test. The material collected in the receiver is used for the appropriate sedimentation test (Section 4.8.2 or 4.8.3).

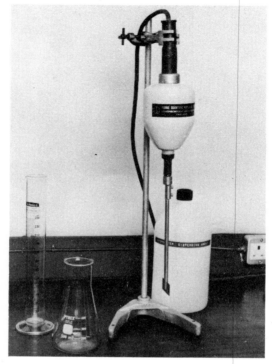

Fig. 4.31   *Vibro-stirrer*

Transfer the material retained on the 63 µm sieve to an evaporating dish, dry and weigh.

Sieve through larger aperture sieves as appropriate, and weigh the amount retained on each sieve, as for simple dry sieving (Section 4.6.1). These masses are used subsequently for the construction of the grading curve in the sand size range.

Any material passing the 63 µm sieve when dry sieved should be added to the sedimentation cylinder.

4.8.2   Pipette Analysis (BS 1377:1975, Test 7 (C))

SCOPE

In this test a special sampling pipette is used to obtain a sample of soil, pretreated as in Section 4.8.1, from a suspension in water at three fixed time intervals. The mass of soil in each sample is determined, from which the percentages of coarse silt, medium silt, fine silt and clay can be calculated.

The British Standard refers to this procedure as the standard or primary method for fine particle analysis. However, the apparatus required is expensive and delicate, and is not suitable for use in a site laboratory running control tests during construction.

APPARATUS

The following apparatus is required in addition to that used for pretreatment.

(1)   Sampling pipette (Fig. 4.32) mounted on a stand with a suitable lowering and raising device (Fig. 4.33). The pipette must be calibrated before use, as described under stage (7) below. It is sometimes referred to as an Andreasen pippette.

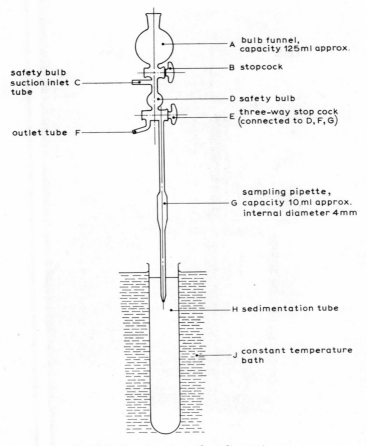

safety bulb
suction inlet C
tube

outlet tube F

A bulb funnel,
   capacity 125 ml approx.

B stopcock

D safety bulb

E three-way stop cock
   (connected to D, F, G)

G sampling pipette,
   capacity 10 ml approx.
   internal diameter 4 mm

H sedimentation tube

J constant temperature
   bath

Fig. 4.32   Sampling pipette for sedimentation test

(2)   Constant temperature bath maintained at 25 °C to ±0.1 °C.
(3)   Two glass sedimentation cylinders, 50 mm diameter, approximately 350 mm long, graduated at 500 ml, with rubber bungs to fit.
(4)   Stop-clock.
(5)   Glass rod about 12 mm diameter and about 400 mm long.

PROCEDURAL STAGES

(1)   Prepare suspension.
(2)   Take samples with pipette.
(3)   Dry and weigh solid matter.
(4)   Take and measure calibration sample.
(5)   Calculate.
(6)   Plot and present results.
(7)   Calibration of pipette (need not be repeated when the internal volume is known).

TEST PROCEDURE

(1)   *Preparation of suspension*

The suspension of pretreated soil passing the 63 μm sieve (obtained as described in Section 4.8.1) is transferred with a glass funnel from the receiver into a 500 ml sedimentation cylinder.

Fig. 4.33   Apparatus for pipette sampling

No soil must be lost in this operation. The water level in the cylinder is made up to the 500 ml calibration mark.

Place the sedimentation cylinder in the constant-temperature bath, set at 25 °C. Into a second sedimentation cylinder place 25 ml of the dispersant solution and make up to the 500 ml graduation mark with distilled water. Place this tube in the same constant-temperature bath.

Insert a rubber bung into each cylinder, and allow them to stand in the constant-temperature bath until they have reached the bath temperature; about 1 h is usually sufficient.

Several sedimentation cylinders (up to six or eight) may be tested at the same time. When all cylinders are in the constant-temperature bath, the water level of the bath should just reach their 500 ml graduation marks.

Remove the sedimentation cylinders in turn, shake thoroughly by inverting end-over-end several times, stir the sediment with the glass rod so that it all goes into suspension, and replace in the bath. Also shake and replace the cylinder containing the dispersant solution.

At the instant when the cylinder containing soil suspension is replaced in the bath, start the stop-clock. The instant of placing the tube in the bath is zero time ($t = 0$) for that sample.

Each cylinder can be timed separately, a stop-clock being used for each, but only one clock is necessary if they are started at regular intervals of, say, 5 min.

(2)   *Sampling with Pipette*

Pipette samples are to be taken at three specified time intervals from zero time for each sample.

Table 4.13.   PIPETTE SAMPLING TIMES (25 °C)

SG of silt and clay fraction	*1st sample*		*2nd sample*		*3rd sample*	
	min	*s*	*min*	*s*	*h*	*min*
2.50	4	30	50	30	7	35
2.55	4	20	49	0	7	21
2.60	4	10	47	30	7	7
2.65	4	5	46	0	6	54
2.70	4	0	44	30	6	42
2.75	3	50	43	30	6	30
2.80	3	40	42	0	6	20
2.85	3	35	41	0	6	10
2.90	3	30	40	0	6	0
2.95	3	25	39	0	5	50
3.00	3	20	38	0	5	41
3.05	3	15	37	0	5	33
3.10	3	10	36	0	5	25
3.15	3	5	35	0	5	18
3.20	3	0	34	30	5	10

Times after shaking of starting sampling operation

These times depend upon the specific gravity of the particles in the suspension, and are given in Table 4.13. The third sampling operation is at about 7 h from zero, so it is convenient to shake up and start a batch of tests first thing in the morning.

Move the sampling pipette on its supporting frame over the sedimentation cylinder. Lower the pipette until the tip just touches the surface of the water in the cylinder, and record the reading R_0 (mm) on the graduated scale. Repeat for each sedimentation tube of the batch, recording each level separately.

About 15 s before a pipette sample is due (Table 4.13), lower the pipette steadily with tap E closed (see Fig. 4.32) into the suspension until the tip is 100 mm below the surface. This will be when the scale reading is equal to $(R_0 + 100)$ mm (or $R_0 - 100$ if the scale reads upwards). The lowering operation should take about 10 s, and must cause no turbulence in the suspension.

Open tap E at the exact sampling time, and draw a sample (volume V_p ml) up into the pipette, until the pipette and the bore in tap E are filled with suspension. A small amount may also be drawn into bulb D. The easiest way of drawing the sample up is by sucking through a length of rubber tubing attached to the inlet C. This operation should take about 10 s. Withdraw the pipette steadily from the suspension, taking about 10 s.

If any suspension has been drawn up into bulb D, wash it away into a beaker down outlet tube F by opening tap E to connect D and F. Allow distilled water to flow from bulb funnel A into D and out through F until no suspension remains.

Place a weighed glass weighing bottle under the end of the pipette and open tap E so that the contents run into the bottle. Wash any suspension on the inside of the pipette into the weighing bottle by allowing distilled water to flow from A via B, D and E into the pipette. The last few drops of suspension can be removed from the end of the pipette by gently blowing through the tube connected to C.

(3)   Determination of dry solid matter

Place the weighing bottle and contents in an oven at 105 °C until the sample is evaporated to dryness.

Cool in a desiccator, and weigh carefully to the nearest 0.001 g or less. Accurate weighing is essential, because the quantity of soil recovered is very small indeed. Check the mass of the weighing bottle again after cleaning out and oven drying.

From the mass of the empty bottle the mass of soil in the sample is determined. This is denoted by m_1, m_2, m_3 for the first, second and third sampling operation, respectively. These masses are used for the calculations described in stage (5).

(4)   Calibration sampling

At any convenient time between the stipulated sampling times, a pipette sample is taken from the sedimentation tube containing dispersant only, exactly as described for the soil suspensions. The time of this sampling need not be recorded.

Transfer the pipette sample to a glass weighing bottle, dry, cool and weigh accurately, exactly as for the soil suspensions. Hence determine the mass of solid material, m_4, in the dispersant sample.

(5)   Calculations

The percentages finer than each sieve size in the gravel (if present) and sand size ranges are calculated as described in the appropriate section under sieving.

Data obtained from the pipette test enable the percentages of coarse silt, medium silt, fine silt and clay present in the pretreated sample (m) to be calculated, as follows. A worked example is given in Fig. 4.34.

The mass of solid material in the whole 500 ml of suspension at any sampling time can be calculated by proportion from the measured mass in the pipette volume, V_p. If the mass in 500 ml is denoted by W_1, W_2, W_3 (corresponding to sample masses m_1, m_2, m_3), then $W_1 = m_1 \times (500/V_p)$ g, and similarly for W_2 and W_3.

The mass of medium silt (particles in the size range 0.02–0.006 mm) is the mass settled out between the first and second pipette samplings, and is therefore equal to the difference between the masses present at these times:

$$\text{mass of medium silt} = W_1 - W_2$$

The percentage of medium silt, P_M, is given by

$$P_M = \frac{W_1 - W_2}{m} \times 100$$

$$= \frac{m_1 - m_2}{m} \times \frac{500}{V_p} \times 100$$

Similarly, the percentage of fine silt, P_F, is given by

$$P_F = \frac{m_2 - m_3}{m} \times \frac{500}{V_p} \times 100$$

The percentage of clay, P_{cy}, is given by

$$P_{cy} = \frac{m_3 - m_4}{m} \times \frac{500}{V_p} \times 100$$

If the total dry mass of soil particles retained on the 63 μm sieve is denoted by m_R, the percentage of sand (and gravel) sizes, P_{SG}, is given by

$$P_{SG} = \frac{m_R}{m} \times 100$$

Mass of sample after pretreatment m = 25.36 g

S.G. = 2.65

Volume of pipette

Mass retained on 63 um sieve m_s = 3.88 g
(all finer than 212 um sieve)

Temperature 25°C

9.656 ml

Pipette sample no. and time	Pipette mass calculation	grams	Mass in 500 ml in suspension	Percentage calculation	Size category
(1) 4 min. 5s	Weighing bottle no. Bottle + sample Bottle Mass of sample m_1	(25) 5.5980 5.3024 0.2956	$0.2956 \times \dfrac{500}{9.656} = 15.307$ $m_1 \times \dfrac{500}{V_p} = W_1$	W_1 15.307 W_2 9.020 $(W_1 - W_2)\ \overline{6.287} \times \dfrac{100}{25.36} = 24.8\%$	Medium silt (20 to 6 μm)
(2) 46 min.	Weighing bottle no. Bottle + sample Bottle Mass of sample m_2	(44) 5.6344 5.4602 0.1742	$0.1742 \times \dfrac{500}{9.656} = 9.020$ $= W_2$	W_2 9.020 W_3 5.717 $(W_2 - W_3\ \overline{3.303} \times \dfrac{100}{25.36} = 13.0\%$	Fine silt (6 to 2 μm)
(3) 6 hours 54 min.	Weighing bottle no. Bottle + sample Bottle Mass of sample m_3	(76) 5.9262 5.8158 0.1104	$0.1104 \times \dfrac{500}{9.656} = 5.717$ $= W_3$	W_3 5.717 W_4 1.336 $(W_3 - W_4\ \overline{4.381} \times \dfrac{100}{25.36} = 17.3\%$	Clay (less than 2 μm)
(4) About 30 min.	Weighing bottle no. Bottle + sample Bottle Mass of sample m_4	(18) 5.4418 5.4160 0.0258	$0.0258 \times \dfrac{500}{9.656} = 1.336$ $= W_4$	$m_s\ \ 3.88 \times \dfrac{100}{25.36} = 15.3\%$	Fine sand (212 to 63 μm)

Total from above   70.4
                      100.0
                      29.6%   Coarse silt
                               (63 to 20 μm)

Cumulative percentages finer than each particle size:

Clay	0.002 mm			17.3%
Fine Silt	0.006 mm	13.0 + 17.3	=	30.3%
Medium Silt	0.02 mm	24.8 + 30.3	=	55.1%
Coarse Silt	0.06 mm	29.6 + 55.1	=	84.7%
Sand and gravel	0.212 mm	15.3 + 84.7	=	100.0%

In this example all the material retained on 63 μm sieve is sand finer than 212 μm

Fig. 4.34 Pipette sedimentation test results and calculations

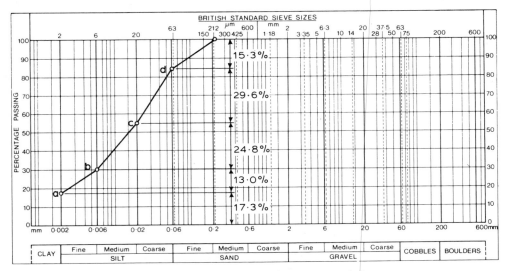

Fig. 4.35   Particle size curve from pipette analysis

The percentage of coarse silt, P_c, is determined by difference:

$$P_c = 100 - (P_{SG} + P_M + P_F + P_{cy})$$

The percentages of each size range may be tabulated, or used as the basis for a particle size distribution curve as described in stage (6).

The sand and gravel fraction retained on the 63 μm sieve is sieved dry through appropriate larger sieves, from which the percentages retained can be derived as indicated in Section 4.6.1.

(6)   Plotting and presentation

The above calculation procedure is based on that given in BS 1377:1975. The particle size curve may be plotted more directly as follows (see Fig. 4.35).

Plot P_{cy} as the percentage corresponding to a particle size of 0.002 mm (point a, Fig. 4.35). Plot ($P_{cy} + P_F$) at 0.006 mm (point b). Plot ($P_{cy} + P_F + P_M$) at 0.02 mm (point c). Plot ($100 - P_{SG}$) at 0.063 mm (point d). Join up these points to give the particle size curve in the silt range. It is not necessary to calculate the percentage of coarse silt in order to draw the curve, but this percentage can be read off the curve as the difference between points c and d.

Add the particle size curve from the sieving test on the material retained on the 63 μm sieve, starting at point d. This then gives the whole grading curve for the pretreated sample of soil.

The result of this test gives readings corresponding only to sizes of 0.002, 0.06, 0.02 mm — that is, the thick ordinates on the grading curve sheet. If required, pipette sampling can be carried out at intermediate times so as to obtain percentages at other equivalent particle diameters. These diameters would have to be computed from the time intervals from zero by use of Equation (4.9) in Section 4.7.3.

(7)   Calibration of pipette

The internal volume (V_p ml) of the sampling pipette is determined as follows.

Clean the pipette thoroughly and dry it. Immerse the nozzle in distilled water, with tap B closed, and open tap E (Fig. 4.32). Using a rubber tube attached to outlet C, suck water into the pipette until it rises above E. Close tap E and remove the pipette from the water. Pour off surplus water in the cavity above E through outlet F into a small beaker. Discharge the water in the pipette and tap E into a glass weighing bottle of known mass, and determine its mass.

The internal volume V_p ml of the pipette and tap is equal to the mass of water in grams, calculated to the nearest 0.01 g.

Make three determinations of the volume, take the average and express the volume V_p to the nearest 0.05 ml.

4.8.3 Hydrometer Analysis (BS 1377:1975, Test 7 (D))

SCOPE

In this method a specific gravity hydrometer of special design is used to measure the density of a soil, pretreated as in Section 4.8.1, in a suspension in water at various intervals of time. From these measurements the distribution of particle sizes in the silt range (60–2 μm) can be assessed. The test is not usually performed if less than 10% of the material passes the 63 μm sieve.

Although this method is referred to in the British Standard as the subsidiary method, it can give results which are sufficiently accurate for most engineering purposes. The techniques are less exacting than those required for the pipette method. The hydrometer method has the additional advantage that it can be performed without much difficulty in a small field laboratory. If a main central laboratory also uses this procedure, the results obtained by both are directly comparable.

Before use every hydrometer must be calibrated, and a calibration curve must be drawn relating 'effective depth' to 'hydrometer reading' (see Section 4.8.4). This calibration must be inserted in the appropriate space in the British Standard nomograph chart. Without it the nomograph chart is useless, a fact which is not made clear in the Standard.

APPARATUS

The following apparatus is required in addition to that used for pretreatment.

(1)   Soil hydrometer of the type illustrated in Fig. 4.36. The British Standard specifies that it shall be calibrated to real 1.000 in pure water at 20 °C. Commercially available hydrometers are calibrated to read *true density* at 20 °C — that is, in pure water at that temperature the reading is 0.9982. However, a correction can be made as described below in stage (3d).
(2)   Two 1000 ml glass measuring cylinders about 330 mm high, marked at 1000 ml, with stoppers or rubber bungs.
(3)   Constant-temperature bath.
(4)   Stop-clock.
(5)   Glass rod about 12 mm diameter and about 400 mm long.

PROCEDURAL STAGES

(1)   Prepare suspension.
(2)   Take hydrometer readings.
(3)   Correct hydrometer readings.
(4)   Calculate.
(5)   Present results.

TEST PROCEDURE

(1)   *Preparation of suspension*

The suspension of pretreated soil passing the 63 μm sieve, obtained as described in Section 4.8.1, is transferred from the receiver into a 1000 ml sedimentation cylinder without losing any soil. The suspension is made up exactly to the 1000 ml mark with distilled water.

Place the sedimentation cylinder in the constant-temperature bath, set at 25 °C. Place a

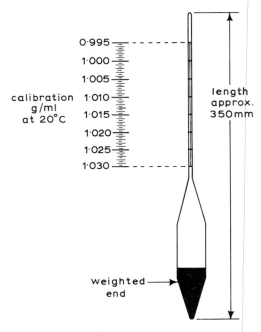

calibration
g/ml
at 20°C

0·995
1·000
1·005
1·010
1·015
1·020
1·025
1·030

length
approx.
350mm

weighted ─────▶
end

Fig. 4.36   Soil hydrometer

second cylinder containing distilled water in the constant-temperature bath; this is for holding the hydrometer between readings.

Allow the cylinders to stand in the bath until they have reached the bath temperature; about 1 h is usually sufficient. Several sedimentation cylinders (up to six or eight) may be tested at the same time. When all cylinders are in the constant-temperature bath, the water level in the bath should just reach the 1000 ml graduation marks.

Insert a rubber bung into the sedimentation cylinder. This must be pushed in sufficiently to obtain a watertight fit, but undue force must not be applied, as otherwise the glass may split and cause serious injury to the hand. The cylinder is then shaken vigorously to obtain a uniform suspension. Stir the sediment with a glass rod so that it all goes into suspension. The cylinder is inverted for a few seconds, and is then stood in the constant-temperature bath without delay. As soon as it is in the upright position, the stop-watch or timer clock is started (zero time, $t = 0$).

(2)   Hydrometer readings

Remove the rubber bung, insert the hydrometer steadily and allow it to float freely (Fig. 4.37). It must not be allowed to bob up and down, or to rotate, when let go. However, a quick rotational twist with the fingers on the end of the stem will dislodge any air bubbles which may adhere to the side.

Readings of the hydrometer are taken, in the manner described in stage (3), at the top of meniscus level at the following times from zero: $\frac{1}{2}$, 1, 2, 4 min.

The hydrometer is then removed slowly, rinsed in distilled water, and placed in the separate cylinder of distilled water in the constant temperature bath.

Insert the hydrometer for further readings at the following times from zero, and remove and place it back in the distilled water cylinder after each reading: 8, 15, 30 min; 1, 2, 4, 8 h; overnight (about 16 h); thereafter (if necessary) twice daily. It is not essential to keep rigidly to these times, provided that the actual time of each reading is recorded on the hydrometer test sheet.

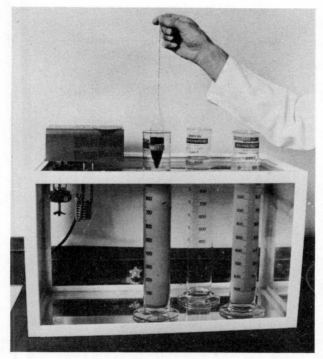

Fig. 4.37   Immersion of hydrometer in sedimentation cylinder

Insertion and withdrawal of the hydrometer into the suspension must be done carefully. Each operation should take about 10 s, and when released the hydrometer should be in its steady floating position.

Disturbance of the suspension, either by the hydrometer or by vibration, must be avoided. If a heater/stirrer unit is fitted to the constant-temperature bath, this must be mounted so that no vibration is transmitted to the sedimentation cylinder.

If six or eight samples are being treated together as a batch, it is best to start them at intervals of 5 min from the start of the first. Between each reading the hydrometer should be dipped into the distilled water cylinder.

The temperature of the suspension should be checked at intervals, but if a reliable constant-temperature bath is used, there should normally be no significant change in temperature throughout the test. A constant temperature of 25 °C is preferable to one of 20 °C, partly because it eliminates the necessity of cooling in all but the hottest climates, and also because it is convenient for the reason given in stage (3).

(3)   *Correction of hydrometer readings*

Each density reading taken on the hydrometer must first be expressed as a hydrometer reading, R'_h, corresponding to the level of the upper rim of the meniscus. This is done by subtracting 1 from the density and moving the decimal point three places to the right (i.e. multiply by 1000). For example, a density of 1.0325 would be a hydrometer reading of $R'_h = 32.5$.

To each reading R'_h must now be applied four corrections, as follows:

(a)   Meniscus correction, C_m.
(b)   Temperature correction, M_t.
(c)   Dispersing agent correction, x.
(d)   Water density correction, C_w.

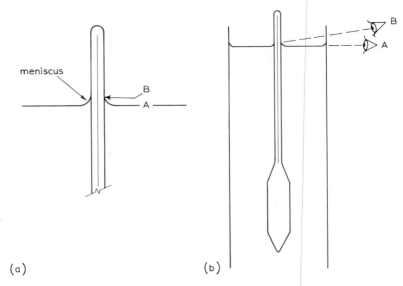

Fig. 4.38   Reading a hydrometer

The first three corrections referred to are explained in the British Standard. The fourth follows from the specification for the calibration of the hydrometer (Head, 1976), but is not explained in the Standard. The way these corrections are obtained is described below.

(a)   *Meniscus correction*   A hydrometer is calibrated to read correctly at the surface of the liquid in which it is immersed (level A in Fig. 4.38a). Since soil suspensions are not transparent enough to permit a reading to be taken at this level, the scale has to be read at the upper rim of the meniscus. This is shown at B in Fig. 4.38(a). It is therefore essential that the meniscus be fully developed, which means that the hydrometer stem must be perfectly clean.

The meniscus correction (C_m) has to be added to R'_h in order to obtain the true reading R_h, because density readings on the stem increase downwards. The correction C_m is a constant for a given hydrometer, and is determined as follows.

The hydrometer is inserted in a 1000 ml cylinder about three-quarters full of distilled water. The plane of the surface of the liquid is seen as an ellipse from just below the surface. The eye is raised until the surface is seen as a straight line, and the scale marking at which this plane intersects the hydrometer stem is noted (reading A in Fig. 4.38b). By looking from just above the plane of the liquid surface, the scale marking at the level of the upper limit of the meniscus is noted (reading B). The difference between the two scale readings, multiplied by 1000, is the meniscus correction:

$$C_m = (B - A) \times 1000$$

For example:

$$\text{If reading } A = 0.9985$$
$$\text{reading } B = 0.9990$$

$$(B - A) = 0.0005$$

$$C_m = +0.5$$

This is a typical value for C_m, but it must be determined for every hydrometer. The true

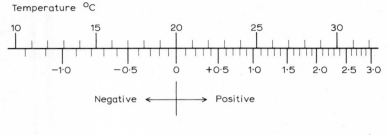

Fig. 4.39   *Temperature correction chart*

hydrometer reading R_h is given by

$$R_h = R'_h + C_m$$

(b)   *Temperature correction*   Hydrometers are usually calibrated at 20 °C. If a test is carried out at a different temperature, both the density of water and the density of the hydrometer (owing to the thermal expansion of glass) will be different. These factors are allowed for in the temperature correction chart (Fig. 4.39), which is reproduced from Fig. 15 of BS 1377:1975.

The value of M_t given on the chart at the appropriate temperature is added to the true hydrometer reading R_h.

Examples:

$$\text{at } 27\,°\text{C}, \ M_t = 1.5$$

$$\text{add 1.5 to } R_h$$

$$\text{at } 17\,°\text{C}, \ M_t = -0.5$$

$$\text{subtract 0.5 from } R_h$$

(c)   *Dispersing agent correction*   The addition of the dispersing agent results in the density of the liquid in which sedimentation takes place being greater than that of water. To determine the correction x, a volume of exactly 50 ml of the dispersing agent solution (i.e. the stock solution) is placed in a weighed bottled or evaporating dish. The water is evaporated by drying in the oven at 105–110 °C, and the mass of dispersing agent m_d g remaining in the container is determined. The correction x to be applied to R_h is given by

$$x = 2m_d$$

This correction is independent of temperature, and is typically 3.5–4.0 for the standard dispersing agent. The value should be checked periodically, and always measured if a non-standard dispersant solution is used.

The x correction is always subtracted from the R_h value.

(d)   *Water density correction*   The 1975 British Standard specifies that the scale of the hydrometer shall be calibrated in g/ml to read 1000 at 20 °C. The density of pure water is exactly 1.000 only at 4°C, at which temperature the density is at its maximum. At higher temperatures it is less than this, as shown in Table 4.10. At 20 °C the density is 0.9982, which would correspond to an R_h value on the hydrometer of -1.8. This is the reading which a normal hydrometer shows when immersed in pure water at 20 °C, because hydrometers are calibrated to read true density in g/ml. There are as yet no hydrometers available which read exactly as required by the British Standard.

To comply with the Standard, therefore, all hydrometer readings must be increased by 1.8

Table 4.14. HYDROMETER READING CORRECTIONS

Temperature (°C)	C_m	M_t	C_w	x	Total correction calculated	Total correction rounded
15	+0.5	−0.75	+1.8	−3.5	−1.95	−2
20	+0.5	0	+1.8	−3.5	−1.20	−1
25	+0.5	+.10	+1.8	−3.5	−0.2	0
30	+0.5	+2.3	+1.8	−3.5	+1.1	+1

when used at 20 °C. At any other temperature the same correction is applied, together with the appropriate M_t correction referred to above.

The fully corrected hydrometer reading R is given by

$$R = R'_h + C_m + M_t - x + 1.8$$

The total of the four corrections at different temperatures is given in Table 4.14. This is based on a meniscus correction $C_m = 0.5$ and a dispersant correction $x = 3.5$.

It is interesting to observe that at 25°C the total correction is -0.2, which for most practical purposes may be taken as zero. Thus, at 25 °C, provided that the standard dispersant is used, together with a standard hydrometer, the observed top of meniscus readings may be taken to be the same as the fully corrected readings R. This is one advantage of conducting the test at 25 °C as the standard constant temperature.

As explained in stage (4), the fully corrected reading R is used only for calculation of percentages of particles smaller than a given size. The value of $R_h = R'_h + C_m$ (i.e. meniscus correction only) is used at all temperatures for computing the particle diameter D, whether by calculation, by using tables or from the nomographic chart, because here the hydrometer is acting as a measuring rod to determine the effective depth at which the density reading applies.

(4) Calculations

(a) *Equivalent particle diameter D* The equivalent particle diameter at a known depth and after a certain time interval from the start of sedimentation can be calculated from the following equation derived in Section 4.7.3:

$$D = 0.005\,531 \sqrt{\frac{\eta H}{(G_s - 1)t}}$$

where D = equivalent particle diameter (mm); η = viscosity of water at test temperature (mPas); H = effective depth (mm); G_s = specific gravity of particles; t = elapsed time (min). Values of η at various temperatures are given in Table 4.10.

This equation can be solved without tedious repetitive calculations by using the nomographic chart, devised by Casagrande (1931), which is given in Fig. 14 in the British Standard and is shown here in Fig. 4.40. An enlarged copy of the chart, mounted on card with a plasticised finish, is commercially available and is easier to read and use than the version in the Standard.

Before the chart can be used, the calibration of the hydrometer used for the test must be added. This consists of a scale of R_h values corresponding to the printed H_r figures at the extreme right-hand side of the chart. The procedure is described in Section 4.8.4.

The nomograph chart is used as follows. The procedure is illustrated by the skeleton diagram in Fig. 4.41.

Find the specific gravity of the soil particles on the G_s scale (1), and the test temperature on

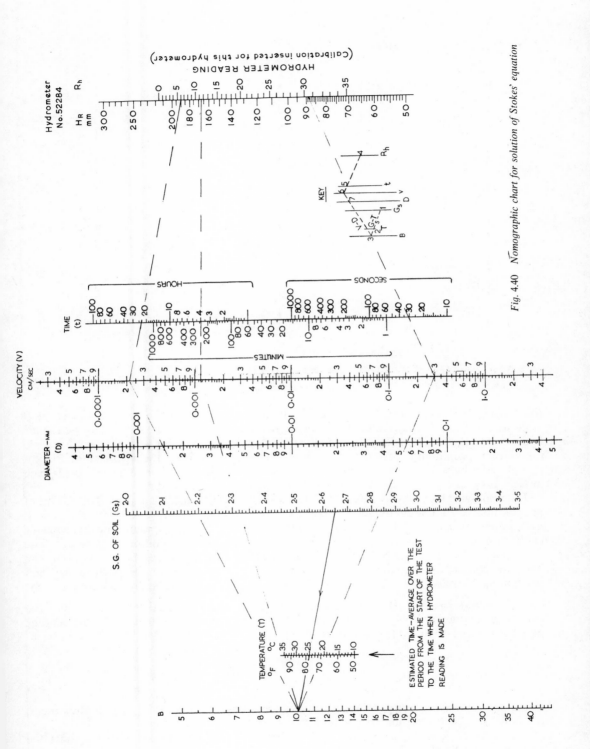

Fig. 4.40 Nomographic chart for solution of Stokes' equation

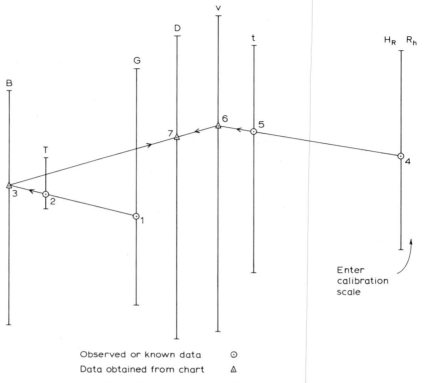

Observed or known data   ☉

Data obtained from chart   △

Fig. 4.41   Use of nomographic chart

the T scale (2). Place a straight-edge between them and extrapolate the line to intersect the B scale at (3). Find the corrected hydrometer reading on the R_h scale for that particular hydrometer (4), using $R_h = R'_h + C_m$. Find the time at which that reading was taken on the t scale (5). Place a straight-edge between them and extrapolate to intersect the v scale at (6). Place a straight-edge between (3) and (6) and read off the diameter where this lines intersects the D scale at (7). This is the particle diameter relevant to the hydrometer reading.

The nomograph chart was devised before the days of electronic calculators. The advent of programmable computers renders the chart obsolete in many laboratories, since the Stokes' Law equation can be solved almost instantaneously with a computer. The hydrometer calibration curve, if linear, can be easily incorporated into the computer program. The percentage calculation (see below) can also be included. It is then necessary to enter only the hydrometer reading and the time in order to obtain immediately the particle size and percentage. If linked to an X–Y plotter, the particle size curve can be drawn automatically.

(b) *Percentage smaller than D*   The percentage, by mass, of particles smaller than the equivalent diameter, D, is denoted by K. This percentage is equivalent to the 'percentage passing' in sieve analysis. It is given by the equation

$$K = \frac{G_s}{m(G_s - 1)} \times R \times 100\%$$

where G_s = specific gravity of soil particles; m = mass of dry soil after pretreatment; R = fully corrected hydrometer reading = $R'_h + C_m + M_t - x + 1.8$.

The value of K is calculated for each hydrometer reading, and is plotted against the corresponding particle size, drawn to a logarithmic scale, exactly as for a grading curve determined by sieving. The same graph sheet is used, the sizes of particles being extended downwards to about 1 μm. Usually the test is terminated at about 2 μm, which is the lower limit of the silt size range. The intersection of the particle size curve with the 2 μm ordinate gives the percentage which is referred to as the 'clay fraction'.

The significance of the correction for the density of water ($C_w = +1.8$) becomes clear at the fine end of the scale. In the early stages of the test, when the hydrometer reading R'_h is 20 or more, this correction amounts to less than 10% of the reading used in the calculation. But towards the end the value of R'_h falls to 5 or less, and this correction is then of the same order of magnitude as the reading. Without this correction the derived value of the clay fraction can be seriously underestimated, and could be only half the correct value.

The above method of calculating percentages applies only when the whole of the sample passes a 2 mm sieve, and is used for pretreatment. If the hydrometer test specimen is obtained by quartering down from a larger sample which has been put through larger aperture sieves, the calculated percentages must be adjusted as described in Section 4.8.5.

(5)   Presentation of results

Calculated percentages finer than each determined size are plotted against the corresponding particle diameter on the same sheet as that used for a sieving analysis. A smooth curve is drawn through the plotted points. If a sieving analysis has also been carried out, a single continuous curve is drawn to give the particle size distribution of the whole sample (see Section 4.8.5).

Details of pretreatment, the size of sample used and the SG of particles used in the calculations are added to the particle size distribution sheet, together with a visual description of the soil.

(6)   Typical results and calculations

A typical set of results obtained from a hydrometer sedimentation test is given in Fig. 4.42, and the sieving data for the dried pretreated portion retained on the 63 μm sieve are given in Fig. 4.43.

The method of calculation of particle diameters, and of percentages finer than each particle size, from the sedimentation test is shown in Fig. 4.42. Normally the nomograph chart would be used for evaluating particle diameters, and the derivation of the first and last readings (30 s and about 24 h), and the readings taken at 4 h, are marked on the nomograph chart in Fig. 4.40.

The percentages corresponding to each particle size are plotted on the particle size sheet in Fig. 4.44, for both the sieving test and the sedimentation test. The masses retained on the sieves are calculated as percentages of mass m, the total dry mass *after* pretreatment. In this example the first point from the sedimentation test does not lie on the smooth curve connecting the sieving curve with the remaining points. This is not uncommon, and the reason for the apparent discrepancy is discussed in Section 4.8.5.

4.84.   Calibration of Hydrometer

The hydrometer must be calibrated in the cylinder in which it is to be used. This is because the cross-sectional area, A, of the cylinder comes into the calibration calculations. In practice, the sectional area varies but little between similar measuring cylinders of one batch; nevertheless each cylinder used should be checked.

To determine the sectional area A, measure the distance, L, in millimetres between two well-spaced graduations (such as 100 and 900 ml) on the cylinder. The volume included between these two marks is 800 ml, so the sectional area, A, is given by

$$A = \frac{800}{L} \times 1000 \, \text{mm}^2$$

Hydrometer no.	52284
Test temperature	25°C
Meniscus correction	$C_m = +0.5$
Temperature correction	$M_t = +1.0$
Dispersant correction	$x = 3.5$

Specific gravity of particles	$G_s = 2.65$
Viscosity of water	$\eta = 0.8909$ mPas
Initial dry mass of soil	59.37
Dry mass after pretreatment	$m = 58.88$ g
Pretreatment loss	$= \overline{0.49}$ g
$\frac{0.49}{59.37} \times 100$	$= 0.83 \%$

(1) Date Time	(2) Elapsed time t minutes	(3) Hydrometer reading R_h'	(4) True reading R_h	(5) Effective depth H_R mm	(6) Fully corrected reading R	(7) Particle diameter D μm	(8) Percentage finer than D K %
1.6.78 0935	0						
	0.5	30	30.5	88.9	29.8	$4.064 \times \sqrt{\dfrac{88.9}{0.5}} = 54.2$	$2.728 \times 29.8 = 81.3\%$
	1	29.5	30	91.0	29.3	$4.064 \times \sqrt{\dfrac{91.0}{1}} = 38.8$	$2.728 \times 29.3 = 79.9\%$
	2	28.5	29	95.1	28.3	$4.064 \times \sqrt{\dfrac{95.1}{2}} = 28.0$	$2.728 \times 28.3 = 77.2\%$
	4	27	27.5	101.2	26.8	$4.064 \times \sqrt{\dfrac{101.2}{4}} = 20.4$	$2.728 \times 26.8 = 73.1\%$
0943	8	24.5	25	111.5	24.3	$4.064 \times \sqrt{\dfrac{111.5}{8}} = 15.2$	$2.728 \times 24.3 = 66.3\%$
0950	15	22.5	23	119.7	22.3	$4.064 \times \sqrt{\dfrac{119.7}{15}} = 11.5$	$2.728 \times 22.3 = 60.8\%$
1005	30	20	20.5	130.0	19.8	$4.064 \times \sqrt{\dfrac{130.0}{30}} = 8.46$	$2.728 \times 19.8 = 54.0\%$
1035	60	17.5	18	140.2	17.3	$4.064 \times \sqrt{\dfrac{140.2}{60}} = 6.21$	$2.728 \times 17.3 = 47.2\%$
1135	120	14.5	15	152.5	14.3	$4.064 \times \sqrt{\dfrac{152.5}{120}} = 4.58$	$2.728 \times 14.3 = 39.0\%$
1335	240	11	11.5	166.9	10.8	$4.064 \times \sqrt{\dfrac{166.9}{240}} = 3.39$	$2.728 \times 10.8 = 29.5\%$
1705 2.6.78	450	8	8.5	179.2	7.8	$4.064 \times \sqrt{\dfrac{179.2}{450}} = 2.56$	$2.728 \times 7.8 = 21.3\%$
0915	1420	5.5	6	189.4	5.3	$4.064 \times \sqrt{\dfrac{189.4}{1420}} = 1.48$	$2.728 \times 5.3 = 14.5\%$

Calculations: Col.(4) $R_h = R_h' + C_m = R_h' + 0.5$

Col.(5) $H_R = 214 - 4.1\,R_h$ (or read off calibration curve)

Col.(6) $R = R_h + M_t - x + 1.8 = R_h + 1.0 - 3.5 + 1.8 = R_h - 0.7$

Col.(7) $D = 0.005531 \sqrt{\dfrac{\eta\,H_R}{(G_s - 1)t}} = 0.005531 \sqrt{\dfrac{0.8909\,H_R}{1.65t}}$ mm $= 4.064 \sqrt{\dfrac{H_R}{t}}$ μm

(or obtain D from nomograph chart)

Col.(8) $K = \dfrac{G_s}{m(G_s - 1)} \times R \times 100 = \dfrac{2.65}{58.88 \times 1.65} \times R \times 100 = 2.728 \times R\%$

Fig. 4.42   *Hydrometer sedimentation test data, results and calculations*

The sides of the cylinder must be parallel, so that the sectional area is constant throughout its length.

On the hydrometer itself the distance from the neck of the bulb to the lowest calibration mark is measured, to the nearest millimetre, with a steel ruler. This is denoted by N in Fig. 4.45. The distances l_1, l_2, etc., from this calibration mark to each of the other major marks are measured as shown, to the nearest millimetre, and tabulated. The distance H corresponding to each reading R_h is given by $(N + l_1)$, $(N + l_2)$, etc.

Initial total dry mass after pretreatment 58.88 g

Dry mass retained on 63 μm sieve   5.92 g

Sieve size μm	Mass retained g	Cumulative mass passing	Percent passing
600	0	58.88	100%
425	1.46	$\dfrac{-1.46}{57.42}$	$\dfrac{57.42}{58.88}$ x 100 = 97.3%
212	1.74	$\dfrac{-1.74}{55.68}$	$\dfrac{55.68}{58.88}$ x 100 = 94.7%
150	0.55	$\dfrac{-0.55}{55.13}$	$\dfrac{55.13}{58.88}$ x 100 = 93.5%
63	1.94	$\dfrac{-1.94}{53.19}$	$\dfrac{53.19}{58.88}$ x 100 = 90.3%
Pass	$\dfrac{0.23}{5.92}$	$\dfrac{-0.23}{52.96}$	

Fig. 4.43   Sieving data and calculations relating to hydrometer test

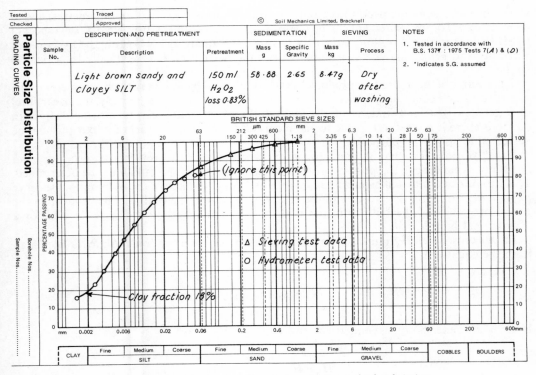

Fig. 4.44   Particle size curve from hydrometer test and related sieving

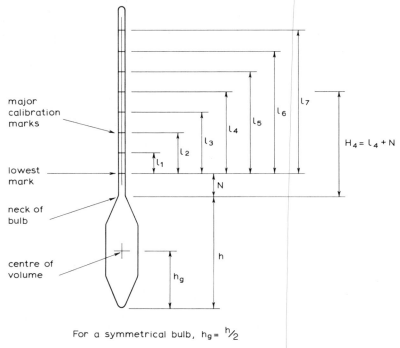

For a symmetrical bulb, $h_g = \frac{h}{2}$

Fig. 4.45   *Measurements for calibration of hydrometer*

The distance from the neck to the bottom of the bulb is measured, in millimetres, and is denoted by h. This measurement can be made either by laying the hydrometer flat on a sheet of paper and projecting down on to the paper using a set-square, or by holding the hydrometer vertically and projecting across with a square to a metre-stick held vertically by a burette stand.

The volume of the hydrometer bulb, V_h, can be measured by weighing the hydrometer to the nearest 0.1 g and equating the mass in grams to its volume in ml. Alternatively, the rise in level of water in a 1000 ml cylinder, initially filled to the 800 ml mark, can be measured. In both methods there is a small error due to the inclusion of the stem or part of it, but this can be neglected for practical purposes.

If the hydrometer bulb is of symmetrical shape, no further measurements are necessary, but if it is not symmetrical, the position of the centre of volume of the bulb must be determined. This can be done with sufficient accuracy by projecting the shape of the bulb on to a sheet of paper and estimating the position of the centre of gravity of the outline. The distance of the centre of gravity of the bulb from the bottom is denoted by h_g (see Fig. 4.45), and for a symmetrical bulb $h_g = \frac{1}{2}h$.

The effective depth H_R (mm), corresponding to each major calibration mark R_h, is calculated from the equation

$$H_R = H_1 + h_g - \frac{V_h}{2A}$$

If the hydrometer bulb is symmetrical, this equation becomes

$$H_R = H_1 + \frac{1}{2}\left(h - \frac{V_h}{A}\right)$$

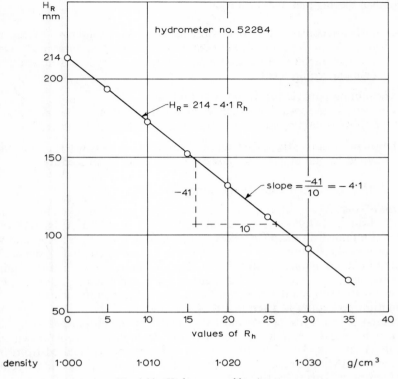

Fig. 4.46   *Hydrometer calibration curve*

Values of H_R are plotted against R_h on ordinary graph paper, and a smooth curve is drawn through the points as shown in Fig. 4.46. The curve usually approximates to a straight line over the range used. This relationship takes into account the effective depth of the suspension at the level being considered at a given time, and allows for the rise of liquid in the cylinder due to displacement by the hydrometer.

By measuring the slope of the calibration line, and reading off its intercepts on the H_R axis, the equation of the calibration line can be written in the form

$$H_R = 214 - 4.1\,R_h$$

which is the equation to the calibration curve shown in Fig. 4.46. This relationship between H_R and R_h can be used in a programmable computer for the calculation of the particle diameter, D, corresponding to each hydrometer reading.

The calibration curve is added to the H_R scale at the extreme right of the nomographic chart as follows. For each of the main hydrometer scale markings (30, 25, 20, etc.) the corresponding value of effective depth H_R is read off from the calibration curve. The R_h value is marked against H_R on the printed scale. (Note that the H_R values printed in the British Standard are given in centimetres, and must be multiplied by 10 to express them in millimetres.) Intermediate values of R_h can be added by subdividing each main division on the scale. When the nomograph chart is used, the values of R_h are referred to, not the H_R scale. This automatically incorporates the hydrometer calibration. If a different hydrometer is used, a new R_h scale must be constructed on the chart.

The R_h scale shown in Fig. 4.40 is derived from the calibration curve in Fig. 4.46.

4.8.5 Combined Sieving and Sedimentation

If the sample under test consists of particles ranging in size from sand size or larger down to silt or clay size, the results of sieving and sedimentation tests must be combined to give a single continuous curve. The method of calculation depends on the type of soil, and the following categories cover most requirements:

(1) Soils containing particles up to but not exceeding 2 mm.
(2) Non-cohesive soils containing particles larger than 2 mm.
(3) Cohesive soils containing gravel or larger sizes, including 'boulder clay'.

Each calculation procedure is described below. They apply equally whether the sedimentation curve was obtained by the pipette analysis or the hydrometer procedure.

SOILS CONTAINING PARTICLES UP TO 2 mm

With this type of soil a representative portion of the whole sample is taken for the pretreatment procedure (Section 4.8.1) followed by sedimentation (Section 4.8.2 or 4.8.3).

The material retained on the 63 μm sieve after washing is dried, resieved and weighed, as described in Section 4.6.4. Percentages passing each sieve are calculated on the basis of the dry mass of soil (m) remaining *after* pretreatment. The fine material washed through the 63 μm sieve is used for the sedimentation test, but each size fraction is calculated as a percentage of the mass (m) of pretreated soil, *not* the mass transferred to the sedimentation cylinder. An example of the calculation peocedure is given in Fig. 4.42.

The calculated percentages can be plotted directly on to the particle size sheet *without further correction*, against the relevant particle size. A smooth curve is drawn through the plotted points, as in Fig. 4.44. The first two or three points calculated from hydrometer readings sometimes do not lie on a smooth curve connecting the sieving and the sedimentation portions of the graph. This is partly because in the early stage of the sedimentation test the assumptions made in the theory based on Stoke's Law (Section 4.7.3) may not be strictly valid. In addition, some of the coarse silt particles are retained on the 63 μm sieve when wet sieved, owing to the effects of surface tension. These particles reappear when dry sieved and unless they are added to the sedimentation cylinder, there is a small deficiency in this size range. The initial readings should be ignored if they do not lie on a smooth curve continued from the sieving curve.

(2) NON-COHESIVE SOILS CONTAINING PARTICLES LARGER THAN 2 mm

These soils require a sample of more than 100 g, the minimum quantity depending upon the maximum size of the particles (see Table 4.4). It is usually necessary to riffle the portion passing the 2 mm sieve to obtain a sample of a size suitable for pretreatment and sedimentation. As far as the sieving test is concerned, calculations of percentages are similar to those described for composite sieving (Section 4.6.2). However, allowance should be made for the fact that part of the sample has been lost in the pretreatment process, as follows.

Let m_1 = dry mass of original sample; m_2 = dry mass passing 2 mm sieve; m_3 = mass of riffled portion of m_2 used for pretreatment, fine sieving and sedimentation; m = dry mass of test sample after pretreatment. The loss due to pretreatment, expressed as a percentage of the riffled mass, m_3, is equal to

$$\frac{m_3 - m}{m_3} \times 100\%$$

If this loss is small, say 1% or less, sieving percentages can be based on the mass m_1 for particle sizes greater than 2 mm, and on the pretreated mass m for particle sizes from 2 mm to 63 μm. The latter are corrected to percentages of the original mass by multiplying by the factor m_2/m_1, as explained in Section 4.6.2. Any errors introduced by this simplification will be insignificant.

If the pretreatment loss is not small, a further correction should be applied. It can be assumed that the percentage of material which would be removed by pretreatment of the unriffled sample, m_2, is the same as that removed from the riffled portion, m_3. The mass removable by pretreatment would therefore be

$$\frac{m_3 - m}{m_3} \times m_2$$

If it can be assumed that particles larger than 2 mm consist of minerals, such as quartz, which are unaffected by pretreatment, this loss would be the same if the whole of the original sample, m_1, were pretreated. The mass of the whole sample remaining after pretreatment (m_0) would be given by

$$m_0 = m_1 - \left(\frac{m_3 - m}{m_3}\right) \times m_2$$

The mass passing the 2 mm sieve would be reduced by the same amount to a corrected value of m_4, where

$$m_4 = m_2 - \left(\frac{m_3 - m}{m_3}\right) \times m_2$$

The mass m_0 should be used in place of m_1 for calculating sieving percentages of particles larger than 2 mm. Percentages retained on sieves below 2 mm should be corrected by the factor m_4/m_0 instead of m_2/m_1.

Calculations illustrating this procedure are given in the upper part of Fig. 4.47.

The sedimentation test is carried out on the material washed through the 63 μm sieve, as described in Section 4.8.2 or 4.8.3. The percentage of each size fraction is first calculated on the basis of the mass of soil after pretreatment (m), not the mass passing the 63 μm sieve. This is then corrected to the percentage of the whole sample by multiplying by the factor m_2/m_1, or m_4/m_0 if the pretreatment loss was significant. The procedures for both hydrometer and sieving tests are indicated in the lower portion of Fig. 4.47.

The complete grading curve is shown in Fig. 4.48.

(3) 'BOULDER CLAY'

The procedure for carrying out a sieve analysis on materials of boulder clay type is described in Section 4.6.7. This enables the grading curve to be plotted down to 63 μm, as shown in Fig. 4.26 and represented by curve (S), portion b–c in Fig. 4.49.

The subsample of material smaller than 2 mm (Section 4.6.7; Selection of sample) is treated as described in method (1) above to obtain a combined sieving and sedimentation curve from 2 mm down to the clay size range. This is shown as curve (F) in Fig. 4.49. It may be assumed that the distribution of particles finer than 63 μm in the fine soil fraction (F) is representative of the original sample, but the distribution from 63 μm up to 2 mm may not be properly representative of the whole.

Curves (F) and (S) are combined by adding the sedimentation curve to the main sieving curve, joining them at 63 μm. If necessary, curve (F) should first be smoothed by rejecting the points derived from the early hydrometer readings, as described in method (1). The sedimentation curve is scaled down by the ratio of percentages passing 63 μm read off from the two curves, as follows.

Let P_1 = percentage passing 63 μm as measured by wet sieving on the total sample; P_2 = percentage smaller than 63 μm on the 'fines' fraction used for sedimentation; y = percentage smaller than any given particle size on the hydrometer test curve; x = percentage smaller than the same particle size on the complete corrected grading curve. Then x (%) is calculated from the equation

Initial dry mass of sample = 2517 g = m_1

Sieve size	Mass retained	Mass passing		Percentage passing	
		Measured	Corrected		
20 mm	0	m_1 = 2517	m_0 = 2482		100
6.3	484	$\dfrac{-484}{2033}$	$\dfrac{-484}{1998}$	$\dfrac{1998}{2482}$ × 100	= 80.5
2	403	$m_2 = \dfrac{-403}{1630}$	$m_4 = \dfrac{-403}{1595}$	$\dfrac{1595}{2482}$ × 100	= 64.3

RIFFLED

m_3 = 91.26   (1538.74)

Pretreated

m = 89.32   (loss 1.94)   Pretreatment loss $= \dfrac{1.94}{91.26}$ × 100 = 2.13%

Equivalent loss from original sample $= \dfrac{2.13}{100}$ × 1630 = 34.7 g

WASH on 63 μm

Retained 51.64   (Sedimentation test 37.68)

∴ m_0 = 2517 − 34.7 = 2482.3 g
and m_4 = 1630 − 34.7 = 1595.3 g

	m = 89.32				
600 μm	21.52	$\dfrac{-21.52}{67.80}$		$\dfrac{67.80}{89.32} \times \dfrac{1595}{2482}$ × 100	= 48.8
212	14.68	$\dfrac{-14.68}{53.12}$		$\dfrac{53.12}{89.32}$ × 64.3	= 38.2
63	15.31	$\dfrac{-15.31}{37.81}$		$\dfrac{37.81}{89.32}$ × 64.3	= 27.2

37.68

| Passing 63 | $\dfrac{0.13}{51.64}$ | add to sedimentation test $\dfrac{0.13}{37.81}$ | | | |

Sedimentation tests

(a) Hydrometer method

Fully corrected hydrometer reading for a given particle diameter, derived as in Fig. 4.42 = R

$P = \dfrac{G_s}{89.32 \, (G_s - 1)}$ × R × 64.3

Example:

R_h = 13.5 at t = 60 minutes

R = 13.5 − 0.7 = 12.8

H_R = 214 − (4.1 × 13.5) = 158.65

D = $0.004064 \sqrt{\dfrac{158.65}{60}}$ = 0.00661 mm

$P = \dfrac{2.67}{89.32 \times 1.67}$ × 12.8 × 64.3 = 14.7%

(b) Pipette method

Percentage finer than each size, calculated as in Fig. 4.34 = p

$P = p \times 64.3$

NOTE: The mass used in the sedimentation calculations is the mass after pretreatment (m = 89.32 g), *not* the mass (37.81 g) transferred to the sedimentation cylinder.

Fig. 4.47   Calculations for combined sieving and sedimentation tests

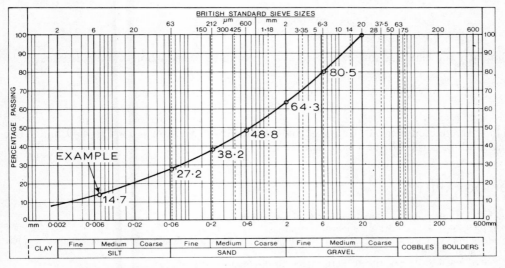

Fig. 4.48 Grading curve from combined sieving and sedimentation test data

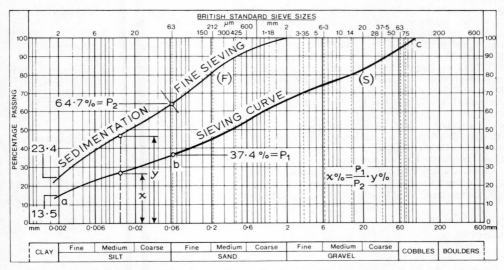

Fig. 4.49 Combination of separate sieving and sedimentation curves ('boulder clay' type of soil)

$$x(\%) = \frac{P_1}{P_2} \times y(\%)$$

As an example, in Fig. 4.49

$$P_1 = 37.4\%$$

$$P_2 = 64.7\%$$

At the particle diameter shown by the dashed line, $y = 46.5\%$.

The percentage ($x\%$) for this diameter on the complete curve is given by

$$x = 64.5 \times \frac{37.4}{64.7} = 26.9\%$$

Similarly the clay fraction, expressed as a percentage of the whole sample, is equal to

$$23.4 \times \frac{37.4}{64.7} = 13.5\%$$

The corrected sedimentation curve is that denoted by the portion a–b in Fig. 4.49, and a–b–c is the grading curve for the whole sample.

BIBLIOGRAPHY

Allen, T. (1974), *Particle Size Measurement*. Chapman & Hall, London

BS 410:1969, 'Test sieves'. British Standards Institution, London

BS Code of Practice CP 2001 (1957), 'Site investigations'. British Standards Institution, London; and document No. 76/11937, draft revision

BS Code of Practice CP 2003 (1959), 'Earthworks'. British Standards Institution, London

BS Code of Practice CP 2004 (1972), 'Foundations'. British Standards Institution, London

Casagrande, A. (1931). *The Hydrometer Method for Mechanical Analysis of Soils and other Granular Materials*. Cambridge, Mass.

Casagrande, A. (1947). 'Classification and identification of soils. Discussion of grain-size classifications and of methods for representing the results of mechanical analysis'. *Proc. Am. Soc. Civ. Eng.*, June 1947

Fuller, W. B. and Thompson, S. E. (1907). 'The laws of proportioning concrete'. *Trans. Am. Soc. Civ. Eng.*, Vol. 59

Glossop, R. and Skempton, A. W. (1945). 'Particle size in silts and sands'. *Proc. Inst. Civ. Eng. Paper* No. 5492

Head, K. H. (1976). 'Particle size analysis for fine-grained soils'. *Ground Engineering*, Vol. 9, No. 7

Kaye, G. W. C. and Laby, T. H. (1973). *Tables of Physical and Chemical Constants*, 14th edition. Longmans, London

Lambe, T. W. (1951). *Soil Testing for Engineers*. Wiley, New York

Neville, A. M. (1975). *Properties of Concrete*, 2nd edition (metric), Chapter 3. Pitman, London

Rothfuchs, G. (1935). 'Particle size distribution of concrete aggregates to obtain maximum density'. *Zement*, Vol. 24, No. 1

Stokes, Sir George G. (1891). *Mathematical and Physical Paper III*. Cambridge University Press.

Transport and Road Research Laboratory (1970). Road Note 29: 'A guide to the structural design for pavements for new roads'. HMSO, London

U.S. Department of the Interior, Bureau of Reclamation (1974). *Earth Manual*, 2nd edition. Test designation E.7, Part C. US Government Printing Office, Washington, DC

West, G. and Dumbleton, M. J. (1972). 'Wet sieving for the particle size distribution of soils'. TRRL Report No. LR 437. Transport and Road Research Laboratory, Crowthorne, Berks.

Chapter 5

Chemical tests

5.1 INTRODUCTION

5.1.1 Scope

The detailed chemical composition of soil is of little interest for civil engineering purposes, but the presence of certain constituents can be very significant. These include organic matter, sulphates, carbonates and chlorides. The pH reaction (acidity or alkalinity) of the groundwater can also be of importance.

   Chemical testing in a soil laboratory is usually limited to routine tests for the determination of the following:

(1)   Acidity or alkalinity (pH value).
(2)   Sulphate content.
(3)   Organic content.
(4)   Carbonate content.
(5)   Chloride content.
(6)   Total dissolved solids (in water).

   Tests for items (1)–(3) are covered by BS 1377:1975, and several methods are described in this chapter. The carbonate content test, item (4), is conveniently carried out with a special self-contained apparatus. For the chloride content of soils, methods based on procedures for concrete aggregates are given here.

   Chemical tests for the presence of other substances would normally be carried out by a specialist chemical testing laboratory, as could the tests referred to above if adequate facilities are not available in the soil laboratory.

5.1.2 Types of Test

The tests described in this chapter are listed in Table 5.1, which includes notes on their use and their limitations.

5.1.3 Relevance and Accuracy of Chemical Tests

In the quantitative chemical analysis of soils by far the biggest source of possible error is in the selection of the test sample. Usually a very small sample of dried soil is required at the outset, and it is essential that this sample be truly representative of the original sample. The proper procedure of mixing, riffling and quartering, described in Section 1.5.5, must be rigidly adhered to. Short cuts in this procedure lead to inconsistent results.

   Results of chemical tests on soils should be regarded as an indication of the order of magnitude of constituents for classification purposes rather than as precise percentages. The British Standard test procedures provide accurate enough results for most soils, but with some

Table 5.1 CHEMICAL TESTS FOR SOILS AND GROUNDWATER

Section reference	Type of test	Procedure	Reference	Uses	Limitations and comments
5.5	pH value	Indicator papers	Supplier's instructions	Simple and quick. Useful for determining approximate pH range for a more sensitive test.	Gives approximate values only.
		Colorimetric (Kuhn's method)	BS Test 11 (B)	Quick field test for soils. Apparatus available as a kit.	Requires colour comparison with chart printed in British Standard. Values given to nearest 0.5.
		Lovibond Comparator	Manufacturer's instructions	Colour comparison with standard coloured discs gives pH to nearest 0.2. Range of indicators available.	
		Electrometric	BS Test 11 (A)	BS 'standard' method. Accurate to 0.1 pH unit or better.	Requires a special electrical apparatus, although low-priced portable battery models are available. Electrodes age slowly, and should be checked periodically with buffer solutions.
5.6	Sulphate content	Total sulphates in soils	BS Test 9 BRE Digest 174	Accurate if performed with care and with proper chemical testing facilities. Gives the total amount of sulphates present, including calcium sulphate, which is insoluble in water.	If the measured sulphate content is greater than 0.5% the water-soluble sulphates should also be measured.
		Water-soluble sulphates in soils (gravimetric)	BS 1377:1967, Test 9 (A)	Accuracy as above. Gives the amount of water-soluble sulphates only, which are those most likely to attack concrete.	
		Water-soluble sulphates in soils (ion exchange)	BS Test 10	Quick, easy	Cannot be used if chloride, nitrate or phosphate ions are present. Requires a special ion-exchange resin which needs reactivating frequently.
		Sulphates in groundwater (ion exchange)	BS Test 10	As above.	As above.
		Sulphates in groundwater (gravimetric)	BS 1377:1967, Test 9 (A)	As for water-soluble sulphates in soils.	

5.7	Organic content	Loss on ignition	*Soil Mechanics for Road Engineers* (**TRRL**, 1952)	Destroys all organic matter. Suitable for sandy soils containing little or no clay or chalk.	High temperature also breaks down certain minerals in clay, and carbonates. Gives error on the high side on clayey and chalky soils. Removes water of crystallisation.
		Peroxide oxidation	*Soil Mechanics for Road Engineers* BS Test 7 (C) and (D), pretreatment stage.	Eliminates organic matter before sedimentation particle size tests.	Has limited action on undecomposed plant remains (e.g. roots and fibres).
		Dichromate oxidation	BS Test 8	Accurate, if proper chemical testing facilities used. Suitable for all soils. Presence of carbon and carbonates does not affect results. Fairly rapid, suitable for small batches.	Presence of chlorides affects results but a correction can be applied if chlorides are measured separately. Their effect can be overcome by adding mercuric sulphate.
5.8	Carbonate content	Calcimeter Collin's modification of Schleiber's apparatus	*Soil Mechanics for Road Engineers* Manufacturer's instructions	Compact, simple, fairly quick. Measures the volume of carbon dioxide evolved.	An approximate method, but accurate enough for most engineering purposes. Atmospheric pressure must be known.
5.9	Chloride content	Reaction with silver nitrate (Volhard's method)	BS 812:1976, Part 4	Titration process requiring proper chemical testing facilities. Designed for concrete aggregates.	Several standardised reagents are required.
		Mohr's titration method	Bowley (1979)	Simpler than Volhard's method. Designed for concrete aggregates.	Both methods require an analytical balance.
5.10.1	Total dissolved solids	Evaporation	BS 2690:1970, Part 9, Section 5	Simple procedure.	Requires very accurate weighing.
5.10.2	Concentration of certain salts	Indicator papers	Manufacturer's instructions	Very simple, quick, inexpensive.	Gives approximate indication only; not for accurate work. Presence of salts other than those being tested might affect readings.

tropical soils there is the possibility that the presence of other constituents could have an undetermined effect on the chemical process for the measurement of a particular substance.

However, this does not mean that care and accuracy are not important. All chemical testing procedures require extreme care, high accuracy and clean working conditions. The atmosphere must be dust-free, and this means that operations such as grinding soil, sieving and mixing of dry soil should not be carried out in the chemical testing area. Smoking should not be permitted.

Processes such as acid extraction require the use of a fume-cabinet and air-extractor fan to take away obnoxious gases and fumes. If a separate laboratory for chemical testing is impracticable, a portion of bench space under an air-extraction hood should be allocated solely to chemical tests so as to minimise the effect of fumes, and to keep glassware and other delicate apparatus away from soils testing equipment.

5.1.4   Practical Aspects

SAFETY

Handling and use of chemicals requires extreme care, and the technician must be aware of the possible hazards involved. Some of the precautions which should be observed are outlined in Section 1.6.8, which should be studied before proceeding with chemical tests.

BATCH TESTING

The apparatus listed for each test is that required for a single analysis. In a commercial laboratory many tests, such as those for the determination of sulphate and chloride content, have to be carried out in batches, typically six at a time. In a specialist chemical laboratory 20 or more tests may be carried out in a batch. Where several tests are run concurrently, additional numbers of items such as beakers, funnels and crucibles are required, but there is no need to duplicate the major items, such as ovens and burettes.

STANDARD SOLUTIONS

Nowadays 'molar' (M) solutions are frequently quoted, but in this book 'normal' (N) solutions are referred to, so as to be consistent with current British Standards. The standard solutions most often used for soil testing are listed in Table 5.3.

All chemicals used should be of analytical reagent (AR) grade. In a few instances, where the highest accuracy is necessary, a calibration check on a 'standard' solution is required as part of the test procedure.

When a litre of standard solution is made up, the constituent should first be dissolved in about three-quarters of that volume of distilled water. This is to allow for the resulting increase in volume. When the constituent is dissolved, further distilled water is added to make 1 litre of solution. Standard solutions of different concentrations are designated by means of a multiplying factor preceding the N; for instance, a 2 N solution requires twice the amount of constituent per litre as in an N solution; a 0.1 N (or N/10) solution requires one-tenth of that amount per litre.

USE OF BURETTE

Volumetric analysis usually requires the use of a burette for the measurement of the volume of a solution used, which is more accurate than the use of a measuring cylinder. For reliable results attention must be paid to the following points:

(1)   The burette must be clean.
(2)   The tap must not leak.

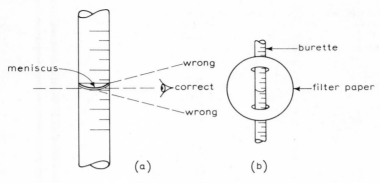

Fig. 5.1   *Reading a burette*

(3)   The tap should be lubricated with only a trace of a lubricant such as Vaseline (*not* silicone grease).

(4)   The jet should be fine enough for the rate of discharge not to exceed about 20 ml/min.

(5)   The burette must be properly clamped to a burette stand so that it is exactly vertical.

Burettes are usually calibrated at intervals of 0.1 ml, and numbered (downwards) every 1 ml. When the liquid level in a burette is being read, observe the bottom of the curve of the dark part of the meniscus, as indicated in Fig. 5.1(a). The eye must be at the same level as the mensicus, to avoid parallax errors.

A sheet of white filter paper slipped around the burette as in Fig. 5.1(b) makes the level of liquid easier to read. Generally, readings to the nearest half division (0.05 ml) are accurate enough for soils testing purposes.

Fill the burette from the open end, using a small funnel. Make sure that any air trapped in the tap or jet is removed by running a little liquid through, before taking the first reading.

PIPETTES

For the delivery of a predetermined volume of liquid in multiples of 5 or 10 ml up to 50 ml, a pipette is convenient, quick and sufficiently accurate for soils testing purposes. A bulb pipette is most often used, but a graduated narrow-tube pipette is required for accurately measuring volumes of less than 1 or 2 ml. They should be used with a suction syringe; acids and other chemical solutions should *never* be drawn into a pipette by mouth suction.

FILTERING UNDER VACUUM

Flasks used for vacuum filtration must be purpose-made of thick glass to withstand the external atmospheric pressure (see Section 1.6.10). When a Buchner funnel is used, the holes are covered with a filter paper which must be of the correct diameter. The stem of the funnel should extend below the level of the side-arm to which the vacuum line is connected.

A conical funnel fitted with a cone of paper is not recommended for filtration by suction.

POLICEMAN

A 'policeman' is a wiper consisting of a rubber sleeve that is slipped on to the end of a stirring rod, covering the end of the rod. A piece of rubber tubing may not be adequate. The policeman is used for wiping the inside of a beaker when a precipitate or solid particles is being transferred to another vessel. It should be made of gum rubber, and should be discarded as soon as it shows signs of going hard.

Table 5.2. ATOMIC WEIGHTS (to three significant figures)

Element	Symbol	Atomic weight
Aluminium	Al	27.0
Barium	Ba	137
Bromine	Br	79.9
Calcium	Ca	40.1
Carbon	C	12.0
Chlorine	Cl	35.5
Chromium	Cr	52.0
Copper	Cu	63.5
Hydrogen	H	1.01
Iron	Fe	55.8
Lead	Pb	207
Magnesium	Mg	24.3
Mercury	Hg	201
Nitrogen	N	14.0
Oxygen	O	16.0
Phosphorus	P	31.0
Potassium	K	39.1
Silicon	Si	28.1
Silver	Ag	108
Sodium	Na	23.0
Sulphur	S	32.1

5.2   DEFINITIONS AND DATA

5.2.1   Definitions

ION   A charged atom, molecule or radical whose migration effects the transport of electricity through an electrolyte.

ELECTROLYTIC DISSOCIATION   The (reversible) breaking down of a substance into ions on dissolution in a suitable liquid.

pH VALUE   The logarithm to base 10 of the reciprocal of the concentration of hydrogen ions in an aqueous solution. It provides a measure of the acidity or alkalinity of the solution.

INDICATOR   A substance whose colour depends (usually) on the acidity or alkalinity of the solution in which it is dissolved. The colour change is often used to indicate the completion of a chemical reaction.

TITRATION   The addition of a solution from a graduated burette to a known volume of a second solution, until the chemical reaction between the two is completed. If the strength of one of the solutions is known, that of the other can be calculated from the volume of liquid added.

EQUIVALENT WEIGHT   The weight of an element or radical which combines with or displaces one gram-atomic weight of hydrogen.

NORMAL SOLUTION   A solution containing the equivalent weight, in grams, of a substance in distilled water to make one litre of solution.

MOLAR SOLUTION   A solution containing the gram-molecular weight (molecular weight in grams) of a substance in distilled water to make up one litre of solution.

5.2.2   Atomic Weights

Atomic weights of elements referred to in this chapter, and their symbols, are given to three significant figures in Table 5.2.

Table 5.3. NORMAL AND MOLAR SOLUTIONS

Normal solution	Constituent	Grams of constituent in 1 litre normal (N) solution	Grams of constituent in 1 litre molar (M) solution
Hydrochloric acid	HCl	36.5	36.5
Sulphuric acid	H_2SO_4	49	98
Nitric acid	HNO_3	63	63
Sodium hydroxide	NaOH	40	40
Silver nitrate	$AgNO_3$	170	170
Potassium dichromate	$K_2Cr_2O_7$	49.035	294

Table 5.4. MAIN CONSTITUENTS OF SEA-WATER

Dissolved salt		Typical percentage by mass
Sodium chloride	NaCl	2.71
Magnesium chloride	$MgCl_2$	0.38
Magnesium sulphate	$MgSO_4$	0.16
Calcium sulphate	$CaSO_4$	0.13
Potassium sulphate	K_2SO_4	0.09
Other	(mainly $CaCO_3$, $MgBr_2$)	0.02
	Total dissolved salts	3.49

5.2.3   Standard Solutions

The content of standard solutions used in the tests described in this chapter are listed as both normal solutions and molar solutions in Table 5.3.

The approximate composition of sea-water, expressed as percentages by mass of dissolved salts, is given in Table 5.4.

5.3   THEORY

5.3.1   Acidity, Alkalinity and pH

THE pH SCALE

All liquids containing water contain at least two kinds of free ions (atoms or groups of atoms) carrying electric charges. These are the hydrogen ions, which are positively charged, and the hydroxyl ions, which are negatively charged. This results from the electrolytic dissociation of some of the water molecules represented by the reversible action

$$H_2O \rightleftharpoons H^+ + OH^-$$

When the numbers of these two kinds of ions are equal, the liquid is said to be neutral.

One litre of pure freshly distilled water contains one ten-millionth of a gram (10^{-7} g) of hydrogen ions (H^+) and the same number of hydroxyl ions (OH^-). The addition of acid to the water increases the concentration of the H^+ ions and decreases the concentration of OH^- ions. The water then gives an acid reaction and the active *acidity* increases in proportion to the increase in the concentration of hydrogen ions.

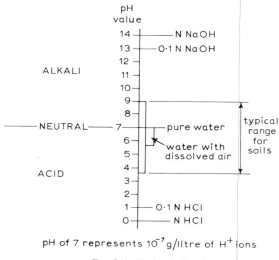

pH of 7 represents 10^{-7} g/litre of H^+ ions

Fig. 5.2   Scale of pH values

The addition of alkali has the opposite effect, and the active alkalinity increases in proportion to the decrease in hydrogen ion concentration.

At a given temperature the product of the concentration of the H^+ and OH^- ions is constant, so if one is known the other can be calculated. It is usual to refer only to the hydrogen ion concentration, which is expressed in grams of active (ionised) hydrogen per litre of liquid. Since these values are minutely small, the concentration is expressed more conveniently on a logarithmic scale, known as the pH scale. The 'p' stands for the mathematical power, and the 'H' for hydrogen ions. The pH value is the logarithm to base 10 of the reciprocal of the hydrogen ion concentration in grams per litre. This means that the pH value is the index, or power of ten, of the hydrogen ion concentration with the negative sign changed to positive. The value can vary with temperature.

Pure distilled water has an H^+ concentration of 10^{-7} g/litre, and its pH value is 7, which is neutral (neither acid nor alkaline). A solution having a pH of less than 7 is acid, and having a pH greater than 7 is alkaline. Since this is a logarithmic scale, a decrease of one unit of the pH scale represents an increase of H^+ ions by a factor of 10; a decrease of two units by a factor of 100; and so on.

The acidity referred to above is the 'active' acidity, which may be described as the intensity of the acidity, and the pH provides a measure of this intensity. The 'total' acidity, or amount of acid present, is a different property, which can be measured quantitatively by titration.

The pH values of some solutions are shown diagrammatically on the pH scale in Fig. 5.2. Freshly distilled or de-ionised water usually shows a slight acid reaction (pH of 6.6–7.0). On exposure to air this can fall to 6.0 or less, owing to rapid absorption of carbon dioxide, which produces acidity.

It is not easy to measure accurately the pH of pure water.

INDICATORS

Certain dyes, known as pH indicators, change colour in a definite manner, according to the acidity or alkalinity of the solution in which they are mixed. This feature is used in the colorimetric methods of determining pH. The best-known indicator is litmus, which is red in an acid solution and blue in an alkaline solution. However, it is not sensitive enough for measurement of pH, because it may require a pH as low as 4.6 to indicate acid, and up to 8.4 to indicate alkali. Indicators used for pH measurement show a complete colour change over a

Table 5.5. INDICATORS FOR SOIL TESTING AND pH MEASUREMENT

Purpose	Indicator	pH range	Lovibond disc reference
General indicators	litmus	<5	
		>8	
	bromophenol blue	2.8–4.6	
	methyl orange	2.8–4.6	
	methyl red	4.4–6.0	
	thymol blue	8.0–9.6	
	phenolphthalein	8.4–10.0	
Indicator papers for pH	full-range	1–14	
	narrow ranges	1–4	
		4–6	
		6–8	
		8–10	
		10–12	
		12–14	
Indicators for use with Lovibond comparator:	full range*	1–13	2/1ZE
Wide-range	Universal*	4–11	2/1P
	B.D.H. Soil*	4–8	2/1N
	B.D.H. 9011	7–11	2/1M
	B.D.H. 1014	10–14	2/1BB
Narrow-range	bromophenol blue	2.8–4.4	2/1B
	bromocresol green	3.6–5.2	2/1C
	bromocresol purple	5.2–6.8	2/1G
	bromothymol blue	6.0–7.6	2/1H
	phenol red	6.8–8.4	2/1J
Double-range	cresol red	1.2–2.8	2/1Y
		7.2–8.8	2/1K
	m-cresol purple	1.0–2.7	2/1W
		7.6–9.2	2/1Z
	thymol blue	1.2–2.8	2/1A
		8.0–9.5	2/1L

* Suitable for many purposes not requiring high accuracy.

small range of pH values. By using a universal indicator and comparing the colour with the BDH colour chart, the pH can be estimated to the nearest 0.5.

A more accurate assessment can be made with the Lovibond Comparator, the procedure for which is described in Section 5.5.3. The colour change of an indicator is identified by comparison with a number of permanent glass colour standards mounted in a rotatable plastics disc. The number marked on the matching standard is the pH value of the solution under test.

There are many different indicators available covering different ranges of pH. Each indicator must be used with the appropriate disc of standard colours. Some indicators have a double range of pH values, for which two different discs are required. Some of the indicators most useful for soils and groundwaters are listed in Table 5.5 together with the pH ranges over which they are applicable. The universal or wide-range indicators are intended to give an approximate pH value only, so that the appropriate narrow-range indicator may be selected for a more accurate assessment.

An indicator is sensitive to colour change only within the limits of its pH range. Beyond these limits there is no further colour change, and another indicator is necessary. There is some overlap between indicators covering adjacent ranges, so that a reading at the limit of one indicator can be confirmed by repeating the test with the next.

INDICATOR PAPERS

Indicator papers are available in booklet form, and consist of strips of absorbent paper impregnated with an indicator. Both universal and narrow-range indicator papers are available. The most useful ranges are given in Table 5.5.

ELECTRICAL CONDUCTIVITY

The electrical conductivity of the hydrogens ions (H^+) in a very dilute solution is almost double that of the hydroxyl ions (OH^-). Electrical conductivity of a solution can therefore be related to its pH, and although the converse is far more complex, this property is made use of in the electrometric method of determining pH, described in Section 5.5.4.

5.3.2 Sulphates

The water-soluble sulphates usually found in soils are sodium sulphate (Na_2SO_4) and magnesium sulphate ($MgSO_4$). Calcium sulphate ($CaSO_4$) is commonly found as gypsum, and is only slightly soluble in water but is readily soluble in dilute hydrochloric acid. Treatment with acid is therefore necessary if the total amount of sulphates is required.

The approximate maximum solubilities in water of the three salts mentioned, expressed in terms of SO_3 per litre at about 20 °C, are as follows:

sodium sulphate (Glauber's salt)	240 g/litre
magnesium sulphate (Epsom salt)	180 g/litre
calcium sulphate (gypsum, or selenite)	1.2 g/litre

In the gravimetric methods described in Section 5.6.5 the dissolved sulphates are precipitated as insoluble barium sulphate as a result of a reaction with barium chloride in slightly acid conditions. The chemical reaction with magnesium and calcium sulphate may be represented by the equation

$$\left.\begin{matrix} Mg \\ Ca \end{matrix}\right\} SO_4 + BaCl_2 \rightarrow Ba\,SO_4 \downarrow + \left.\begin{matrix} Mg \\ Ca \end{matrix}\right\} Cl_2$$

(soluble)   (soluble)   (insoluble, precipitated)   (soluble)

The reaction is similar with sodium sulphate, but the equation is

$$Na_2SO_4 + BaCl_2 \rightarrow BaSO_4 \downarrow + 2NaCl$$

The precipitate of barium sulphate is filtered out, dried and weighed. From the atomic weights the equivalent percentage of SO_3 in the original sample is calculated, as follows.
Molecular weight of barium sulphate ($BaSO_4$):

Element	Atomic weight (approx.)	No. of atoms	
Ba	137		$\times 1 = 137$
S	32		$\times 1 = 32$
O	16		$\times 4 = 64$
			_____
		Molecular weight	233

Weight of SO_3:

Element	Atomic weight (approx.)	No. of atoms
S	32	$\times 1 = 32$
O	16	$\times 3 = 48$
		80

Thus the mass of SO_3 will be $80/233 = 0.343$ times the mass of barium sulphate precipitated. If the mass of precipitate is m_4 and the mass of soil used is m_3, the percentage of SO_3 in the soil used is given by

$$\frac{m_4 \times 0.343}{m_3} \times 100\% = 34.3 \times \frac{m_4}{m_3} \underline{}$$

If the mass m_3 was taken not from the original sample, but from the fraction passing a 2 mm sieve, the calculated percentage must be multiplied by m_2/m_1 to convert it to a percentage of the original sample, where $m_1 = $ mass of selected sample before sieving; $m_2 = $ mass of sample passing the 2 mm sieve.

It is customary to express sulphates in terms of SO_3. The BRE Digest 174 (1975) and other references which give practical advice on concrete in sulphate-bearing soils all base their recommendations on SO_3 content.

The addition of bromine water during the acid extraction stage oxidises any extraneous metallic compounds which may be present and which could otherwise contaminate the barium sulphate. These oxides are insoluble in an alkaline solution, and the addition of ammonia causes them to be precipitated out, so that they can be removed before the reaction with barium chloride.

During the final filtration the presence of remaining soluble chlorides is indicated by turbidity when a drop of the washings is tested with silver nitrate solution:

$$CaCl_2 + 2AgNO_3 \rightarrow 2AgCl\downarrow + Ca(NO_3)_2$$

$$BaCl_2 + 2AgNO_3 \rightarrow 2AgCl\downarrow + Ba(NO_3)_2$$
$$\text{(white precipitate}$$
$$\text{if chloride}$$
$$\text{is present)}$$

Washings must continue until no turbidity is indicated; otherwise the chlorides will be included in the final weighing.

5.3.3 Organic Matter

Organic matter contains carbon, which may occur in complex chain compounds with hydrogen, oxygen, nitrogen and other elements. In a test these compounds are broken down in various ways, depending on the process used.

IGNITION TEST

In the ignition test the carbon burns to combine with oxygen forming carbon dioxide, which is driven off:

$$C + O_2 \rightarrow CO_2 \uparrow$$
(in complex organic compounds)

The other constituents of the organic compounds break down, and most are also lost as gases.

PEROXIDE TEST

In the peroxide test hydrogen peroxide (H_2O_2) releases nascent oxygen which vigorously oxidises most of the organic matter present:

$$H_2O_2 \rightarrow H_2O + O \uparrow$$

In each process the mass of organic matter present is assumed to be equal to the mass lost, which is expressed as a percentage of the dry mass of the soil.

DICHROMATE OXIDATION

In the dichromate oxidation method it is assumed that the organic matter in soil contains 58% of carbon by mass, and that approximately 77% of that carbon is oxidised by the action of potassium dichromate. These factors are taken into account in the equation for the determination of organic matter content:

$$\text{percentage organic matter} = \frac{0.67 \times V}{m_3}$$

where V is the volume of potassium dichromate used to oxidise the organic matter in the soil, of initial mass m_3. It is measured by titration with ferrous sulphate.
V is calculated from the equation

$$V = 10.5 \times \left(1 - \frac{y}{x}\right)$$

where y = total volume if ferrous sulphate used in the test; x = volume used in standardisation test.

If the mass of soil used, m_3, is taken from the mass m_2 passing a 10 mm sieve from an initial mass m_1, the calculated percentage must be multiplied by m_2/m_1 to give the organic matter content as a percentage of the whole original sample.

5.3.4 Carbonates

When hydrochloric acid is allowed to react with a carbonate, such as calcium carbonate (which is the main form of carbonates in soils), the chloride is formed and carbon dioxide is evolved:

$$CaCO_3 + 2HCl \rightarrow CaCl_2 + H_2O + CO_2 \uparrow$$

If the carbon dioxide evolved is collected and its volume measured, its mass can be determined if its temperature and pressure are known. The mass of carbonates in the treated soil sample can then be calculated, and expressed as a percentage of the original dry mass.

Small changes of temperature and pressure can result in significant changes in volume of a given amount of gas. One of the practical difficulties of this type of test is the maintenance of a known steady temperature and pressure. The apparatus described in Section 5.8.1 (Fig. 5.12) was developed to fulfil these conditions.

The mass of 100 ml of carbon dioxide collected in this apparatus depends on the volume of hydrochloric acid used for the test, as well as the temperature and pressure. At the standard conditions (20 °C and 760 mmHg), if 20 ml of acid is used, the mass of 100 ml of carbon dioxide is 200 mg. If the atmospheric pressure is higher, or the temperature lower, the mass of gas will be greater, and vice versa. The mass of 100 ml of carbon dioxide (W_2) can be obtained either from a table, or by using the special slide-rule provided with the apparatus.

$$W_1 = \text{mass of solid sample tested (g)}$$
$$W_2 = \text{mass of 100 ml of } CO_2 \text{ (mg)}$$
$$V_g = \text{volume of } CO_2 \text{ evolved (ml)}$$

$$\text{mass of } CO_2 \text{ evolved} = \frac{W_2}{100} \times V_g \quad \text{mg}$$

$$= \frac{W_2 \times V_g}{100\,000} \quad \text{g}$$

Expressed as a percentage of W_1,

$$CO_2 \text{ content} = \frac{W_2 \times V_g}{100\,000 \times W} \times 100\%$$

or

$$\text{carbonate content (as } CO_2) = \frac{W_2 \times V_g}{1000\,W_1}$$

In a simplified procedure (Section 5.8.2) the volume of acid used is first determined from the prevailing pressure and temperature conditions, so that no subsequent correction is necessary.

5.3.5 Chlorides

The methods for the determination of chloride content described in Section 5.9 depend upon the exchange reaction which takes place between silver nitrate and the chloride salt in solution. For sodium chloride this is represented by the equation

$$NaCl + AgNO_3 \rightarrow NaNO_3 + AgCl\downarrow$$
$$\text{(soluble)} \quad \text{(soluble)} \quad \text{(soluble)} \quad \text{(precipitate)}$$

In the BS 812 test (Volhard's method; Section 5.9.2) an excess of silver nitrate is used to precipitate the chloride, and the quantity of the unreacted portion of silver nitrate is determined from a titration with potassium thiocyanate. Silver thiocyanate is precipitated, until all the silver has been used up. In the acidified solution the next few drops of potassium thiocyanate react with the ferric alum indicator to produce ferric thiocyanate, which gives the permanent brown colour and indicates that the end point has been reached.

In Mohr's method (Section 5.9.3) the silver nitrate is added to the chloride in a neutral solution which also contains potassium chromate. The silver has a far greater affinity for the chloride than for the chromate, so the above reaction takes place until all the chloride is combined as silver chloride. The next few drops of silver nitrate react with the potassium chromate, producing a red colour which remains even after the titration flask is swirled, indicating that the end point has been reached:

$$2AgNO_3 + K_2CrO_4 \rightarrow 2KNO_3 + 2AgCrO_4$$
$$\text{(red)}$$

Provided that the same intensity of colour is observed in the test solution as in the 'blank', the difference between the volumes of silver nitrate used in the two flasks is the volume required to react with the chloride.

5.4   APPLICATIONS

5.4.1   pH Value

Excessive acidity or alkalinity of the groundwater in soils can have detrimental effects on concrete buried in the ground. Even a moderate degree of acidity can cause corrosion of metals. Measurement of the pH value of the groundwater reveals these potential dangers so that remedial measures can be taken.

In the stabilisation of soils for roads, some resinous materials are unsuitable for alkaline soils, yet may be satisfactory with neutral or slightly acid soils.

As well as being used for the above purposes, the pH value is usually determined whenever the sulphate content is measured.

5.4.2   Sulphate Content

Groundwater containing dissolved sulphates can attack concrete, and other materials containing cement, placed in the ground or on the surface. A reaction takes place between the sulphates and the aluminate compounds in cement, causing crystallisation of complex compounds. The expansion which accompanies crystallisation induces internal stresses in the concrete, which results in mechanical disintegration.

Measurement of the sulphate content enables the ground conditions to be classified according to potential sulphate attack. Appropriate precautionary measures, such as the use of sulphate-resisting cement or of a richer, denser concrete mix, can be taken during construction.

Sulphates in soil can also cause disintegration of precast members, such as slabs and concrete pipes, and can lead to corrsoion of metal pipes placed in contact with the soil.

The soluble sulphates (sodium and magnesium) are much more aggressive to concrete than calcium sulphate, which is relatively insoluble in water. Therefore if the predominating sulphate present in the soil is calcium sulphate, the test for total sulphates (Section 5.6.2) is likely to give a pessimistic indication of the danger due to sulphates. If the total sulphate content exceeds 0.5% the sulphate content of a 1:1 soil–water extract should be determined (Section 5.6.3). Because of its low solubility (Section 5.3.2) calcium sulphate will show a sulphate content in the aqueous extract of not more than 1.2 g/litre (0.12%). A sulphate content in excess of this figure in the soil–water extract indicates the presence of other and more harmful salts.

Although the solubility of calcium sulphate is low, appreciable quantities can be dissolved away in the long term if the groundwater can be continually replenished.

Further information on sulphates in soils is given in Building Research Establishment Digest 174 (1975), which replaces the earlier BRS Digest No. 90. The classification of soil and groundwater according to sulphate content is reproduced in Table 5.6, together with an indication of the type of cement which may be required for buried concrete. Reference should be made to BRE Digest 174 for details of the requirements for concrete mixes.

Table 5.6. CLASSIFICATION OF SOIL AND GROUNDWATER ACCORDING TO SULPHATE CONTENT

| | Concentration of sulphates expressed as SO_3 | | | |
| | In soil | | | |
Class	Total SO_3 (%)	SO_3 in 2:1 aqueous extract (g/litre*)	In groundwater SO_3 (g/litre*)	Type of cement recommended for buried concrete
1	<0.2	–	<0.3	Ordinary Portland or Portland-blastfurnace
2	0.2–0.5	–	0.3–1.2	Ordinary Portland or Portland-blastfurnace Sulphate-resisting Portland Supersulphated
3	0.5–1.0	1.9–3.1	1.2–2.5	Sulphate-resisting Portland or Supersulphated
4	1.0–2.0	3.1–5.6	2.5–5.0	Sulphate-resisting Portland or Supersulphated
5	>2	>5.6	>5.0	As for Class 4, but with the addition of adequate protective coatings of inert materials

* 1 g/litre = 100 parts per 100 000.
Reproduced from BRE Digest 174, 'Concrete in sulphate bearing soils and ground waters', by permission of the Controller HMSO. Crown Copyright

5.4.3 Organic Matter Content

Organic matter in soil is derived from a wide variety of animal and plant remains, so there can be a great variety of organic compounds. They all can have undesirable effects on the engineering behaviour of soils. These can be summarised as follows:

(1)   Bearing capacity is reduced.
(2)   Compressibility is increased.
(3)   Swelling and shrinkage potential due to changes in moisture content is increased.
(4)   The presence of gas in the voids can lead to large immediate settlements, and can affect the derivation of consolidation coefficients from laboratory tests.
(5)   The gas can also give misleading shear strength values derived from total stress tests.
(6)   The presence of organic matter (e.g. in peat) is usually associated with acidity (low pH), and sometimes with the presence of sulphates. Detrimental effects on foundations could result if precautions are not taken.
(7)   Organic matter is detrimental in soils used for stabilisation for roads.

A measure of the organic content of soils is necessary in order to make allowance for these effects.

5.4.4 Carbonate Content

Knowledge of the carbonate content of soils is useful for the following reasons:

(1)   Carbonate content can be used as an index to assess the quality of chalk as a foundation material. A high carbonate content means a low clay mineral content, and usually indicates a relatively high strength.
(2)   In cemented soils and soft sedimentary rocks the carbonate content can indicate the degree of cementing.
(3)   In roads construction chalky subgrades are susceptible to frost action.
(4)   The carbonate content of chalk or limestone is an indication of its suitability for the manufacture of cement.

5.4.5   Chloride Content

The chloride content is most often used as an indication of whether or not the groundwater is sea-water, or whether the soil has been affected by sea-water. In some coastal situations, notably in the Middle East, the concentration of sodium chloride in the groundwater can be very much higher than that in sea-water. High concentrations can also be present in soils and permeable rocks not now directly in contact with sea-water.

Aqueous solutions of chlorides cause corrosion of iron and steel, including steel reinforcement in concrete. If the presence and concentration of chlorides is known, suitable preventative measures can be taken in the design and construction of buried or underwater reinforced concrete structures.

5.4.6   Corrosion of Metals

Adverse ground conditions can cause corrosion of metals buried in the ground as well as initiating attack on concrete. Steel and cast-iron pipes, and steel sheet piles and tie-bars, are perhaps the most common examples of buried metalwork. In addition, steel bars in reinforced concrete members may become exposed if the surrounding concrete is attacked.

Many factors contribute to the corrosion of iron and steel in the ground, but simple chemical tests can often indicate whether corrosion is likely to develop. An acid environment (low pH) is always potentially aggressive, but corrosion of iron and steel can take place in neutral or alkaline conditions if sulphate-reducing bacteria are also present. These bacteria flourish under anaerobic conditions (i.e. where oxygen is absent), and their presence is indicated by the presence of sulphides as well as sulphate conditions (i.e. in the presence of oxygen). The presence of chlorides can accelerate the corrosion process, even in alkaline (high pH) conditions.

Tests for the presence of sulphides, aggressive bacteria, and other corrosive agents apart from those covered in this chapter, require the facilities of a specialist chemical testing laboratory.

5.5   TESTS FOR pH

5.5.1   Indicator Papers

To determine the pH of water, simply dip a strip of the indicator paper into the water and lay it on a white tile or similar white non-absorbent surface. After 30 s compare its colour with the colour chart on the packet or dispenser. The number on the colour which matches the test strip most closely is the pH value of the water.

If the approximate pH of the water is not known to begin with, use a 'universal' or wide-range paper first. Then select a narrow-range paper appropriate to the approximate value indicated. It should be possible to assess the pH to 0.5 unit with a narrow-range paper.

If the water is turbid, it is better to place a drop of water on one side of the paper and observe the same spot on the reverse side for comparison with the colour chart.

To test the pH of a soil, place a quantity of the soil in a test-tube, add distilled water and shake vigorously until all the soil is in suspension. Dip the test paper in the water (if clear), or place a spot of the water on the paper, and observe the pH value as explained above. The quantity of soil used should be such that the ratio of water to soil by volume is about 5 for clay, 3 for silt and 2 for sand.

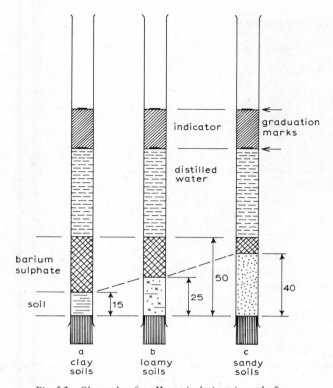

barium
sulphate

soil

indicator

distilled
water

graduation
marks

a
clay
soils

b
loamy
soils

c
sandy
soils

Fig. 5.3   Glass tubes for pH test (colorimetric method)

5.5.2   Colorimetric (Kuhn's Method) (BS 1377:1975, Tests 11 (B))

This is referred to in the British Standard as a subsidiary method for determining the pH value of soil, and is intended primarily for use in the field. The colour of the indicator is compared with the colours on a printed colour chart such as that given in Fig. 19 of BS 1377:1975. The method was first reported by S. Kuhn in 1930.

APPARATUS

(1)   A number of glass tubes, fitted with rubber bungs at each end. The tubes are about 200 mm long and 13 mm internal diameter, with two graduation marks at 115 mm and 140 mm from one end (see Fig. 5.3).
(2)   Wooden rack for the tubes.
(3)   Chattaway spatula with blade about 130 mm long and 10 mm wide.
(4)   Wash bottle with distilled water.
(5)   Colour chart.

REAGENTS

(1)(a) Indicator solution, obtainable specially for soil pH tests.
  (b) Alternatively, an indicator solution may be made up using bromothymol blue, methyl red, thymol blue, and sodium hydroxide, as follows. Weigh out accurately 0.15 g bromothymol blue, 0.063 g methyl red and 0.013 g of thymol blue and transfer them to a 1000 ml beaker. Add 500 ml of distilled water. The beaker is heated tently and its contents stirred with a glass rod until the indicators have dissolved. The colour of the mixture of indicators in the beaker is adjusted by carefully adding drops of approxi-

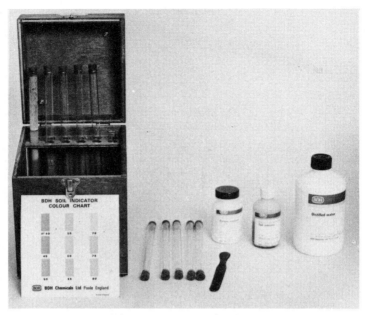

Fig. 5.4   Colorimetric apparatus for pH measurement

mately 0.1 N sodium hydroxide solution until the colour approximately matches the colour corresponding to pH 7.0 on the colour chart. The mixture is allowed to cool, and is diluted to 1 litre with distilled water. The indicator should be stored in a stoppered bottle.

(2)   Barium sulphate of soil-testing grade.

The above apparatus and reagents are commercially available in a portable kit known as the BDH Soil Testing Outfit, as shown in Fig. 5.4.

PROCEDURE

(1)   Prepare an air-dried sample of 20–25 g of soil which passes a 3.35 mm sieve. Particles larger than that should first be broken down, and the main sample subdivided by riffling.
(2)   Place a stopper in the end of one of the glass tubes furthest from the graduation marks. Place the soil in the tube to a depth of (a) 15 mm for clay soil, (b) 25 mm for loamy (silty) soil, (c) 40 mm for sandy soil, as indicated in Fig. 5.3.
(3)   Add barium sulphate on top of the soil so that the combined depth of soil and barium sulphate is 50 mm.
(4)   Add distilled water to the tube up to the first graduation mark.
(5)   Add indicator up to the second graduation mark.
(6)   Place a rubber bung in the open end of the tube and shake the tube vigorously until all the soil and barium sulphate are in suspension.
(7)   Place the tube in the rack to allow the solids to settle. The barium sulphate (which is insoluble in water) accelerates the settlement of the clay particles, which would otherwise remain as a turbid suspension, and leaves a clear-coloured supernatant liquid above the sediment.
(8)   Compare the colour of the supernatant liquid in the tube with the colour chart. Record the pH value of the colour which matches most closely.
(9)   Report the pH value to the nearest 0.5 pH unit, stating that the colorimetric method was used.

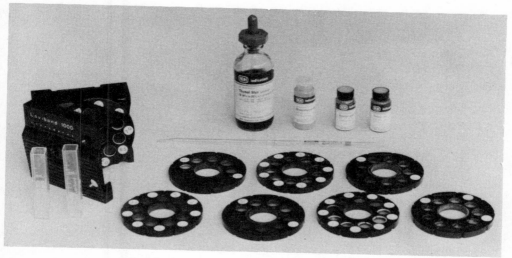

Fig. 5.5 Lovibond Comparator for pH measurement

If at stage (7) the suspension clears only very slowly, too little water was added. Part of the suspension can be poured out, the remainder diluted with more water and more indicator, and the tube reshaken. However, as much soil as possible (consistent with obtaining a clear solution) should be used, because too little soil gives an unreliable result.

5.5.3 Lovibond Comparator

This procedure can be used either with a full-range indicator, to obtain an approximate measure of pH (to the nearest unit), or with one of the many available narrow-range indicators, with which the pH can be read to 0.2 of a unit. It is intended for use with groundwater or with aqueous solutions.

APPARATUS

The Lovibond Comparator consists of a plastics holder in which two small test-tubes can be placed, and to which a rotatable comparator colour disc can be fitted (Fig. 5.5). A special pipette is provided.

There are over 30 different comparator discs available, and almost as many indicators. Sixteen discs with appropriate indicators are listed in Table 5.5. One of the three wide-range indicators marked with an asterisk would be sufficient for many purposes for which the greatest accuracy is not required.

PROCEDURE

(1)   If the water or solution to be tested contains solid particles in suspension, allow the solids to settle and filter if necessary.
(2)   Fill both test tubes up to the 10 ml mark with the water or solution to be tested.
(3)   Using the pipette provided, add the appropriate quantity of the selected indicator to the right-hand tube only. Do not immerse the tip of the pipette below the surface of the liquid. For most indicators 0.5 ml is used, but some require 0.2 or 0.1 ml. Follow the supplier's instructions.
(4)   Carefully mix the indicator into the liquid, using a clean glass stirring rod.
(5)   Insert the appropriate comparator disc in the recess provided. Hold the device so that the

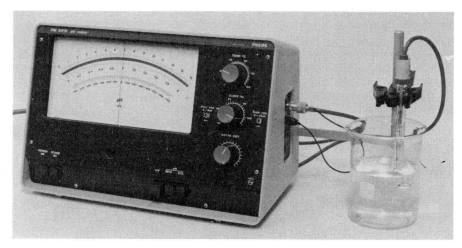

Fig. 5.6   Electric pH meter

tubes may be viewed against north daylight or a source of white light (not a tungsten electric light bulb).

(6)   Rotate the comparator disc until the nearest colour match is obtained between the sample tube and the colour disc.
(7)   Read the pH number which appears in the indicator recess.
(8)   Report the result to the nearest whole number if a wide-range or universal indicator is used, or to the nearest 0.2 of a unit if a narrow-range indicator is used. State that the Lovibond Comparator method was used.
(9)   Wash the tubes and pipette after use.

5.5.4   Electrometric Method (BS1377:1975, Test 11 (A)

The electrometric determination of the pH value of a soil suspension, or of groundwater, can be the most accurate method, and gives a direct reading to 0.05 pH unit, or with some instruments to 0.02 pH unit. However, the equipment is relatively expensive, and if the electrodes are not maintained in perfect condition, the readings can be unreliable.

This type of instrument is intended for use in a main laboratory where pH tests are carried out regularly. It is less suitable for intermittent use in a site laboratory.

PRINCIPLE OF OPERATION

The operation of an electrical pH meter is based on the principle that the solution to be tested can be considered as an electrolyte of a voltaic cell. One electrode, known as the reference electrode, remains at a constant voltage with respect to the solution and is unaffected by changes of pH. The voltage of the other electrode is affected by the conductivity, and indirectly by the pH, of the test solution, and the complex relationship between pH and voltage can be determined. In most instruments the voltage indicator is calibrated to read directly in pH units. The reference electrode most commonly used is the saturated calomel type. The other electrode may be of various types, of which the glass type is considered to be the most reliable. This consists of a thin-walled glass bulb, of a special kind of glass, enclosing a suitable electrolyte and electrode.

APPARATUS

(1)   Electric pH meter of the type referred to above. A typical instrument is shown in Fig. 5.6.
(2)   Three 100 ml glass beakers with cover-glasses and stirring rods.

(3)   Two 500 ml volumetric flasks.
(4)   Wash-bottle and distilled water.
(5)   Pestle and mortar.
(6)   3.35 mm aperture BS sieve.
(7)   Balance reading to 0.001 g.

REAGENTS

(1)   Buffer solution of pH 4.0.
(2)   Buffer solution of pH 9.2.
         These are obtainable as powders ready for solution in water as directed by the manufacturers. Alternatively, they can be made up as follows:
         Solution of pH 4.0: Dissolve 5.106 g of potassium hydrogen phthalate in distilled water and make up to 500 ml.
         Solution of pH 9.2: Dissolve 9.54 g of sodium tetraborate (borax) in distilled water and make up to 500 ml.
(3)   Potassium chloride, saturated solution (for maintenance of calomel electrode).

PROCEDURE

(1)   The soil sample is crushed to pass through a 3.35 mm sieve, and reduced by riffling to produce a sample of 30–35 g.
(2)   Weigh 30 g of soil and place in a 100 ml beaker.
(3)   Add 75 ml of distilled water, stir for a few minutes.
(4)   Allow to stand overnight. Stir again immediately before testing.
(5)   After calibrating the pH meter (see below) wash the electrodes with distilled water and immerse them in the suspension.
(6)   Take two or three readings of pH, stirring briefly between each reading, when the meter reaches equilibrium. These readings should agree within ± 0.05 pH unit. About 1 min may be required to reach a constant value.
(7)   Remove the electrodes and wash with distilled water.
(8)   Recheck the calibration, using one of the buffer solutions. If out of adjustment by more than 0.05 pH unit, reset the instrument and repeat the test until consistent readings are obtained.
(9)   Leave the electrodes standing in distilled water when the instrument is not in use.
(10)   If the pH of groundwater is to be measured, place about 80 ml in the beaker and follow stages (5)–(9).

The above procedure is a general guide but the detailed instructions provided by the manufacturer should be followed carefully.

REPORT RESULTS

Report the pH value of the soil suspension, or groundwater, to the nearest 0.1 pH unit, stating that the electrometric method was used.

CALIBRATION OF pH METER

The detailed procedure for calibration is given in the manufacturer's instructions accompanying each instrument and must be compensated for temperature. Calibration should be carried out at regular intervals, and the instrument, especially the electrodes, should be inspected frequently. The main stages for calibration are as follows.

Wash the electrodes with distilled water.
Set the electrical controls as directed.
Immerse the electrodes and a thermometer in the buffer solution of pH 4.0.

Adjust the buffer controls so that the pH scale reads 4.00. If the temperature of the solution is not 20 °C, use the correction tables or graphs to determine the required corrected value.

Wash the electrodes with distilled water.

Immerse them in the buffer solution of pH 9.2, as a check on the reading before use.

Wash the electrodes with distilled water before using in the soil suspension or groundwater sample.

5.6   SULPHATE CONTENT TESTS

5.6.1   Scope of Tests

In this section methods of test for the determination of the following are described:

Total (acid-soluble) sulphates in soils, expressed as a percentage.
Water-soluble sulphates in soils, expressed as a percentage.
Sulphates in groundwater, expressed in grams per litre (in some references parts per 100 000, or parts per million (p.p.m.)).

There are two methods of analysis given in the British Standard for determining sulphates in soils and groundwater. These are:

(1)   The gravimetric method, in which the sulphates are precipitated as insoluble barium sulphate, which is collected and weighed.
(2)   A more rapid method using an ion-exchange column, which involves titration against a standardised sodium hydroxide solution.

For sulphate tests on soils, whether for total sulphates or for water-soluble sulphates, and whichever type of test procedure is used, it is first necessary to obtain a liquid extract containing the sulphates in solution. The gravimetric method of analysis of the extract is the same whether an acid extract or a water extract is obtained, and this procedure can also be used for the analysis of water samples.

The ion-exchange procedure can be used either for groundwater or for the water-soluble extract from soils, but not for the acid extract.

The ways in which these procedures are related to each other, and to the results obtained, are shown diagrammatically in Fig. 5.7.

For most of these tests the size of soil sample actually tested is very small, in some instances only about 2 g. It is, therefore, essential that the test sample be correctly prepared so as to be properly representative of the initial sample.

Another method which is available for determination of sulphate content is the analysis of the density of the precipitated insoluble sulphates by optical means, using a turbidimetric analysis. A nephelometer is an instrument of this type which can be calibrated to give a rapid indication of the amount of sulphates present if the concentration is below a certain level. Further details are given in the paper by Bowley (1979). The equipment is expensive, and the accuracy is much lower than that obtainable by the standard methods, but it enables large numbers of tests to be carried out quickly.

5.6.2   Total Sulphates in Soils — Preparation of Acid Extract (BS 1377:1975, Test 9)

This section describes the procedure for the extraction of a solution of acid-soluble sulphates in soil. For most practical purposes this includes all naturally occurring sulphates. The analysis of the extract is described in Section 5.6.5.

The procedure can be divided into two main parts:

(A)   Physical preparation of soil sample.
(B)   Chemical treatment to prepare acid extract.

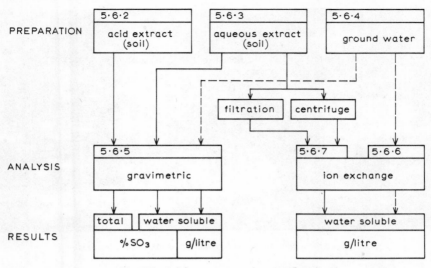

Fig. 5.7   *Sulphate test procedures — flow diagram*

APPARATUS

This list does not include the apparatus required for the analysis of the acid extract.

(1)   Analytical balance weighing to 0.001 g or better.
(2)   Glass weighing bottle and stopper.
(3)   Two conical beakers, 500 ml, with cover-glasses.
(4)   Conical beaker, 250 ml, with cover-glass.
(5)   Glass filter funnel, 100 mm diameter.
(6)   Desiccator and desiccant.
(7)   Two glass stirring rods about 200 mm long.
(8)   Filter papers: Whatman No. 541, 110 mm diameter;
                 Whatman No. 44, or Barcham Green No. 800.
(9)   Wash-bottle, rubber tubing.
(10) Drying oven, set at 80 °C.
(11) Test sieves, 2 mm and 425 μm aperture, with receivers.
(12) Riffle-box, small size, 300 cm^3 capacity.
(13) Pestle and mortar.

The above items are included in Fig. 5.8.

REAGENTS

This list does not include the reagents needed for the analysis of the extract.

(1)   Dilute hydrochloric acid (10% HCl). Add 100 ml of concentrated hydrochloric acid (SG 1.18) to about 800 ml of distilled water, then add more distilled water to make 1 litre of solution.
(2)   Dilute ammonia solution. Take 500 ml of ammonia (SG 0.880) and add distilled water to make 1 litre of solution.
(3)   Bromine water. Add 6 ml of liquid bromine to 500 ml of distilled water, and shake.

All reagents must be of analytical reagent grade.

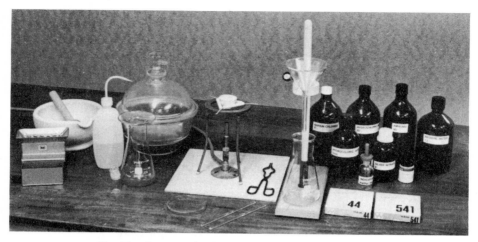

Fig. 5.8   Apparatus for determination of total sulphates in soil

PROCEDURE

(*A*)   *Preparation of sample*

This test requires only a very small sample (about 2 g), but the preparation of the sample to be properly representative can be a lengthy procedure.

The actual mass of sample used (stage 10) depends on the amount of sulphate present, and ideally should produce a precipitate of barium sulphate weighing about 0.2 g.

(1)   The bulk sample should be oven dried at a temperature not exceeding 80 °C, to avoid loss of water of crystallisation from any gypsum which may be present.

(2)   After cooling in a desiccator, the sample is weighed to an accuracy of 0.1% (m_1).

(3)   Sieve on a 2 mm sieve, breaking down all particles *other than stones* to pass the sieve. All stones are rejected unless they consist of lumps of gypsum, which should be removed by hand, crushed to pass the sieve and added to the portion passing the sieve.

(4)   Weigh the sieved material to an accuracy of 0.1% (m_2). Take care to ensure that there is no loss of fines during preparation and subsequent operations.

(5)   Reduce the sieved sample by successive riffling to a sample weighing about 100 g.

(6)   Pulverise this sample with pestle and mortar so that it passes a 425 μm sieve.

(7)   Subdivide the sample again until a sample of about 10 g is obtained. Mix thoroughly, and avoid segregation during riffling.

(8)   Place the riffled sample in a weighed glass weighing bottle, and dry overnight in the oven at 75–80 °C. Make successive weighings to check that the sample is dry.

(9)   Cool in the desiccator, and weigh to the nearest 0.001 g (m_5).

(10)   Transfer about 2 g of material (more if the sulphate concentration is low) into a 500 ml conical glass beaker (beaker A) without losing any fines. Weigh the weighing bottle again (m_6), and calculate the mass of soil transferred (m_3) by difference; i.e. $m_3 = m_5 - m_6$.

(*B*)   *Preparation of acid extract*

(11)   Add 200 ml of the 10% hydrochloric acid to the soil in beaker A. If effervescence occurs, take care to ensure that no material is lost.

(12)   Cover with a clock-glass, and boil gently for 4–5 min. Rinse the underside of the clock-glass with a little distilled water.

(13)   While the solution continues to boil, add 3 ml of the bromine water.

(14)   Add ammonia solution slowly from a burette to the boiling solution with constant stirring until the sesquioxides are precipitated and the liquid smells slightly of ammonia.

(15) Filter through a 110 mm diameter Whatman No. 541 filter paper on a conical filter funnel into a second 500 ml conical beaker (B).
(16) When beaker A has been drained and filtration has stopped, carefully remove the filter paper and contents and transfer them, without loss, back into beaker A.
(17) Add 20 ml of the 10% hydrochloric acid to beaker A, and stir until all the sesquioxides are dissolved. Add more acid if required, but no more than is necessary.
(18) Take the filter paper from the beaker and wash with distilled water until all traces of yellow colouration have disappeared. Collect the washings in beaker A. Reject the filter paper.
(19) Bring the contents of beaker A to the boil.
(20) Add ammonia solution as before (stage 14) to re-precipitate the sesquioxides.
(21) Filter the contents of beaker A through a Whatman No. 541 filter paper (as for stage 15) into beaker B.
(22) The contents of beaker B are now ready for the gravimetric analysis described in Section 5.6.5.

5.6.3 Water-soluble Sulphates in Soils — Preparation of Aqueous Extract (BS 1377:1975, Test 10)

This section gives the procedure for obtaining a 1:1 soil–water extract from a soil sample. Two procedures for measuring the sulphate content of the aqueous extract are described in Sections 5.6.5 and 5.6.7. The quantity of extract used for test depends upon which procedure is followed.

It may sometimes be difficult to prepare a 1:1 soil–water extract, especially if a centrifuge is not available, on account of the small quantity of water which can be filtered off. Under these circumstances it would be easier to prepare a 1:2 soil–water extract, but the appropriate correction must be applied when calculating the results.

The procedure consists of two main parts:

(A) Physical preparation of soil sample.
(B) Preparation of aqueous extract.

APPARATUS

Items (1)–(13) of the apparatus listed in Section 5.6.2, with the addition of the following (see Fig. 5.9):

(14) Mechanical shaker, or stirrer, capable of keeping 50 g of soil in continuous suspension in 50 ml of water.
(15) Watch-glass 75 mm diameter.
(16) Extraction bottle of 100 ml capacity.
(17) Centrifuge and centrifuge tubes.
(18) Conical beaker 250 ml.
(19) Pipettes, 25 ml and 50 ml.
(20) Buchner funnel and vacuum filtration flask.
(21) Filter papers, Whatman Nos. 44 and 50, or Barcham Green Nos. 800 and 975.
(22) Vacuum line and tubing.

PROCEDURE

(A) Preparation of sample

Stages (1)–(9). The soil sample is prepared from the initial bulk sample in the same way as for the total sulphates test (Section 5.6.2), except that a riffled sample of about 60 g instead of 100 g is prepared at stage (5), and stage (7) is omitted.
(10) From the dried sample, weigh out 50.00 g of soil on to a watch-glass.

(11)   Transfer the soil to a clean, dry extraction bottle of about 100 ml capacity. If a centrifuge is available, transfer to a centrifuge tube instead.

(B)   Preparation of water extract

(12)   Add exactly 50 ml of distilled water to the extraction bottle or centrifuge tube, using the 50 ml pipette, and stopper tightly.
(13)   Place in the shaker and agitate for 30 min.

The final preparation stages (14–16, or 17–19, or 20–22) depend upon whether a gravimetric analysis (Section 5.6.5) or the ion-exchange method (Section 5.6.7) is to be followed, and for the latter whether filtration or a centrifuge is used.

Gravimetric method

(14)   Filter the soil suspension through a Whatman No. 50 or Barcham Green No. 975 filter paper into a clean, dry filter flask, using a Buchner funnel. Wash the soil with distilled water to make a total volume of about 250 ml, which should contain no soil particles.
(15)   Transfer the extract to a clean, dry 500 ml conical beaker.
(16)   The extract is now ready for the gravimetric analysis described in Section 5.6.5.

Ion-exchange method (filtration)

(17)   After stage (13) filter the soil suspension through a Whatman No. 50 or Barcham Green No. 975 filter paper into a clean, dry filter flask, using a Buchner funnel. Do *not* add any additional water, and do not wash the soil remaining on the filter paper.
(18)   Transfer exactly 25 ml of the water extract to a clean, dry 250 ml conical beaker, using the 25 ml pipette.
(19)   The extract is now ready for the ion-exchange method of analysis described in Section 5.6.7.

Ion-exchange method (centrifuge)

(20)   After stage (13) centrifuge the suspension so that the solids settle out, leaving a clear, supernatant liquid above them.
(21)   Draw off exactly 25 ml of the supernatant liquid into the pipette and run into a 250 ml conical beaker.
(22)   The extract is now ready for the ion-exchange method of analysis described in Section 5.6.7.

5.6.4   Sulphates in Groundwater — Preparation of Water Sample

This section describes the simple procedure required for preparing a groundwater sample for the analysis of the sulphate content, which is expressed as grams per litre. The methods of analysis are described in Section 5.6.5 (gravimetric method) or Section 5.6.6 (ion-exchange procedure).

APPARATUS

Items (3), (4), (5), (8) of the apparatus listed in Section 5.6.2.

PROCEDURE

(1)   A sample of groundwater of at least 500 ml should be collected in a clean bottle. The method of sampling is given in BRE Digest 174.
(2)   The water sample is filtered through a Whatman No. 44 or Barcham Green No. 800 filter paper to remove any particles in suspension.

Continue with either stage (3) or stage (4), depending on which procedure is to be used.

(3) *Gravimetric analysis* (Section 5.6.5). Transfer exactly 50 ml of the filtered water sample to a clean, dry 500 ml conical beaker, and add approximately 100 ml of distilled water.

(4) *-on-exchange method* (Section 5.6.6). After stage (2) transfer exactly 100 ml of the sample to a clean dry 250 ml conical beaker.

The water sample is now ready for the analysis described in Section 5.6.5 or Section 5.6.6.

5.6.5 Sulphate Analysis on Prepared Solution — Gravimetric Method (BS 1377:1975, Test 9, and BS 1377:1967, Test 9 (A))

The gravimetric procedure for the determination of the sulphate content of a liquid extract from soil (either an acid extract, or a 1:1 soil–water extract), and of a sample of groundwater, is described below. The analysis is the same in all cases, but the method of calculation differs slightly, depending on the purpose of the test.

APPARATUS

The following are additional to the apparatus required for preparation of the extract (items 1–22 of Sections 5.6.2 and 5.6.3). They are included in Fig. 5.8, except item (2).

(1) Porcelain or silica ignition crucible.
(2) Electric muffle furnace capable of maintaining 800 °C.
(3) Bunsen burner, tripod and pipeclay triangle (if muffle furnace not available).
(4) Glass stirring rod fitted with rubber policeman.
(5) Filter papers, Whatman No. 44 or Barcham Green No. 800.

REAGENTS

The following are additional to the reagents required for the preparation of the extract (items 1–3 of Section 5.6.2).

(4) Barium chloride, 5% solution. Dissolve 50 g of barium chloride in 1 litre of distilled water.
NOTE: BARIUM CHLORIDE IS POISONOUS.
(5) Concentrated hydrochloric acid (SG 1.18).
(6) Concentrated sulphuric acid (SG 1.84).
(7) Indicators: methyl red, or blue litmus paper.
(8) Silver nitrate, 5% solution.

All reagents must be of analytical reagent grade.

PROCEDURAL STAGES

The test is carried out on the prepared solution in the conical flask, and follows on from the stage previously described, as under:

Total sulphates in soil — from stage (22) of Section 5.6.2.
Water-soluble sulphates in soil — from stage (16) of Section 5.6.3.
Sulphates in groundwater — from stage (3) of Section 5.6.4.

The procedural stages given below are divided into six parts:

(A) Prepare precipitate.
(B) Collect precipitate.
(C) Ignite precipitate.
(D) Weigh.

(E)   Calculate.
(F)   Report results.

PROCEDURE

(*A*)   *Preparation of precipitate*

(1)   Add two drops of methyl red indicator solution to the solution in the flask, and add dilute (10%) hydrochloric acid to acidify it (red coloration), plus a slight excess. An indicator solution is preferable to a paper indicator (as given in the British Standard) because it is easier to ensure thorough mixing. Methyl red is more sensitive than litmus.
(2)   Heat to boiling point.
(3)   Add 25 ml of the 5% barium chloride solution, drop by drop, while stirring the solution in the beaker.
(4)   Cover the solution and keep hot, but not boiling, for at least 1 h. This digestion period is necessary to enable the precipitate to form in particles large enough to be retained by filtration.
(5)   Allow the suspension to settle, and add a drop or two of barium chloride solution to the supernatant liquid.
(6)   If any slight cloudiness is observed as the drops enter, precipitation is incomplete and stages (2)–(4) should be repeated.

(*D*)   *Collection of precipitate*

(7)   Transfer the precipitate with extreme care to a Whatman No. 44 or Barcham Green No. 800 filter paper in the glass funnel placed over a receiver (e.g. beaker A), and filter. Wash several times with hot, distilled water until the washings are free from chloride.
(8)   To check for the presence of chloride, test a drop of the filtrate with a little of the silver nitrate solution. If there is no turbidity, the washings are free from chloride. Use a rubber policeman to remove the barium sulphate adhering to the walls of the beaker.
(9)   Transfer the filter paper and precipitate carefully to a porcelain or silica crucible, which has previously been ignited and weighed to 0.001 g (m_7).

(*C*)   *Ignition of precipitate*

If a muffle furnace is available, follow stages (10) and (11), omit stages (12*) and (13*), and continue from stage (14). Otherwise, continue from stage (12).

(10)   Place the crucible and contents in an electric muffle furnace at room temperature.
(11)   Raise the temperature of the furnace slowly to red heat (800 °C), and maintain that temperature for 15 min. The filter paper should char slowly, not inflame.
(12*) Dry the filter paper slowly at first over a small bunsen flame. Do not allow the filter paper to inflame, but let it char slowly.
(13*) Ignite by heating to red heat, and maintain this heat for 15 min.
(14)   Allow to cool in a desiccator for 30 min.
(15)   A footnote in the British Standard gives a procedure for treatment with acids after ignition, to eliminate any sulphide introduced by reduction of some of the barium sulphate by the carbon of the filter paper. However, it states that this effect may be ignored for most practical purposes, so this treatment need not be part of the routine procedure.

(*D*)   *Weighing residue*

(16)   Weigh the crucible and contents to 0.001 g (m_8).
(17)   Calculate the mass of precipitated (m_4) by difference:

$$m_4 = m_8 - m_7$$

(E)   *Calculations*

(18)   *Total sulphates* (Section 5.6.2). The percentage of total (acid-soluble) sulphates, expressed as SO_3, in the original soil sample is calculated from the equation

$$SO_3 \ (\%) = \frac{34.3 \times m_2 \times m_4}{m_1 \times m_3}$$

(19)   *Water-soluble sulphates* (Section 5.6.3). The percentage of water-soluble sulphates, expressed as SO_3, in the original soil sample is calculated from the equation

$$SO_3 \ (\%) = \frac{0.686 \times m_2 \times m_4}{m_1}$$

This is based on the mass of soil used (m_3) being exactly 50 g. If the mass differs from this, the equation given in stage (18) must be used instead.

(20)   The water-soluble sulphates may alternatively be expressed as grams per litre in the 1:1 aqueous extract, as follows:

$$\text{sulphates } (SO_3) \text{ in aqueous extract} = 6.86 \times m_4 \ \text{g/litre}$$

(21)   *Sulphates in groundwater* (Section 5.6.4). The concentration of sulphates, expressed as SO_3, in the 50 ml groundwater sample is given by

$$SO_3 = 6.86 \times m_4 \ \text{g/litre}$$

(F)   *Report results*

(22)   The sulphate content of soil, whether total (acid-soluble) or water-soluble, is reported as the percentage of SO_3 to the nearest 0.01%. It is important to state whether this relates to total sulphates or water-soluble sulphates. The sulphate content of groundwater, or of an aqueous solution, is reported as SO_3 in grams per litre, to the nearest 0.01 g/litre.

5.6.6   Sulphates in Groundwater (Ion-exchange Column) (BS 1377:1975, Test 10)

This method is quicker and easier than the gravimetric procedure given above. However, it cannot be used if the groundwater contains chloride, nitrate or phosphate ions. If these or other anions are present, the gravimetric procedure (Sections 5.6.4 and 5.6.5) must be used instead.

APPARATUS

(1)   Glass tube 400 mm long and 10 mm diameter, with swan-neck outlet, as detailed in Fig. 5.10(a) and shown on the left of Fig. 5.9.
(2)   Constant-head device, made from a round-bottomed flask, as detailed in Fig. 5.10. If this is not available, the tap on a burette can be adjusted to provide a steady rate of flow into item (1), as shown in Fig. 5.9.
(3)   Buchner funnel, 100 ml diameter.
(4)   Filter flask, about 500 ml capacity, for the funnel.
(5)   Filter papers, Whatman No. 44 or Barcham Green No. 800, 100 mm diameter.
(6)   250 ml beaker.

Fig. 5.9   Apparatus for determination of water-soluble sulphates in soil and groundwater

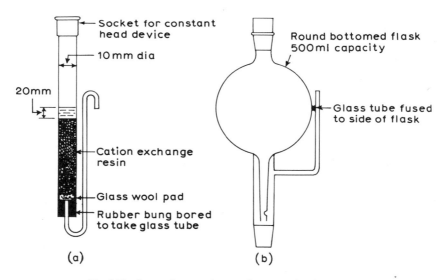

Fig. 5.10   Ion-exchange column and constant-head apparatus

REAGENTS

(1)   A strongly acidic cationic exchange resin. Suitable types are Zeo-Karb 225 or Amberlite IR-120.

(2)   Sodium hydroxide solution, approximately 0.1 N. Dissolve 2 g of sodium hydroxide in 500 ml of distilled water. Determine the exact normality of this solution by titration with standardised hydrochloric acid.

      If 50 ml of sodium hydroxide solution is just neutralised by a volume V_a ml of 0.1 N hydrochloric acid, the normality, n, of the sodium hydroxide solution is given by

$$n = \frac{V_a}{50} \times \frac{N}{10}$$

Keep the solution in an airtight plastics container.

(3)   Hydrochloric acid, approximately 4 N. Add 360 ml of concentrated hydrochloric acid (SG 1.18) to about 800 ml of distilled water, then add more distilled water to make 1 litre.

(4)   Silver nitrate solution. Dissolve 0.5 g of silver nitrate in 100 ml of distilled water. Store in an amber-coloured glass bottle.

(5)   Nitric acid, approximately 1 N solution. Add 15 ml of concentrated nitric acid (SG 1.42) to about 60 ml of distilled water, then add more distilled water to make 100 ml.

(6)   Indicator, such as screened methyl orange, which gives a distinct colour change in the range pH 4 to pH 5.

PROCEDURAL STAGES

(1)   Prepare ion-exchange column.
(2)   Prepare groundwater sample.
(3)   Boil.
(4)   Pass through ion-exchange column.
(5)   Add indicator.
(6)   Titrate.
(7)   Calculate.
(8)   Report result.

TEST PROCEDURE

(1)   *Preparation of ion-exchange column*

Take a quantity of ion-exchange resin sufficient to half-fill the column, place in a beaker and stir with distilled water. Empty the suspension of resin in water into the column. When the resin has settled, drain off surplus water so that about 20 mm depth of water remains above the resin (see Fig. 5.10a).

The resin is activated by leaching with 100 ml of 4 N hydrochloric acid followed by washing with distilled water. If the constant-head device (Fig. 5.10b) is used, place the acid in the round-bottomed flask, replace the stopper and leave the acid to pass through the column. Rinse the flask with distilled water and leave the water to percolate through the column until the liquid coming out shows no turbidity when tested with about 1 ml of silver nitrate solution acidified with a few drops of nitric acid.

If the constant-head device is not available, add the acid and water in increments but allow each increment to drain away before adding the next. Alternatively, use a burette as a substitute for the constant head device.

After four successive determinations of sulphate content, the column must be regenerated as described above.

(2)   *Preparation of sample*

As described in Section 5.6.4.

(3)   *Boiling*

Heat the water (from stage 4 of Section 5.6.4) to boiling point, and boil gently for 5 min, taking care that no water is lost. Leave to cool.

(4)   *Use of ion-exchange column*

Place a 500 ml conical beaker under the outlet of the ion-exchange column. Pass the water through the column, followed by rinsing with two 75 ml increments of distilled water. Collect the water and washings together in the conical beaker.

(5)   *Addition of indicator*

Add indicator to the collected liquid to impart sufficient colour for the detection of the end point of the titration.

(6)   *Titration*

Titrate the liquid against the standardised sodium hydroxide solution. Record the volume of sodium hydroxide solution required to just neutralise the liquid to the nearest 0.05 ml (V). This has occurred when the solution retains a yellow colour after swirling the flask.

(7)   *Calculation*

The sulphate content expressed as SO_3 of the groundwater is calculated from the equation

$$SO_3 = 0.4\, nV \text{ g/litre}$$

where n is the normality of the sodium hydroxide solution (see under 'Reagents'.)

(8)   *Result*

The sulphate content of the groundwater, expressed as SO_3, is reported in grams per litre to the nearest 0.01 g/litre, as determined by the ion-exchange column method.

5.6.7   Sulphates in Aqueous Soil Extract (Ion-exchange Method) (BS 1377:1975, Test 10)

This method is quicker and easier than the gravimetric method, but it cannot be used if the soil contains chloride, nitrate or phosphate ions, or other anions.

The procedure is similar to that described for groundwater samples in Section 5.6.6, except that a 1:1 soil–water extract must first be prepared as described in Section 5.6.3. The same apparatus and reagents as referred to in both these sections are required.

PROCEDURE

(1)   *Preparation of ion-exchange column.* As Stage 1 of Section 5.6.6.
(2)   *Preparation of aqueous extract.* As described in Section 5.6.3. To the 25 ml of aqueous extract in the 250 ml beaker from Stage 19 or 22, add distilled water to make up to 100 ml. Stages (3)–(6) are similar to stages (3)–(6) in Section 5.6.6.
(7)   Calculation. The sulphate content, expressed as SO_3, of the soil–water extract, is calculated from the equation

$$SO_3 = 1.6\, nV \text{ g/litre}$$

where n is the normality of the sodium hydroxide solution.
(8)   *Results.* The sulphate content of the soil–water extract, expressed as SO_3, is reported in grams per litre to the nearest 0.01 g/litre, as determined by the ion-exchange column method. The percentage of material removed from the soil sample by sieving on the 2 mm sieve is also reported.

5.7 ORGANIC CONTENT TESTS

5.7.1 Scope of Tests

Three different procedures are given for the determination of the organic content of soils:

Loss on ignition.
Oxidation with hydrogen peroxide.
Dichromate oxidation.

The dichromate method is the one given in BS 1377:1975 as the standard procedure for soils. Limitations of the other methods are summarised in Table 5.1.

5.7.2 Loss on Ignition

This method is suitable for sandy soils which contain little or no clay or chalky material. It is based on the procedure given in *Soil Mechanics for Road Engineers* (TRRL, 1952).

APPARATUS

(1) Balance accurate to 0.001 g.
(2) Silica or porcelain crucible, about 30 ml capacity.
(3a) Muffle furnace
     or
(3b) Meker gas burner, pipeclay triable and tripod.
(4) Drying oven, 105–110 °C, and desiccator.

PROCEDURE

(1) Prepare a representative sample of about 20 g of oven-dried soil, cooled in the desiccator.
(2) Clean, dry and weigh the crucible to an accuracy of 0.001 g (m_1).
(3) Place the dried soil in the crucible and weigh soil and crucible to 0.001 g (m_2).
(4) Place the crucible in the muffle furnace and heat to 800 °C.
(5) Alternatively, if a muffle furnace is not available, heat the crucible to red heat over the Meker burner. Take care to avoid loss of soil due to the currents of burning gas.
(6) Allow the crucible to cool, and weigh to 0.001 g (m_3).

CALCULATION

Calculate the ignition loss from the equation

$$\text{loss on ignition} = \frac{m_2 - m_3}{m_2 - m_1} \times 100\%$$

RESULT

Report the result as the organic matter content by loss on ignition as a percentage to the nearest 0.1%.

5.7.3 Peroxide Oxidation Method

This method is used as part of the pretreatment of soils before a fine particle size analysis, to eliminate colloidal organic matter (Section 4.8.1). However, hydrogen peroxide has only a limited action on undecomposed plant remains, such as roots and fibres.

APPARATUS

(1) Balance accurate to 0.001 g.
(2) Porcelain evaporating dish, 140 mm diameter.
(3) Glass stirring rod.
(4) Thermometer, 0–100 °C.
(5) Buchner funnel and filter flask.
(6) Source of vacuum.
(7) Filter papers, Whatman No. 50.
(8) Beaker, 400 ml.
(9) Drying oven 105–110 °C and desiccator.

REAGENT

Hydrogen peroxide, 6% (20 volume) solution.

PROCEDURE

(1) The original soil sample is oven dried and cooled in the desiccator, and riffled to give a representative sample of 50–100 g passing a 2 mm sieve. Weigh the sample to 0.01 g (m_1) and place in a clean dry wide-mouth conical flask.
(2) Add 150 ml of hydrogen peroxide and stir gently with a glass rod. Cover, and allow to stand overnight.
(3) Heat gently to a temperature of about 60 °C, stirring to release bubbles of gas. Avoid frothing over.
(4) Allow the reaction to continue until gas is no longer evolved at a very rapid rate.
(5) Boil the mixture, to reduce the volume to about 50 ml and to decompose excess peroxide.
(6) When cool, add more peroxide if necessary to complete the oxidation, and repeat stages (4) and (5). With very organic soils this process may take one or two days.
(7) Filter through a Whatman No. 50 filter paper, using a Buchner funnel and vacuum flask. Wash thoroughly with distilled water.
(8) Transfer the soil to a weighed and dried glass evaporating dish (mass m_2).
(9) Dry the dish and contents in the oven at 105–110 °C, to constant mass.
(10) Weigh dish and contents (m_3).
(11) Calculate the loss due to hydrogen peroxide treatment from the equation

$$\text{loss} = \frac{m_3 - m_2}{m_1} \times 100\%$$

(12) Report the result as a percentage to the nearest 0.1% as the organic matter content determined by hydrogen peroxide treatment.

5.7.4 Dichromate Oxidation Method (BS 1377:1975, Test 8)

This method was first introduced by Walkley and Black in 1934, and was revised by Walkley in 1935. It has been found to give reproducible results. The accuracy is not absolute but is sufficient for most engineering purposes.

Soils containing sulphides or chlorides have been found to give high results by this method. These substances, if present, can be removed at the sample preparation stage by the appropriate chemical treatment, as described under 'Procedure'.

APPARATUS

(1) Balance accurate to 0.001 g.
(2) Two 1 litre volumetric flasks.

Fig. 5.11   Apparatus for organics content test (dichromate oxidation)

(3)   Two 25 ml burettes, graduated to 0.1 ml, and burette stands.
(4)   10 ml pipette with rubber teat.
(5)   1 ml pipette with rubber teat.
(6)   Two 500 ml conical flasks.
(7)   Graduated measuring cylinders, 200 ml and 20 ml.
(8)   Glass weighing bottle.
(9)   Sieves, 10 mm and 425 μm aperture.
(10) Pestle and mortar.
(11) Wash-bottle with distilled water.
(12) Drying oven 105–110 °C and desiccator.
(13) Riffle box, small (300 cm³).

The main items are shown in Fig. 5.11.

REAGENTS

(1)   Potassium dichromate, 1 N solution. Dissolve 49.035 g of potassium dichromate in distilled water to make 1 litre of solution.
(2)   Sulphuric acid, 0.5 N solution. Add 14 ml of concentrated sulphuric acid to distilled water to make 1 litre of solution.
(3)   Sulphuric acid, concentrated (SG 1.84).
(4)   Ferrous sulphate, approximately 0.5 N solution. Dissolve 140 g of ferrous sulphate in 0.5 N sulphuric acid to make 1 litre of solution. This solution is unstable in air and should be kept tightly stoppered. It should be standardised against the dichromate solution weekly.
(5)   Orthophosphoric acid, 85% (SG 1.70–1.75).
(6)   Indicator solution. Dissolve 0.25 g of sodium diphenylamine sulphonate in 100 ml of distilled water.

PROCEDURAL STAGES

The procedure described below is divided into five parts:

(A)   Standardise ferrous sulphate solution.
(B)   Prepare sample.
(C)   Test for organic matter.
(D)   Calculate.
(E)   Report result.

PROCEDURE

(A)   *Standardisation of ferrous sulphate solution*

(1)   Run 10 ml of the 1 N potassium dichromate solution from the burette into a 500 ml conical flask.
(2)   Very carefully add 20 ml concentrated sulphuric acid. This will generate heat. Swirl the mixture and allow to cool, standing on a heat-insulating surface or pad. Protect from draughts.
(3)   Add 200 ml of distilled water.
(4)   Add 10 ml of orthophosphoric acid and 1 ml of indicator, and mix thoroughly.
(5)   Add ferrous sulphate from the burette in increments of 0.5 ml, swirling the flask, until the colour changes from blue to green.
(6)   Add a further 0.5 ml of potassium dichromate, changing the colour back to blue.
(7)   Add ferrous sulphate drop by drop, with continued swirling, until the colour of the solution changes from blue to green after the addition of a single drop. Record the total volume of ferrous sulphate used, x, to the nearest 0.05 ml.

(B)   *Preparation of sample*

Only a small sample is required for the test, but it must be correctly prepared to be representative.

(8)   The bulk sample is oven dried at 105–110 °C, cooled in a desiccator and weighed to an accuracy of 0.1% (m_1).
(9)   Sieve on a 10 mm sieve and crush all particles, other than stones, so as to pass the sieve.
(10)   Weigh the mass passing to an accuracy of 0.1% (m_2). Take care not to lose any fines.
(11)   Reduce the sample by successive riffling to a sample of about 100 g.
(12)   (This stage applies only if the soil contains sulphides, which can give a falsely high result.) To eliminate sulphides, add dilute sulphuric acid (2 N) until no further evolution of hydrogen sulphide (H_2S) occurs, then wash with water. Oven dry before proceeding to stage (14).
(13)   (This stage applies only if the soil contains chlorides, which can give a falsely high result.) To remove chlorides, wash the soil with distilled water until no turbidity is observed when a drop of the wash-water is tested with silver nitrate solution. Oven dry before proceeding further.
(14)   Pulverise this sample with pestle and mortar so that it passes a 425 μm sieve.
(15)   Subdivide further until a sample weighing about 5 g is obtained. Mix thoroughly, and avoid segregation.
(16)   Place the sample in a glass weighing bottle and dry in an oven at 105–110 °C.
(17)   Cool in a desiccator, and weigh to 0.001 g (m_5).
(18)   Transfer a small quantity for the test to a clean dry 500 ml conical flask, without losing any fines.
      The amount to be transferred will range from about 5 g for a soil low in organic matter, to as little as 0.2 g for a very peaty soil. Experience will indicate the most suitable size; alternatively, several sizes of sample should be tested, and the most suitable one taken as giving the correct result. (See stage 27.)

(19) Weigh the weighing bottle again (m_6), and calculate the mass of sample transferred (m_3) by difference:

$$m_3 = m_5 - m_6$$

(C) *Testing for organic matter*

(20) Run 20 ml of 1 N potassium dichromate into the conical flask from the burette.
(21) *Very carefully* add 20 ml of concentrated sulphuric acid from a measuring cylinder.
(22) Swirl the mixture for 1 min and stand it on an asbestos or wood mat for 30 min while it cools and to allow oxidation of the organic matter to proceed. Protect the flask from draughts.
(23) Add 200 ml of distilled water.
(24) Add 10 ml of orthophosphoric acid and 1 ml of indicator, and shake thoroughly. If the indicator is absorbed by the soil add a further 1 ml of indicator, and shake.
(25) Add ferrous sulphate from the bottle in increments of 0.5 ml, and swirl the flask, until the colour of the solution changes from blue to green.
(26) Add a further 0.5 ml of potassium dichromate, changing the colour back to blue.
(27) Add ferrous sulphate drop by drop from a burette, with continued swirling, until the colour changes from blue to green after the addition of a single drop. Record the total volume of ferrous sulphate used, y, to the nearest 0.05 ml. This volume will be between 5 ml and 8 ml if the correct amount of material was taken for test in stage (18). If several sizes of sample were tested, accept that result which gives a value of y within this range.

(D) *Calculations*

(28) Calculate the total volume, V ml, of potassium dichromate used to oxidise the organic matter in the soil from the equation

$$V = 10.5 \times \left(1 - \frac{y}{x}\right)$$

where y = total volume of ferrous sulphate used in the test (stage 27); x = total volume used in standardisation test (stage 7). Calculation the percentage organic matter present in the oven-dry soil sample from the equation

$$\text{percentage organic matter content} = \frac{0.67 \times m_2 \times V}{m_1 \times m_3}$$

where m_1 = mass of sample before sieving (stage 8); m_2 = mass of sample passing 10 mm sieve (stage 10); m_3 = mass of soil used in the test (stage 19).

(E) *Results*

(30) Report the result as the percentage organic matter content, determined by the dichromate oxidation method, to the nearest 0.1% of the original mass of oven-dry soil.

5.8 CARBONATE CONTENT

5.8.1 Collins Calcimeter — Standard method

This procedure for measuring the carbonate content of soils was first published by S. H. Collins in 1906. The apparatus which bears his name was developed from that known as Scheibler's apparatus. Collins also devised a special slide-rule to simplify the necessary calculations.

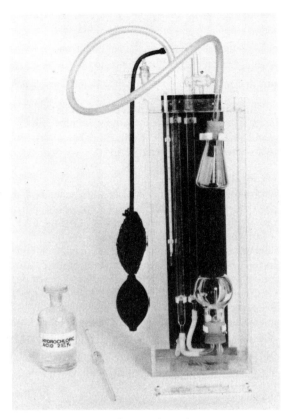

Fig. 5.12   Collins Calcimeter for carbonate content test

In this test a weighed amount of soil is treated with hydrochloric acid. The volume of carbon dioxide given off is measured, and is corrected for temperature and atmospheric pressure. The carbonate content of the soil is calculated from the corrected volume of carbon dioxide. The method is accurate enough for most engineering purposes.

APPARATUS

(1)   Collins Calcimeter. This consists of several components (which are listed below) in a clear Perspex tank which can be filled with water. This is to maintain a uniform temperature in all components throughout the duration of the test. The results are very sensitive to temperature.

The apparatus is illustrated in Fig. 5.12, and is shown diagrammatically in Fig. 5.13. It consists of

Measuring cylinder for hydrochloric acid (A) (preferably of clear plastics)
reaction flask (B)
graduated burette reading to $0.1 \, cm^3$ (E)
levelling tube (F)
reservoir flask (R)
hand pressure bulb (D)
tap (G)
two-way tap (H) (see Fig. 5.14)
outlet (C)

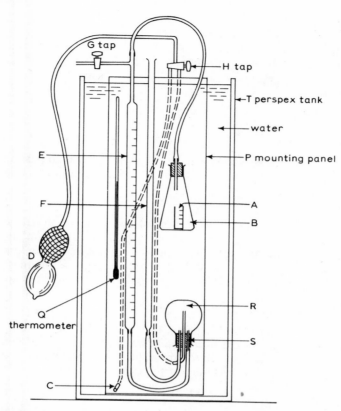

Fig. 5.13   *Diagrammatic arrangement of Collins Calcimeter (Courtesy Macfarlane Robson Ltd.)*

     rectangular Perspex tank (T)
     mounting panel (P)
     thermometer, 0–40 °C reading to 0.2 °C (Q)

(2)   Special slide-rule, provided with the Calcimeter (Fig. 5.15).
(3)   Barometer, or means of ascertaining local barometric pressure.
(4)   Analytical balance reading to 0.001 g.
(5)   Drying oven and equipment.

REAGENT

Hydrochloric acid solution, 25% v/v. Dilute one volume of concentrated hydrochloric acid
with three volumes of distilled water.

PROCEDURAL STAGES

(1)   Prepare sample.
(2)   Weigh sample.
(3)   Prepare Calcimeter apparatus.
(4)   Insert acid.
(5)   Connect reaction flask.
(6)   Agitate water.
(7)   Adjust burette to zero.
(8)   Mix acid and soil.
(9)   Agitate water.

(10) Equalise burette.
(11) Repeat stages (7)–(10) until steady.
(12) Apply temperature correction.
(13) Read or obtain atmospheric pressure.
(14) Calculate.
(15) Report results.

TEST PROCEDURE

(1) *Preparation of sample*

A representative sample of oven-dried soil is broken down so as to pass a 6.3 mm sieve. It is then riffled to about 50 g, and is oven dried and cooled in a desiccator.

(2) *Weighing sample*

The sample for test is weighed to 0.001 g (W_1) and placed in the reaction flask (B, Fig. 5.13). The quantity required is such as to evolve about 10–25 ml of carbon dioxide gas. As an approximate guide, this amount will range from 0.1 or 0.2 g for a pure limestone or chalk to 20 g for a relatively non-calcareous soil. If the probable carbonate content is not known, one or two preliminary tests should be carried out first so as to establish the approximate range.

(3) *Preparation of calcimeter*

Fill the Perspex tank (T) with water at room temperature to within about 25 mm from the top. Use boiled or distilled water if the local mains water is hard. Before connecting the hand pressure bulb (D), and with tap G open to atmosphere, pour water into the levelling tube (F), using a funnel, so as to half-fill the reservoir flask (R) with water.

(4) *Insertion of acid*

Add hydrochloric acid (25% v/v) to the graduated cylinder (A) from a pipette or burette, then carefully place it in the reaction flask (B). The amount of acid required will range from 10 ml for soils containing little calcareous matter to 15 ml for highly calcareous soils. Take care not to spill any acid on to the soil sample.

(5) *Connection of reaction flask*

Connect the reaction flask to the apparatus by inserting the rubber stopper (S). Place the flask under the water in the tank and secure it into position with the spring clip on the mounting board.

(6) *Agitation of water*

Set tap H so as to connect the hand pressure bulb to the tube (C) (the D–C position, Fig. 5.14a). Gently blow air through the water in the tank by squeezing the hand bulb so as to obtain a uniform temperature. Record the temperature (T_1 °C) when it is steady.

(7) *Adjustment of burette*

Open tap G and set tap H to the D–R position (Fig. 5.14b). Gently squeeze the bulb with one hand until the water level in burette E exactly coincides with the zero mark, then close tap G with the other hand. The water level in tube F should be at the same level as that in E. Release the pressure bulb, and the water level in tube F will fall. The level in E will also drop slightly.

(8) *Mixing acid and soil*

Carefully remove flask B from the water-bath, keeping it connected to the burette (E). Tilt the

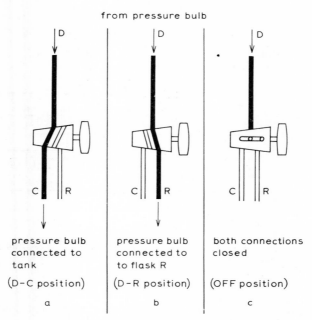

Fig. 5.14   *Two-way tap on Collins Calcimeter*

flask so that the acid in the cylinder (A) spills over the soil sample. Shake the flask well (being careful not to break or crack cylinder A), and replace it in the water-bath.

(9)   *Agitation of water*

Set tap H to the D–C position, and gently blow air through the water in the tank until a uniform steady temperature (T_2 °C) is established.

(10)   *Equalisation of water levels*

Set tap H to the D–R position. Gently squeeze the pressure bulb until the water levels in tubes E and F are equal. Then close tap H (Fig. 5.14c). Read the volume indicated in the burette (E).

(11)   *Check reaction complete*

Remove reaction flask B without disconnecting it, shake, return to the water-bath, and adjust the levels in tubes E and F (stages 7–10). Repeat this process until there is no further increase in the volume indicated by burette E. Record the final reading as the volume of CO_2 evolved (V_g ml).

(12)   *Temperature correction*

If the temperature at the end of the reaction (T_2) differs from that at the beginning (T_1), a correction to the measured volume is necessary. For each 0.2 °C *rise* in temperature from T_1 to T_2, *subtract* 0.1 ml from the observed volume. If the temperature falls, for each 0.2 °C *fall* from T_1 to T_2, *add* 0.1 ml. (This correction is based on the fact that 136 ml of air expands 0.1 ml for a temperature of 0.2 °C. The volume of flask B is 150 ml, and the volume of acid used is 14 ml, giving an air volume of 136 ml.)

(13)   *Barometric pressure*

Read the barometric pressure at the time of the test. If an accurate barometer is not available,

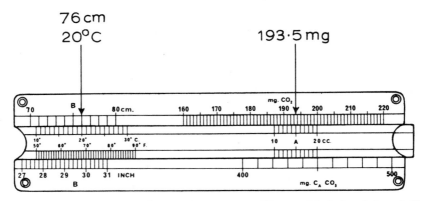

Fig. 5.15   *Slide rule for calculating carbonate content (Courtesy Macfarlane Robson Ltd.)*

the local meteorological office can give this information, or it may be estimated approximately from the day's newspaper weather report.

(14)   *Calculations*

The carbonate content (expressed as CO_2) of the sample can be calculated either: (a) using the special slide-rule which is available for use with the calcimeter, or (b) using tables.

(a)   Using the slide-rule, set the mean temperature $T_m = \frac{1}{2}(T_1 + T_2)$ °C on the left-hand side of the slide opposite the barometric pressure (cmHg). The mass of CO_2 (W_2 mg) equivalent to 100 ml of gas measured in the apparatus is found on the upper scale opposite the volume of acid used (V_a) on the scale at the right-hand end of the slide.

The carbonate content is calculated from the equation

$$\text{carbonate \% (as } CO_2) = \frac{W_2 V_g}{1000 \, W_1}$$

The slide-rule is shown in Fig. 5.15 and is set for a pressure of 76 cmHg at a temperature of 20 °C. If 15 cm³ of acid was used (mark A), the mass of CO_2 is read off as 193.5 mg (say 194 mg).

(b)   The mass (W_2) of 100 ml of CO_2 may alternatively be obtained from Table 5.7 on the same line as the mean temperature T_m and underneath the volume of acid used, V_a. This mass must then be corrected for: (i) barometric pressure, from the right-hand side of the table; (ii) volume of gas evolved, from below the main table.

The equation given in (a) above is used with the corrected value of W_2 to calculate the percentage carbonate as CO_2.

(15)   *Results*

Report the result as the percentage carbonates in the soil sample, expressed as CO_2, to two significant figures.

5.8.2   Collins Calcimeter — Simplified method

This procedure is less accurate than the standard method described above, but it gives a direct reading of carbonate content without any calculations.

This is achieved by first reading the temperature of the water-bath and observing the

Table 5.7. DATA FOR COLLINS CALCIMETER TEST

T_m (°C)	Mass W_2 (uncorrected) Volume of acid used, V_a (ml)								Corrections for pressure Barometric pressure (mm Hg)					Corrections for volume V_g (ml)			
	10	12	14	16	18	20	30	40	740	750	760	770	780	0–9	10–19	20–29	30–40
12	198	201	204	208	211	214	230	249	−6	−3	0	+3	+6	+1	0	0	−1
14	196	199	202	204	207	210	226	244	−5	−3	0	+3	+5	+1	0	0	−1
16	193	196	198	201	203	206	221	238	−5	−3	0	+3	+5	+2	0	0	−2
18	191	193	196	198	201	203	216	233	−5	−3	0	+3	+5	+2	+1	−1	−2
20	188	190	193	195	198	200	212	228	−5	−3	0	+3	+5	+3	+1	−1	−3
22	186	188	190	192	194	196	208	223	−5	−3	0	+3	+5	+3	+1	−1	−3
24	183	185	187	189	191	193	204	218	−5	−3	0	+3	+5	+5	+3	−3	−5
26	181	183	185	186	188	190	200	213	−5	−3	0	+3	+5	+8	+4	−4	−8
28	178	180	182	183	185	187	196	208	−5	−3	0	+3	+5				
30	176	178	179	181	182	184	192	203	−5	−3	0	+3	+5				

Notes:
1. Data are void for a 150 ml reaction flask (B).
2. The main section of the table gives the mass of 100 ml CO_2 (mg) based upon a measured volume of 20 ml of CO_2 at 760 mmHg pressure, and at mean temperatures between 12 and 30°C (T_m).
3. The correction sections show values to be added or subtracted for different pressures and measured volumes of CO_2.

Example: $V_a = 20$ ml, $T_m = 22$°C, barometer reading 750 mmHg, $V_g = 10$ ml. Therefore 100 ml of CO_2 weighs: $196 − 3 + 1 = 194$ mg $= W_2$. If the dry mass of sample used was 2.51 g $= W_1$,

$$\text{carbonate content \% (as } CO_2) = \frac{194 \times 10}{1000 \times 2.51} = 0.77\%$$

* Based on data supplied by Macfarlane Robson Ltd., Blaydon-on-Tyne, Durham, manufacturers of the Collins' Calcimeter and slide-rule.

Table 5.8. DATA FOR SIMPLIFIED PROCEDURE Collins Calcimeter Test
(a)   Volume (ml) of acid to be used

Temperature (°C)	Barometric pressure (mmHg)						
	730	740	750	760	770	780	790
12	19	17	15	13	11	9	7
14	21	19	17	15	13	11	9
16	23	21	19	17	15	13	11
18	25	23	21	19	17	15	13
20	27	25	23	21	19	17	15
22	29	27	25	23	21	19	17

Note: Where the quantity of acid indicated exceeds 15 ml, then half the stated quantity of 50% (v/v) HCl should be placed in tube A and the same amount of distilled water in flask B with the sample.

(b)   Conversion to carbonate content

Mass of soil (g)	Carbonate content represented by each cm^3 of CO_2 (%)
0.2	1.0
2.0	0.1
20.0	0.01

barometric pressure. The volume of hydrochloric acid (25% v/v) required for the test is dependent on these two readings, and is read off from Table 5.8(a). This volume of acid is placed in the measuring cylinder (A, Fig. 5.13). The mass of soil used should be 0.2 g for limestone or chalk, 20.0 g for soil with little carbonate content or 2.0 g for an intermediate material.

The test is carried out exactly as described in Section 5.8.1. The volume of carbon dioxide evolved is read from burette E. This is converted to carbonate content of the soil (%) as indicated in Table 5.8(b).

Where the quantity of acid indicated in Table 5.8(a) exceeds 15 ml, use half the stated quantity of 50% (v/v) hydrochloric acid (instead of 25% HCl) in measuring cylinder A and the same amount of distilled water in flask B with the sample.

5.9   CHLORIDE CONTENT

5.9.1   Qualitative Test

A quick test can be made to indicate the presence of chlorides. Take about 5 ml of filtered groundwater, or of 1:1 soil–water extract, in a test-tube. If this is highly alkaline (pH 12–14), add a few drops of nitric acid to acidify it. Add a few drops of 1% silver nitrate solution. If there is appreciable turbidity, it indicates that chloride is present in a measurable quantity.

Either of the following tests can be used for a quantitative determination of chloride content. Mohr's method (Section 5.9.3) is simpler to carry out than Volhard's method (Section 5.9.2), but both methods require careful observation and accurate weighing.

5.9.2   Chloride Content — Volhard's Method

This procedure is based on the method given in BS 812: Part 4: 1976, 'Methods for sampling and testing mineral aggregates, sands and fillers', for the determination of the amount of water-soluble chloride salts present in concrete aggregates. It can be equally applied to soils. The principle used is that due to Volhard, in which an excess of silver nitrate solution is added to the acidified chloride solution and the unreacted portion is back-titrated with potassium thiocyanate, with ferric alum used as an indicator.

APPARATUS

(1) Analytical balance accurate to 0.1 mg.
(2) Two 1000 ml volumetric flasks.
(3) Graduated measuring cylinders, 10 ml and 500 ml.
(4) Pipettes, 100 ml and 25 ml.
(5) Two 50 ml burettes and burette stands.
(6) At least four conical flasks, 250 ml.
(7) Wash-bottle and distilled water.
(8) Amber-coloured glass reagent bottle.
(9) Three plastics bottles with wide mouth and screw tops, 1–1.25 litre capacity.
(10) Filtration funnel.
(11) Filter papers, Whatman Nos. 541 and 44.
(12) Ashless tablet.
(13) Drying oven, desiccator.

REAGENTS

(1) Silver nitrate, 0.1 N solution. Dry about 20 g of silver nitrate in the oven at 110 °C for 1–2 h, and cool in the desiccator. Weigh out 16.987 g of the dried silver nitrate, dissolve in distilled water, and make up to exactly 1000 ml with distilled water in the volumetric flask. Store the solution in the amber-coloured bottle away from sunlight.
(2) Potassium thiocyanate, approximately 0.1 N solution. Dissolve 10.5 g of potassium thiocyanate in distilled water and make up to 1000 ml in a volumetric flask. This solution will be a little stronger than 0.1 N, and the exact normality is determined as described in steps (1)–(5) of the procedure below.
(3) Nitric acid, approximately 6 N. Dilute 100 ml of nitric acid (SG 1.42) with distilled water to 250 ml, and boil until it is colourless.
(4) 3,5,5-Trimethylhexan-1-ol.
(5) Ferric alum indicator solution. Make a saturated solution of ferric ammonium sulphate (about 40%) in water to which a few drops of 6 N nitric acid have been added.

PROCEDURAL STAGES

The procedure described below is divided into five main sections:

(A) Standardise potassium thiocyanate solution.
(B) Prepare sample.
(C) Prepare solution and titrate.
(D) Calculate.
(E) Report result.

PROCEDURE

(A) Standardisation of potassium thiocyanate solution

(1) Use a 25 ml pipette to place 25 ml of 0.1 N silver nitrate solution into a 250 ml conical flask.
(2) Add 5 ml of 6 N nitric acid and 1 ml of ferric alum indicator solution.
(3) Add potassium thiocyanate solution from a burette until the colour first permanently changes from colourless to pale brown.

(4)   Note the volume of thiocyanate solution used (V_1 ml).
(5)   Calculate the normality (T) of the potassium thiocyanate solution from the equation

$$T = \frac{2.5}{V_1}$$

(B)   *Preparation of sample*

(6)   Three samples of soil each of about 0.5 kg and containing particles not larger than 2 mm are prepared by subdividing the original sample in the usual way. Each sample is oven dried at 105–110 °C, and cooled in the desiccator.

(C)   *Chemical treatment and titration*

The chloride content is determined on each of the three samples separately, using the procedures outlined below.

(7)   Take about 500 g of the dried soil and weigh to 0.1 g (m). Place in the wide-mouth screw-top plastics bottle.
(8)   Add 500 ml of distilled water and allow to stand for 24 h with occasional shaking.
(9)   If the supernatant liquid is discoloured owing to suspended silt or clay, it should be filtered through a Whatman No. 541 filter paper. It may sometimes be necessary to use a Whatman No. 44 filter paper if the suspension is very fine, with the addition of a quarter of an ashless tablet to the filter funnel.
(10)   Use a pipette to draw off 100 ml of the supernatant liquid, and place it in a 250 ml conical flask.
(11)   Add a little nitric acid to the conical flask until the pH is between 6 and 7, followed by 0.1 N silver nitrate solution from a burette untill all the chloride has been precipitated.
(12)   Then add 10 ml excess of silver nitrate solution. Record the total volume (V_2 ml) of solution added.
(13)   Add 2 ml of 3,5,5-trimethylhexan-1-ol and agitate the solution vigorously to coagulate the precipitate.
(14)   Add 5 ml of ferric alum indicator solution.
(15)   Add standardised potassium thiocyanate solution from a burette until the first colour change occurs (colourless to pale brown). Note the volume (V_3 ml) of potassium thiocyanate solution used.

(D)   *Calculation*

(16)   The percentage chloride salts ($C\%$) in the soil, expressed as equivalent sodium chloride content, for each of the three samples is calculated from the equation

$$C = (V_2 - 10\,V_3\,T) \times \frac{2.925}{m}$$

       Calculate the average of the three separate results.

(E)   *Results*

(17)   The average of the three determinations is reported to the nearest 0.01% as the chloride salt content (expressed as equivalent sodium chloride content). The three individual results are also reported.
(18)   If the soil contains particles larger than 2 mm, follow stages (6)–(17) but using about 1 kg of soil instead of 500 g. Particles larger than 20 mm should first be removed.

5.9.3 Chloride Content — Mohr's Method

This method is based on Mohr's procedure, and is outlined by M. J. Bowley (1979). The test solution and a blank for comparison are each titrated with 0.02 N silver nitrate solution, potassium chromate being used as an indicator.

APPARATUS

(1)   Analytical balance accurate to 0.1 mg.
(2)   Theee conical flasks, 250 ml.
(3)   Burette, 50 ml, and burette stand.
(4)   Filtration funnel.
(5)   Filter papers, Whatman Nos. 541 and 44.
(6)   Ashless tablets.
(7)   Glass bottles, about 500 ml, with stopper.
(8)   Pipette, 25 ml.
(9)   Drying oven, desiccator.

REAGENTS

(1)   Sulphuric acid, 0.01 N solution.
(2)   Silver nitrate, 0.02 N solution.
(3)   Potassium chromate saturated solution.
(4)   Indicator papers, narrow-range (pH 6.0–7.0).

PROCEDURAL STAGES

The procedure described below is divided into five main sections:

(A)   Prepare sample.
(B)   Prepare solutions.
(C)   Titrate.
(D)   Calculate.
(E)   Report result.

PROCEDURE

(A)   *Preparation of sample*

(1)   From the original sample prepare a representative sample of about 1 kg containing particles not larger than 20 mm by subdividing in the usual way.
(2)   After oven drying at 105–110 °C and cooling, crush the larger particles so as to pass a 600 μm sieve.
(3)   Dry the crushed material in the oven, cool in the desiccator and weigh out exactly 100 g of the dried soil. Place it in a 500 ml bottle.

(B)   *Preparation of solutions*

(4)   To the soil sample add 200 ml of distilled water. Insert the stopper and shake frequently over a period of 24 h so as to dissolve all the water-soluble chlorides. Do not apply heat.
(5)   Filter a portion of the liquid, if necessary, to remove any solids in suspension, and transfer 25 ml by means of a pipette into a 250 ml conical flask.
(6)   Acidify with 0.02 N sulphuric acid, just sufficiently to bring the pH to between 6 and 7, checking with a narrow-range pH paper.
(7)   Add two drops of saturated potassium chromate solution. Add the same to two similar conical flasks each containing 25 ml distilled water. These are for colour comparison and blank determination.

(C)   Titration

(8)   Titrate the blank with 0.02 N silver nitrate solution until a blood-red tinge is just obtained and remains permanent.

(9)   Titrate the test solution in the same way, and record the volume (*V* ml) of silver nitrate solution used.

(D)   Calculation

(10)   Calculate the percentage of soluble chloride (*C*%) as the equivalent sodium chloride content from the equation

$$C\% = 1.17 \times V$$

If the mass of dry soil is *m* g instead of 100 g, this equation becomes

$$C\% = 1.17 \times \frac{V}{m} \times 100$$

Notes

(11)   If the chloride content is very high, it is more convenient to use a 0.1 N solution of silver nitrate instead of 0.02 N in steps (8) and (9). The calculation given in step (10) then becomes

$$C\% = 5.85 \times V$$

if 100 g of dry soil is used or

$$C\% = 5.85 \times \frac{V}{m} \times 100$$

if a mass *m* g of dry soil is used.

(12)   The volume of water extract taken for the titration analysis is given in step (5) as 25 ml, but this is suitable only when the concentration of chlorides is fairly low. If the concentration is high, take a lesser quantity, say 5 ml (measured accurately with a pipette), and dilute with distilled water to make up to 25 ml for a working solution. The aim should be to use approximately 10 ml of silver nitrate solution during titration. An amount greatly in excess of this is wasteful, and using very much less (say less than 5 ml) reduces the accuracy of the measurement by burette.

If 5 ml of extract is used instead of 25 ml, the calculations for chloride content given in step (10) must be amended by a factor of 5, so that

$$C\% = 5.85 \, V$$

or

$$C\% = 5.85 \times \frac{V}{m} \times 100$$

Similarly, if any other volume (*x* ml) of extract is used, the calculation must be proportionately adjusted by multiplying by the factor 25/*x*.

(E)   *Report results*

(13)   The result is reported as a percentage to two significant figures as the soluble chloride salt content, expressed as equivalent sodium chloride content.

5.10   MISCELLANEOUS

5.10.1   Scope

The tests described below are general procedures which have not been covered in the foregoing sections. The first is a test for the determination of the total solids dissolved in groundwater, irrespective of the actual substances. The second is an outline of the use of indicator papers for assessing the approximate concentration of certain dissolved salts.

5.10.2   Total Dissolved Solids

This test is based on that given in Section 5 of BS 2690: Part 9: 1970, 'Methods of testing water used in industry'.

The result obtained by this method may not be exact, expecially if ammonium salts are present, but it is sufficiently accurate for practical purposes where an indication of the amount of dissolved salts is required.

APPARATUS

(1)   Buchner funnel and flask.
(2)   Vacuum supply.
(3)   Filter papers.
(4)   Evaporating dish.
(5)   Drying oven, 180 °C.
(6)   Volumetric flask.

PROCEDURE

(1)   Filter the sample of groundwater, using the Buchner funnel and flask so as to remove any suspended solids.
(2)   Collect a known volume (V ml) of filtered water in a volumetric flask. The volume of water tested should be such as to yield between 2.5 mg and 1000 mg of dissolved solids.
(3)   Heat the evaporating dish to 180 °C for 30 min, allow to cool in a desiccator and weigh to 0.5 mg (m_1).
(4)   Pour a portion of the filtered water into the dish and evaporate over a bunsen burner (or electric hot-plate) on a boiling water-bath. This must be done in a clean atmosphere to prevent contamination with airborne solids.
(5)   Add further portions of the water samples to the dish as evaporation proceeds. When the flask is empty, rinse it twice with 10 ml of distilled water and add these to the evaporating dish.
(6)   Allow to evaporate to dryness and wipe the outside of the dish.
(7)   Heat the dish and contents in an oven at 180 °C for 1 h.
(8)   Allow to cool in a desiccator, and weigh (m_2).
(9)   Repeat stages (7) and (8), but heating for 30 min, until the difference between successive weighings does not exceed 1 mg.

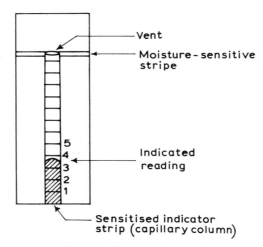

Fig. 5.16   *Typical indicator strip*

CALCULATIONS

If m_1 and m_2 are measured in grams, the mass of residue m is equal to $(m_2 - m_1) \times 1000$ mg. The total dissolved solids (TDS) dried at 180 °C is given by the equation

$$TDS = \frac{m}{V} \times 1000 \quad \text{p.p.m.}$$

REPORT RESULT

The result expressed in parts per million is reported to three significant figures as the total dissolved solids in the water sample when dried at 180 °C.

If it was not possible to filter the sample free of all traces of turbidity, this fact should be reported.

5.10.3   Use of Indicator Papers

Litmus papers, blue and red, for indicating acidity or alkalinity are the most familiar type of indicator papers used in chemical testing. Indicator papers for the measurement of pH are described in Section 5.5.1. Sensitised papers of a more complex type have become available in recent years for the rapid quantitative determination of many substances present in solution in water. The most important of these in soils work are sulphate content and chloride content. One brand of these indicating papers is known as 'Quantabs'. They are available in sealed bottles containing 50 test strips.

The test strip carries a scale marked from 0 to 10, at the top end of which is a thin coloured horizontal stripe (see Fig. 5.16). The test strip is immersed in the solution to be tested, and held there for a few seconds until the horizontal indicator stripe turns colour. This confirms that water has made its way to the top of the test strip by capillary action, so that the whole length of the strip has been wetted. The length of the strip, from the bottom upwards, which has turned colour after the lapse of 30 s is dependent on the concentration of the substance in the solution. The reading on the scale where the colour-change ends is referred to a table supplied with the indicator strips, from which the percentage (e.g. of sulphates) is read off.

Full instructions are provided by the manufacturer.

These indicators are not intended to be a substitute for chemical analysis, but they are useful for providing an approximate indication. They can be used in the first instance to assess whether or not a full analysis is necessary.

BIBLIOGRAPHY

Bowley, M. J. (1979). 'Analysis of sulphate-bearing soils'. Building Research Establishment Current Paper CP 2/79, March 1979. Building Research Establishment, Garston, Watford

BS 812: Part 4: 1976, 'Methods for sampling and testing mineral aggregates, sands and fillers'. British Standards Institution, London

BS 2690: Part 9: 1970, 'Methods of testing water used in industry'. British Standards Institution, London

Building Research Establishment (1975). Digest No. 174: 'Concrete in sulphate-bearing soils and groundwaters'. Building Research Establishment, Garston, Watford

Collins, S. H. (1960). 'Scheibler's apparatus for the determination of carbonic acid in carbonates; an improved construction and use for accurate analysis'. J. Soc. Chem. Ind., Vol. 25

Cumming, A. C. and Kay, S. A. (1948). Quantitative Chemical Analysis. Gurney and Jackson, London

Gutt, W. H. and Harrison, W. H. (1977). 'Chemical resistance of concrete'. BRE Current Paper CP 23/77, May 1977. Building Research Establishment, Garston, Watford. (Reprint from Concrete 1977, Vol. 11, No. 5)

Kuhn, S. (1930). 'Eine neue kolorimetrische Schnellmethode zur Bestimmung des pH von Boden'. Z. Pflernahr. Dung., Vol. 18A

Transport and Road Research Laboratory (1952). Soil Mechanics for Road Engineers, Chapter 5. HMSO, London

Vogel, A. I. (1961), A Textbook of Qualitative Inorganic Analysis, 3rd edition. Longmans, London

Walkley, A. and Black, I. A. (1934). 'An examination of the Degtjareff method of determining soil organic matter, and a proposed modification of the chromic acid titration method'. Soil Sci., Vol. 37(1)

Wilson, C. L. and Wilson, D. W. (1959). Comprehensive Analytical Chemistry. Elsevier, Amsterdam

Chapter 6

Compaction tests

6.1 INTRODUCTION

6.1.1 Scope

Many civil engineering projects require the use of soils as 'fill' material. Whenever soil is placed as an engineering fill, it is nearly always necessary to compact it to a dense state, so as to obtain satisfactory engineering properties which would not be achieved with loosely placed material. Compaction on site is usually effected by mechanical means such as rolling, ramming or vibrating. Control of the degree of compaction is necessary to achieve a satisfactory result at reasonable cost. Laboratory compaction tests provide the basis for control procedures used on site.

Compaction tests furnish the following basic data for soils:

(1)   The relationship between dry density and moisture content for a given degree of compactive effort.
(2)   The moisture content for the most efficient compaction — that is, at which the maximum dry density is achieved under that compactive effort.
(3)   The value of the maximum dry density so achieved.

Item (1) is expressed as a graphical relationship from which items (2) and (3) can be derived. The latter are the moisture and density criteria, against which the compacted fill can be judged if *in situ* measurements of moisture content and density are made.

There are several different standard laboratory compaction tests. The test selected for use as the basis for comparison will depend upon the nature of the works, the type of soil and the type of compaction equipment used on site. This chapter describes the tests accepted in Britain as standard practice, and two tests of American origin which have special applications.

Tests which are carried out on site to determine the density and other characteristics of the compacted fill are not described here.

6.1.2 Development of Test Procedures

A test to provide data on the compaction characteristics of soil was first introduced by R. R. Proctor in the USA in 1933, in order to determine a satisfactory state of compaction for soils being used in the construction of large dams, and to provide a means for controlling the degree of compaction during construction. The test made use of a hand rammer and a cylindrical mould with a volume of $1/30\,\text{ft}^3$, and became known as the standard 'Proctor' compaction test (Proctor, 1933; Taylor, 1948). The test now known as the British Standard 'ordinary' compaction test is very similar, and even today the mould used is often referred to as the 'Proctor' mould.

At that time it was believed that the Proctor test represented in the laboratory the state of compaction which could be reasonably achieved in the field. But with the subsequent introduction of heavier earth-moving and compaction machinery, especially for the construction of large dams, higher densities became obtainable in practice. A laboratory test using increased energy of compaction was then necessary to reproduce these higher compacted densities, so a test was introduced which used a heavier rammer with the same mould. This intensified procedure became known as the 'modified AAHSHO' test. It is very similar to the British Standard 'heavy' compaction test used today.

Granular soils, especially gravels, are most effectively compacted by vibration. A laboratory test using a vibrating hammer was introduced into British Standards in 1967 to establish the compaction characteristics for these conditions. Because particles up to coarse gravel size are necessary so as to represent these materials as closely as possible in the test, a larger mould (the CBR mould) is used. This procedure is known in this country as the British Standard vibrating hammer compaction test.

Dry densities measured on compacted soils *in situ* are still often expressed as a percentage of the 'Proctor maximum dry density' (or simply 'Proctor density'). This percentage is called the 'relative compaction' of the soil. However, when the dry density required on site is greater than the BS 'ordinary' ('Proctor') maximum dry density, the field density is more usually related to the BS 'heavy' compaction curve, rather than quoting values of relative compaction in excess of 100%.

6.2 DEFINITIONS

COMPACTION   The process of packing soil particles more closely together, usually by mechanical means, thus increasing the dry density of the soil.

OPTIMUM MOISTURE CONTENT (OMC)   The moisture content of a soil at which a specified amount of compaction will produce the maximum dry density.

MAXIMUM DRY DENSITY   The dry density obtained using a specified amount of compaction at the optimum moisture content.

RELATIVE COMPACTION   The percentage ratio between the dry density of the soil and its maximum dry density, determined from a specified laboratory compaction test.

DRY DENSITY–MOISTURE CONTENT RELATIONSHIP   The relationship between dry density and moisture content of a soil under a given compactive effort.

PERCENTAGE AIR VOIDS (V_a)   The volume of air voids in a soil expressed as a percentage of the total volume of the soil.

AIR VOIDS LINE   A line showing the dry density–moisture content relationship for a soil containing a constant percentage of air voids.

SATURATION LINE (ZERO AIR VOIDS LINE)   The line showing the dry density–moisture content relationship for a soil containing no air voids.

6.3 THEORY

6.3.1 Process of Compaction

Compaction of soil is the process by which the solid soil particles are packed more closely together by mechanical means, thus increasing the dry density (Markwick, 1944). It is achieved through the reduction of the air voids in the soil, with little or no reduction in the water content. This process must not be confused with consolidation, in which water is squeezed out under the action of a continuous static load. The air voids cannot be eliminated altogether by compaction, but with proper control they can be reduced to a minimum. The effect of the amount of water present in a fine-grained soil on its compaction characteristics, when subjected to a given compactive effort, is discussed below.

At low moisture content the soil grains are surrounded by a thin film of water, which tends to keep the grains apart even when compacted. (Fig. 6.1a). The finer the soil grains, the more

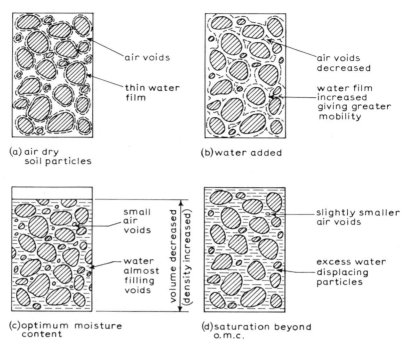

Fig. 6.1   Representation of compaction of soil grains

significant is this effect. If the moisture content is increased, the additional water enables the grains to be more easily compacted together (Fig. 6.1b). Some of the air is displaced and the dry density is increased. The addition of more water, up to a certain point, enables more air to be expelled during compaction. At that point the soil grains become as closely packed together as they can be (i.e. the dry density is at the maximum) under the application of this compactive effort (Fig. 6.1c). When the amount of water exceeds that required to achieve this condition, the excess water begins to push the particles apart (Fig. 6.1d), so that the dry density is reduced. At higher moisture contents little or no more air is displaced by compaction, and the resulting dry density continues to decrease.

If at each stage the compacted dry density is calculated, and plotted against moisture content, a graph similar to curve A in Fig. 6.2 is obtained. This graph is the 'moisture–density relationship' curve. The moisture content at which the greatest value of dry density is reached for the given amount of compaction is the optimum moisture content (OMC), and the corresponding dry density is the maximum dry density. At this moisture content the soil can be compacted most efficiently under the given compactive effort. The relationship between bulk (wet) density and moisture content is shown by the dotted curve (W) in Fig. 6.2. This type of plot is not generally used, except perhaps as a guide during a compaction test before the moisture contents are measured.

A typical compaction curve obtained from the British Standard 'ordinary' compaction test (Section 6.5.2) is shown in Fig. 6.3 as curve A. If a heavier degree of compaction, corresponding to the Standard 'heavy' compaction test (Section 6.5.3), is applied at each moisture content, higher values of density, and therefore of dry density, will be obtained. The resulting moisture–density relationship will be a graph such as curve B in Fig. 6.3. The maximum dry density is greater, but the optimum moisture content at which this occurs is lower, than in the 'ordinary' test.

Every different degree of compaction on a particular soil results in a different compaction curve, each with unique values of optimum moisture content and maximum dry density. For

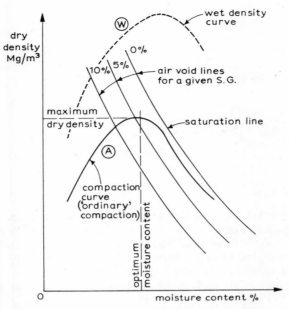

Fig. 6.2   *Dry density–moisture content relationship for soils*

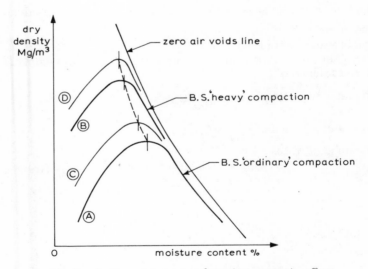

Fig. 6.3   *Dry density–moisture curves for various compactive efforts*

instance a compaction test similar to the British Standard 'ordinary' test but using, say, 50 blows per layer instead of 27 would give a graph similar to that shown by curve C in Fig. 6.3. A test similar to the Standard 'Heavy' compaction test but using a greater number of blows would give a graph similar to curve D. It can be seen that increasing the compactive effort increases the maximum dry density, but decreases the optimum moisture content.

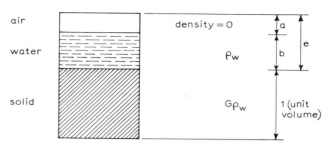

Fig. 6.4   Representation of soil with air voids

6.3.2   Air Voids Lines

A compaction curve is not complete without the addition of air voids lines. An air voids line is a (curved) line showing the dry density–moisture content relation for soil containing a constant percentage of air voids. A set of air voids lines can be drawn from calculated data if the SG of the soil grains is known, and three are indicated in Fig. 6.2. The derivation of the equation relating dry density to moisture content for a given percentage of air voids, V_a, is given below. Note that V_a is the volume of air voids in the soil expressed as a percentage of the *total* volume of soil, as in BS 1377:1975, Section 1.2.1 (26), and not as a percentage of the voids. V_a is *not* the same as $(100 - S)$, where S is the saturation expressed as a percentage of the total voids.

If all the air voids are removed, so that the total voids between solid particles are filled with water, the soil reaches the fully saturated condition. The equation relating the saturated dry density to moisture content, from which the zero air voids line can be drawn, can be derived by putting V_a equal to zero.

The notation is the same as that used in Section 3.3.2, with the addition of the following:

Volume of air voids $= a$
Volume of water in voids $= b$ for a unit volume of solids (see Fig. 6.4)
Volume of air voids, expressed as a percentage of the *total* volume of the soil $= V_a$
Moisture content $= w\%$

Therefore, mass of water present $= \dfrac{w}{100}(G_s\rho_w)$

Therefore, volume of water $= \dfrac{w}{100} \times G_s = b$

Total volume of soil $= V_s$
$\qquad\qquad = (\text{volume of solid}) + (\text{volume of water}) + (\text{volume of air})$
$\qquad\qquad = 1 + b + a$

But

$$V_a = \frac{a}{V_s} \times 100 \tag{6.1}$$

$$= \frac{a}{1 + b + a} \times 100 \tag{6.2}$$

Dry density $= \dfrac{\text{dry mass}}{\text{total volume}}$

that is,

$$\rho_D = \frac{G_s \times \rho_w \times 1}{V_s} \tag{6.3}$$

From Eq. (6.1)

$$V_s = a \times \frac{100}{V_a}$$

Substitute in Eq. (6.3)

$$\rho_D = \frac{G_s \rho_w}{a \times \dfrac{100}{V_a}} \tag{6.4}$$

From Eq. (6.2)

$$\frac{V_a}{100}(1+b+a) = a$$

Therefore,

$$a = \frac{\dfrac{V_a}{100}(1+b)}{1 - \dfrac{V_a}{100}} \tag{6.5}$$

Substitute for a and b in Eq. (6.4):

$$\rho_D = \frac{G_s \rho_w}{\left(\dfrac{1+\dfrac{wG_s}{100}}{1-\dfrac{V_a}{100}}\right)} = \frac{G_s\left(1-\dfrac{V_a}{100}\right)}{1+\dfrac{wG_s}{100}}\,\rho_w$$

that is,

$$\rho_D = \frac{1-\dfrac{V_a}{100}}{\dfrac{1}{G_s}+\dfrac{w}{100}}\,\rho_w \tag{6.6}$$

For the fully saturated condition (no air voids), $V_a = 0$. Therefore,

$$\rho_{D(\text{sat})} = \frac{1}{\dfrac{1}{G_s}+\dfrac{w}{100}}\,\rho_w \tag{6.7}$$

This equation defines the zero air voids line, or the saturation line. It is impossible for a point on a compaction curve (in terms of dry density) to lie to the right of this line, whatever degree of compactive effort is applied.

Curves for 0, 5 and 10% air voids (i.e. $V_a = 0, 5, 10\%$) are shown in Fig. 6.2. These curves are

Table 6.1. DATA FOR CONSTRUCTING AIR VOIDS LINES: dry densities (Mg/m^3) corresponding to various moisture contents for soils of different specific gravities

Moisture content, w (%)	Air voids, V$_a$ (%)	Specific gravity of particles, G$_s$				
		2.60	2.65	2.70	2.75	2.80
0	0	2.60	2.65	2.70	2.75	2.80
	5	2.47	2.52	2.57	2.61	2.66
	10	2.34	2.39	2.43	2.48	2.52
5	0	2.30	2.34	2.38	2.42	2.46
	5	2.19	2.22	2.26	2.30	2.33
	10	2.07	2.11	2.14	2.18	2.21
10	0	2.06	2.09	2.13	2.16	2.19
	5	1.96	1.99	2.02	2.05	2.08
	10	1.86	1.89	1.91	1.94	1.97
15	0	1.87	1.90	1.92	1.95	1.97
	5	1.78	1.80	1.83	1.85	1.87
	10	1.68	1.71	1.73	1.75	1.77
20	0	1.71	1.73	1.75	1.77	1.79
	5	1.63	1.65	1.67	1.69	1.71
	10	1.54	1.56	1.58	1.60	1.62
25	0	1.58	1.59	1.61	1.63	1.65
	5	1.50	1.51	1.53	1.55	1.56
	10	1.42	1.43	1.45	1.47	1.48
30	0	1.46	1.48	1.49	1.51	1.52
	5	1.39	1.40	1.42	1.43	1.45
	10	1.31	1.33	1.34	1.36	1.37
35	0	1.36	1.37	1.39	1.40	1.41
	5	1.29	1.31	1.32	1.33	1.34
	10	1.23	1.24	1.25	1.26	1.27

defined only by the specific gravity of the soil grains. Sets of standard curves can be drawn up for various SGs, so that the set applicable to a particular soil can be selected, either by use of the data given in Table 6.1 or direct from Eqs. (6.6) and (6.7). The air voids lines do not apply to the wet density curve (W) in Fig. 6.2.

6.3.3 Compactive Efforts

The methods used for various standard types of compaction test are summarised in Table 6.2. Data relating to the obsolescent imperial size 'Proctor' mould and rammers are shown in italics.

The mechanical energy applied in each type of test, in terms of the work done in operating the rammer, is derived below.

BS 'ORDINARY' TEST

$$(2.5 \, \text{kg}) \times \frac{(300 \, \text{mm})}{1000} \times 27 \times 3 = 60.75 \, \text{kg m}$$

$$= 60.75 \times 9.81 \, \text{Nm} = 596 \, \text{J}$$

(kg m $\times$ 9.81 = newton metres = joules).

Table 6.2. COMPACTION PROCEDURES

Type of test (and BS 1377:1975 Test No.)	Container	Rammer		No. of layers	Blows per layer	Refer to Section
		mass (kg)	drop (mm)			
'Ordinary' compaction (Test 12)	**Old 'Proctor'**	**2.5**	**305**	**3**	**25**	–
	BS mould	2.5	300	3	27	6.5.2
	CBR mould	2.5	300	3	62	6.5.6
'Heavy' compaction (Test 13)	**Old 'Proctor'**	**4.5**	**457**	**5**	**25**	–
	BS mould	4.5	450	5	27	6.5.3
	CBR mould	4.5	450	5	62	6.5.6
Vibrating hammer (Test 14)	CBR mould	32 to 41*	(vibration)	3	(1 min)	6.5.4
Dietert	2 inch diameter	8.14	50.8	2 ends	10 each end	6.62

* Downward force to be applied.

Volume of soil used $= 1000 \, \text{cm}^3 = 0.001 \, \text{m}^3$. Therefore,

$$\text{work done per unit volume of soil} = \frac{596}{1000} \, \text{J/cm}^3 = 596 \, \text{kJ/m}^3$$

BS 'HEAVY' TEST

$$4.5 \times \frac{450}{1000} \times 27 \times 5 \times 9.81 = 2682 \, \text{J, or } 2682 \, \text{kJ/m}^3$$

'ORDINARY' COMPACTION IN CBR (CALIFORNIA BEARING RATIO) MOULD

$$\text{Volume} = 2305 \, \text{cm}^3$$

$$2.5 \times 0.3 \times 62 \times 3 \times 9.81 = 1368 \, \text{J}$$

$$\frac{1368}{2305} \times 1000 = 594 \, \text{kJ/m}^3$$

'HEAVY' COMPACTION IN CBR MOULD

$$4.5 \times 0.45 \times 62 \times 5 \times 9.81 = 6158 \, \text{J}$$

$$\frac{6158}{2300} \times 1000 = 2672 \, \text{kJ/m}^3$$

The calculations verify that for the 'ordinary' compaction tests, whether carried out with the Proctor mould or the CBR mould, the compactive energy per unit volume of soil is about the same.

For the 'heavy' test the energy is similar with both procedures. The energy applied per unit volume in the 'heavy' test is 4.5 times as much as that used in the 'ordinary' test ($2682/596 = 4.5$ exactly).

VIBRATING HAMMER COMPACTION (see Section 6.5.4)

Assuming a 600 W motor, and that 50% of the electrical input is converted to mechanical energy, half of which is absorbed by the soil sample (the other half being taken mainly by the operator),

$$\text{energy applied to sample} = 600 \times \tfrac{1}{2} \times \tfrac{1}{2} \times 60 \times 3 \text{ J}$$

$$= 27\,000 \text{ J}$$

$$\frac{27\,000}{2300} \times 1000 = 11\,739 \text{ kJ/m}^3$$

The ratio of the calculated energy applied by the vibrating hammer to that applied by the 'heavy' compaction test is $11\,739/2672 = 4.39$, which is about the same as the ratio (4.5) of the 'heavy' to the 'ordinary' compactive effort.

DIETERT COMPACTION (See Section 6.6.2)

$$8.14 \times \frac{50.8}{1000} \times 10 \times 2 \times 9.81 = 81.13 \text{ J}$$

Volume of 2 inch depth of 2 inch diameter mould

$$= \frac{\pi}{4} \times (50.8)^2 \times \frac{50.8}{1000} = 103 \text{ cm}^3$$

$$\frac{81.13 \times 1000}{103} = 788 \text{ kJ/m}^3$$

The Dietert compaction procedure appears to provide about 30% more effort than the British Standard 'ordinary' procedure. However in using the much smaller Dietert mould the relative effects of side friction are more significant than with the standard 1000 cm^3 mould, and undoubtedly some of the compactive effort is lost in overcoming this fraction. This effect is less for cohesive soils than for silty or sandy soils. An empirical relationship between liquid limit and the number of blows required to make the test equivalent to the BS 'ordinary' compaction is shown in Fig. 6.26.

6.3.4 Effect of Stone Content

In the standard laboratory compaction tests only the fraction of soil passing a 20 mm sieve is used. Particles larger than 20 mm which are removed before test may consist of gravel, fragments of rock, shale, brick or other hard material, and are collectively referred to below as 'stones'. The soil actually tested is called the 'matrix' material.

The density achieved on site for the total material cannot be compared directly with the results of laboratory compaction tests on the matrix material only. If the matrix material is compacted to reach a particular density, the presence of stones will give the total material a higher density, because the stones have a greater density than the matrix material they displace. The resulting *in situ* density of the whole material can be calculated from the equation derived below, provided that (a) the proportion of stones in the total material is known; (b) this proportion is not large (i.e. not more than about 25% of the total by dry mass), so that the stones are distributed within the matrix in such a way that they are not in contact with each other (Maddison, 1944; McLeod, 1970).

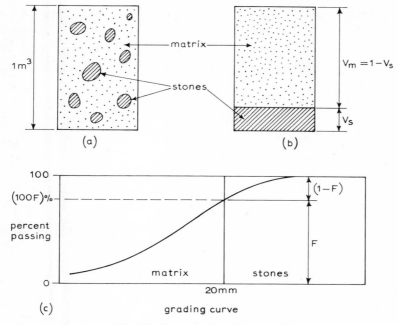

Fig. 6.5 *Representation of stoney soil*

Table 6.3. SYMBOLS FOR STONE CONTENT EQUATIONS

Soil properties	Symbols used		
	Matrix material	*Stones*	*Total material*
Dry density	ρ_{mD}		$\rho_D = ?$
Specific gravity	G_s	G_t	
Volume	V_m	V_s	1
Mass in a unit volume of soil	m_m	m_s	$(m_m + m_s)$

In practice, the presence of stones requires additional compactive effort to achieve the same degree of compaction of the matrix as when the matrix is compacted alone. However, this effect is not great for small percentages of stones, and does not affect these calculations. If the percentage of stones is quite large, there may not be sufficient matrix materials completely to fill the voids between the stones, and this could be an unsatisfactory fill material for many purposes.

A unit volume of 'stony' soil is represented diagrammatically in Fig. 6.5(a), and the stones are imagined to be fused together in one piece occupying a volume V_s, as in Fig. 6.5(b). The idealised grading curve of the whole material is shown in Fig. 6.5(c). The proportion of material finer than 20 mm, expressed as a decimal fraction, is denoted by F.

The symbols used in the expressions below are summarised in Table 6.3. Four relationships can be written in the form of equations, as follows.

The total mass in a unit volume is equal to ρ_D, therefore

$$\rho_D = m_m + m_s \tag{6.8}$$

The mass of matrix materials is equal to its density multiplied by its volume — that is,

$$m_m = (1 - V_s)\rho_{mD} \tag{6.9}$$

The mass of stones is equal to the volume of solid material multiplied by the density of that material — that is,

$$m_s = V_s G_t \rho_w \tag{6.10}$$

From the grading curve, the fraction of the matrix material to the whole is equal to the ratio of its dry mass to the total dry mass — that is,

$$F = \frac{m_m}{m_m + m_s} \tag{6.11}$$

From these equations the relationship between the dry density of the material containing stones, ρ_D, and the dry density of the matrix material measured in the laboratory, ρ_{mD}, can be derived, and is as follows:

$$\rho_D = \frac{G_t \rho_w}{(1 - F) + F\left(\dfrac{G_t \rho_w}{\rho_{mD}}\right)} \tag{6.12}$$

Using customary SI units, and putting $\rho_w = 1 \text{ Mg/m}^3$, this equation becomes

$$\rho_D = \left[\frac{G_t}{(1 - F)\rho_{mD} + FG_t}\right]\rho_{mD} \tag{6.13}$$

This is the theoretical dry density to be expected *in situ*, derived from the dry density, ρ_{mD}, of the matrix material measured in the laboratory.

The overall moisture content of the total material will differ from that of the matrix, owing to the presence of the stones. The stones themselves may absorb a certain amount of moisture, which will be removed by the normal oven drying procedure. Let w_m = moisture content of matrix, and w_s = moisture content (absorbed moisture) of stones. This absorbed moisture does not alter the volume of the stones. Moisture contents are expressed as decimal fractions. Other notation is as before.

Mass of water contained in matrix

$$= w_m m_m = w_m F(m_m + m_s)$$

$$= w_m F\rho_D$$

Mass of water contained in stones

$$= w_s m_s = w_s(1 - F)\rho_D$$

Therefore, total mass of water contained in unit volume of combined material

$$= W = w_m F + w_s(1 - F)\rho_D$$

Moisture content of total material

$$= w = \frac{W}{\text{dry mass}} = \frac{W}{\rho_D}$$

Therefore,

$$w = Fw_m + (1 - F)w_s \tag{6.14}$$

If the stones contains no absorbed water (e.g. if they consist of pieces of quartz gravel), the value of w_s is zero and w is simply equal to $F \times w_m$.

6.4 APPLICATIONS

6.4.1 Objectives of Proper Compaction

Soils may be used as fill for many purposes, the most usual being:

(1)   To refill an excavation, or a void adjacent to a structure (such as behind a retaining wall).
(2)   To provide made-up ground to support a structure.
(3)   As a sub-base for a road, railway or airfield runway.
(4)   As a structure in itself, such as an embankment or earth dam, including reinforced earth.

Compaction, by increasing the density, improves the engineering properties of soils. The most significant improvements, and the resultant effects on the mass of fill as a whole, are summarised in Table 6.4.

Table 6.4. EFFECTS OF PROPER COMPACTION OF SOILS

Improvement	Effect on mass of fill
Higher shear strength	Greater stability
Lower compressibility	Less settlement under static load
Higher CBR value	Less deformation under repeated loads
Lower permeability	Less tendency to absorb water
Lower frost susceptibility	Less likelihood of frost heave

6.4.2 Construction Control

The relationship between dry density and moisture content for soil subjected to a given compactive effort, established by laboratory compaction tests, provides reference data for the specification and control of soil placed as fill. On many projects the laboratory compaction tests are supplemented by field compaction trials by using the actual placing and compacting equipment which is to be employed for construction (Williams, 1949).

Sometimes it is necessary to adjust the natural moisture content of a soil to a value at which it can be most effectively compacted, or at which it has the highest strength. The required moisture content, and the dry density to be achieved, can be assessed on the basis of the dry density — moisture content relationship derived from laboratory compaction tests on samples taken from the borrow area.

While compaction in situ needs to be of a sufficient degree to obtain the required density, it is equally important not to overcompact fine-grained soils. Overcompaction not only is wasteful of effort, but should be avoided because overcompacted soil, if not confined by overburden, can readily absorb water, resulting in swelling, lower shear strength and greater compressibility. Toes and sides of embankments are particularly sensitive to this effect.

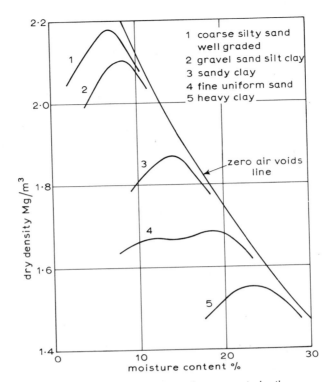

Fig. 6.6   Compaction curves for some typical soils

6.4.3   Design Parameters

When the compaction characteristics of a soil are known, it is possible to prepare samples in the laboratory at the same dry density and moisture content as that likely to be attained after compaction in the field. These samples can be subjected to laboratory tests for the determination of their shear strength, compressibility and other engineering properties. Design parameters derived from these tests enable the stability, deformation and other characteristics of the fill to be assessed. They can also provide the basis for the initial design of an embankment or earth dam.

More elaborate tests can be carried out on compacted samples to measure the changes of pore pressure due to changing conditions of applied stress. During construction, pore pressures can be monitored so as to ensure that they do not at any time exceed certain limiting values established by the tests.

A specification for compacted fill may require a certain 'relative compaction' (measured in terms of dry density) to be achieved, within specified limits of moisture content. More usually a specification defines the maximum air voids permitted in the compacted soil within the required dry density range. For this reason it is necessary to determine the SG of soil particles so that air voids lines can be added to the compaction test graphs.

6.4.4   Types of Compaction Curve

The form of compaction curve for five typical materials is shown in Fig. 6.6. For ease of comparison they have been related to a common zero air voids line by adjusting the curves to the same specific gravity. These curves relate to British Standard 'ordinary' compactive effort.

In general, clay soils, and well-graded sandy or silty soils, show a clearly defined peak to the compaction curve. Uniformly graded soils, consisting of a narrow range of particle sizes, give a flatter compaction curve from which the optimum condition is not easy to define. A 'double peak' is often obtained from uniformly graded fine sands. For these materials the moisture content for optimum compaction is less critical than for those soils giving a steeper compaction curve.

6.5 STANDARD COMPACTION TEST PROCEDURES

6.5.1 Types of Test

The tests described in this section are included in BS 1377:1975 as standard tests for the determination of the moisture–density relationship of soils.

These tests (with BS Test No.) are as follows:

(1) BS 'ordinary' compaction test ('Proctor' test) (Test 12) (Section 6.5.2).
(2) BS 'heavy' compaction test (Test 13) (Section 6.5.3).
(3) Compaction using vibrating hammer (Test 14) (Section 6.5.4).
(4) Compaction of soils containing large particles in CBR mould (Tests 12 and 13) (Section 6.5.6).

The use of a mechanical compaction apparatus as an alternative to hand compaction (Section 6.5.7) is included, and so is the British Standard procedure for the verification of the suitability of a vibrating hammer for test (3) (Section 6.5.5). Reference is also made to compaction by the application of a static load (Section 6.5.8), but this is not a test in itself.

It is important to refer to the test designation in full when reporting results or when quoting the tests; for instance, 'BS 1377:1975, Test 12'. This is because the tests were designated by different Test Nos. in the 1967 and 1961 editions of BS 1377.

6.5.2 'Ordinary' Compaction Test (2.5 kg rammer method) (BS 1377:1975, Test 12)

This test is still often referred to as the 'Proctor' test, and sometimes as the 'moisture–density relationship' test. It is suitable for soils containing particles no larger than 20 mm. The detailed procedure depends on whether or not the granular material is susceptible to crushing during compaction.

If the soil contains particles larger than 20 mm, refer to Section 6.5.6.

APPARATUS

(1) Cylindrical metal mould, internal dimensions 105 mm diameter and 115.5 mm high. This gives a volume of 1000 cm³. The mould is fitted with a detachable baseplate and removable extension collar (see Fig. 6.7).
(2) Metal rammer with 50 mm diameter face, weighing 2.5 kg, sliding freely in a tube which controls the height of drop to 300 mm (see Fig. 6.8).
(3) Measuring cylinder, 200 ml or 500 ml.
(4) 20 mm British Standard sieve and receiver.
(5) Large metal tray, say 600 × 500 × 80 mm deep.
(6) Balance, 10 kg capacity reading to 1 g.
(7) Jacking apparatus for extracting compacted material from the mould.
(8) Small tools: palette knife; steel straight-edge, 300 mm long; steel rule; scoop or garden trowel.
(9) Drying oven, 105–110 °C, and other equipment for moisture content determination.

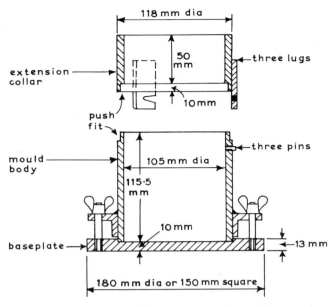

Fig. 6.7   British Standard compaction mould

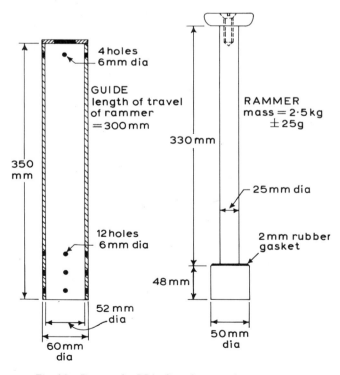

Fig. 6.8   Rammer for BS 'ordinary' compaction test

Fig. 6.9   Equipment for compaction tests

Compaction test equipment is shown in Fig. 6.9. If a mechanical compaction apparatus is available, refer to Section 6.5.7. The procedure described below is the same in principle whether compaction is effected by the hand rammer or by the machine.

PROCEDURAL STAGES

(1)   Prepare apparatus.
(2)   Obtain soil sample.
(3)   (A) Prepare test sample *or* (B) prepare five batches.
(4)   Compact soil into mould.
(5)   Trim off.
(6)   Weigh.
(7)   Remove soil from mould.
(8)   Measure moisture content.
(9)   Break up and remix.
(10) Repeat stages (4)–(8) *either* (A) after adding more water to sample *or* (B) using next batch (total of at least five compactions).
(11) Calculate.
(12) Plot graph.
(13) Read off optimum values.
(14) Report results.

TEST PROCEDURE

(1) *Prepare apparatus.*   See that the mould, extension collar and baseplate are clean and dry. Weigh the mould body to the nearest 1 g (m_1). Measure its internal diameter (D mm) and length (L mm) in several places to 0.1 mm using vernier calipers, and calculate the mean dimensions. Calculate the internal volume of the mould (V cm^3) from the equation

$$V = \frac{\pi \times D^2 \times L}{4000}$$

The present standard mould is designed to give a volume $V = 1000\,cm^3$, but this may change slightly with wear. The old 'Proctor' mould was 4 inches in diameter and gave a volume of $1/30\,ft^3$ or $944\,cm^3$. It is essential to know which type of mould is being used, so that the correct compactive effort can be applied. Extension collars and baseplates are not interchangeable between the present standard equipment and the old. The standard rammer differs little from the older type, having a drop of 300 mm compared with 12 inches (304.8 mm).

Check that the lugs or clamps hold the extension collar and baseplate securely to the mould, and assemble them together. A wipe with a slightly oily cloth on the internal surfaces will assist removal of soil afterwards. A disc of thin filter paper may be placed on the baseplate for the same purpose.

Check the rammer to ensure that it falls freely through the correct height of drop, and that the lifting knob is secure.

(2) *Obtain soil sample.* The original bulk sample is air dried and weighed (W_1). If it contains particles larger than 20 mm, these are removed by passing the sample through a 20 mm sieve. The material retained on the 20 mm sieve is weighed (W_2).

The mass of material required for test, and the preparation of the sample, depends on whether the soil particles are susceptible to crushing. If they are not (condition A), the mass W_2 required is about 5 kg. If they are (condition B), the mass W_2 should be at least 15 kg. Procedures (A) and (B) are described separately below. The compaction procedure is the same for both.

(A)   Soil particles not susceptible to crushing

(3) *Prepare test sample.* Riffle the sieved material so as to obtain a representative sample of about 5 kg. Add a suitable amount of water, and mix thoroughly. The amount of water depends on the soil type, as suggested below:

Sandy and gravelly soils: 4–6% (200—300 ml of water to 5 kg of soil).
Cohesive walls: About 8–10% below plastic limit — i.e. (PL − 10) to (PL − 8). For example, a soil of PL = 20 should be made up to a moisture content of 10–12% (add 500–600 ml of water to 5 kg of soil).

These quantities may be taken as a general guide, but a suitable amount is best judged by experience.

Thorough mixing in of the water is essential. With clay soils the mixed sample should be stored overnight in a sealed container before proceeding with the test.

(4) *Compact into mould.* Place the mould assembly on a solid base, such as a concrete floor or plinth. A resilient base may result in inadequate compaction.

Add loose soil to the mould so that it is about half-filled. Compact the soil by applying 27 blows of the rammer dropping from the controlled height of 300 mm (Fig. 6.10). If the old type of mould ($944\,cm^3$) is being used, the number of blows should be 25 instead of 27.

Take care to see that the rammer is properly in place before releasing. The hand which holds the tube must be kept well clear of the handle of the falling rammer. Do not attempt to grab the lifting knob before the rammer has come to rest; a finger or thumb trapped between knob and tube can sustain a nasty injury.

The first few blows of the rammer, which are applied to soil in a very loose state, should be applied in a systematic manner to ensure the most efficient compaction and maximum reproducibility of results. The sequence shown in Fig. 6.11(a) should be followed for the first four blows. By this means the effort dissipated in displacing loose material is kept to a minimum. After that the rammer should be moved progressively around the edge of the mould between successive blows, as indicated in Fig. 6.11(b), so that the blows are uniformly distributed over the whole are. Soil must not be allowed to collect inside the tube of the rammer, because this will impede the free fall of the rammer. Make sure that the end of the tube is resting on the soil surface, and does not catch on the edge of the mould, before releasing the

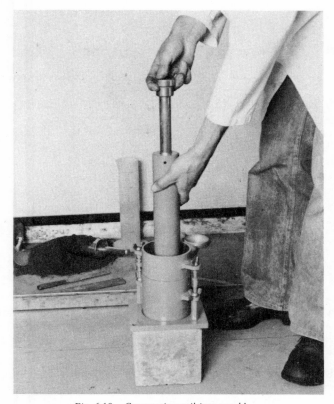

Fig. 6.10   *Compacting soil into mould*

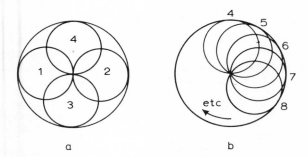

Fig. 6.11   *Sequence of blows using hand rammer*

rammer. The guide tube must be held vertically. Place the tube gently on the soil surface; the rammer does the compaction, not the tube.

If the correct amount of soil has been used, the compacted surface should be at about one-third of the height of the mould body — that is, about 75 mm below the top of the mould body, or 125 mm below the top of the extension collar. If the level differs significantly (by more than, say, 5 mm) from this, remove the soil, break it up, mix it with the remainder of the prepared material and start this stage again.

Lightly scarify the surface of the compacted soil with the tip of a spatula or point of a knife. Place a second, approximately equal, layer of soil in the mould, and compact with 27 blows as before (or 25 blows if using the old type of mould). Repeat with a third layer, which should then

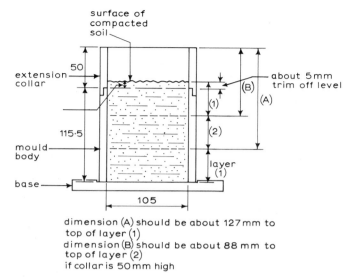

dimension (A) should be about 127mm to
top of layer (1)
dimension (B) should be about 88 mm to
top of layer (2)
if collar is 50mm high

Fig. 6.12   Soil in mould after compaction

bring the compacted surface in the extension collar to about 6 mm above the level of the mould body (see Fig. 6.12). If the soil level is higher than this, the result will be inaccurate, so the soil should be removed, broken up and remixed, and the test repeated with slightly less soil in each layer.

(5) *Trim off.*   Remove the extension collar carefully. Cut away the excess soil and level off to the top of the mould, checking with the straight-edge. Any small cavities resulting from removal of stones at the surface should be filled with fine material.

(6) *Weigh.*   Remove the baseplate carefully, and trim the soil at the lower end of the mould if necessary. Weigh soil and mould to the nearest 1 g (m_2).
   The British Standard procedure does not call for the removal of the baseplate before weighing. If the soil is granular and will not hold together well, the baseplate is best left on. In this case the mass m_1 refers to the mould with baseplate. If the soil is cohesive enough to hold together, it is preferable not to include the baseplate in the weighings, because the mould with baseplate weighs substantially more than the soil it contains.

(7) *Remove soil.*   Fit the mould on to the extruder and jack out the soil (Fig. 6.13). Alternatively, remove the soil by hand, but this can be difficult with gravelly soils containing a clay binder. Break up the sample on the tray.

(8) *Measure moisture content.*   Take up to three representative samples in moisture content containers for measurement of moisture content, using the standard procedure described in Section 2.5.2. This must be done immediately, before the soil begins to dry out. The average of the three measurements is denoted by $w\%$.
   Alternatively, moisture content samples may be taken, one from each layer, as the soil is placed in the mould for compaction.

(9) *Break up and remix.*   Break up the material on the tray, by rubbing through a 20 mm sieve if necessary, and mix with the remainder of the prepared sample. Add an increment of water, approximately as follows:

Sandy and gravelly soils: 1–2% (50–100 ml of water to 5 kg of soil).
Cohesive soils: 2–4% (100–200 ml of water to 5 kg of soil).

Mix in the water thoroughly.

Fig. 6.13   Jacking soil out of mould

(10)   *Repeat with added water.*   Repeat stages (4)–(9) for each increment of water added, so that at least five compactions are made. The range of moisture contents should be such that the optimum moisture content (at which the dry density is maximum) is within that range. If necessary to define the optimum value clearly, carry out one or more additional tests at suitable moisture contents. Keep a running plot of density against moisture content so as to see when the optimum condition has been passed.

Above a certain moisture content the material may be extremely difficult to compact. For instance, a granular soil may by then contain excessive free water, or a clay soil may be very soft and sticky. In either event the optimum condition has been passed and there is no point in proceeding further.

(B)   Soil particles susceptible to crushing

If the sample contains granular material such as soft limestone, sandstone or other minerals likely to be broken down by the action of the rammer (stage 4), separate batches should be prepared for compaction at each moisture content. The above stages are amended as follows.

(3)   *Prepare batches for test.*   From the sieved sample take five or more representative samples each of about 2.5 kg. Mix each batch with a different amount of water to give a suitable range of moisture contents. The moisture content of the first batch could be as suggested in stage (3) above. Additional increments of water should be added to the other batches as suggested in stage (9) above. The range of moisture contents should be as stated in stage (10) above.

Thorough mixing in of the water is essential. With clay soils each batch should be sealed and stored overnight before proceeding with the test.

Stages (4)–(8).   Treat the first batch as described in stages (4)–(8) above.

(9)   The sample remains may be discarded when it is established that no further tests are needed.

Compaction Test Work sheet

B.S. 1377:1975	Test 12/~~Test 13/Test 14~~	Location: Easthampstead		Loc No. 1998
No. of layers 3	Rammer 2·5 kg	Soil description Brown sandy clay with a little fine gravel		Sample No. 27/4
Blows per layer 27	Drop 300 mm	Sample type Bulk bag	Operator C.B.A.	Date started 10.3.78
Compacted by hand/machine		Sample preparation Air dried and riffled		
Proctor/CBR/cylinder no. 7		No. of separate batches 5	Special techniques Separate batches used	

DENSITY   Volume of cylinder (V) 1002 cm³

Measurement no.			(1)	(2)	(3)	(4)	(5)
Cylinder & soil	A	g	3786	3907	3999	3962	3908
Cylinder	B	g	1917	1917	1917	1917	1917
Wet soil	A−B	g	1869	1990	2082	2045	1991
Wet density	ρ Mg/m³		1·865	1·986	2·078	2·041	1·987

MOISTURE CONTENT

Container no.		64	44	18		
Wet soil & container	g	104·12	97·48	89·67		
Dry soil & container	g	96·02	90·42	82·76		
Container	g	9·36	16·58	9·51		
Dry soil	g	88·66	73·84	73·25		
Moisture loss	g	8·10	7·06	6·91		
Moisture content $W_{1,2,3}$	%	9·35	9·56	9·44		
AVERAGE MOISTURE	%	9·45	12·55	15·95	18·71	21·47
DRY DENSITY ρ_D Mg/m³		1·704	1·765	1·792	1·719	1·636

$$\rho = \frac{A-B}{V}$$

$$W = \frac{W_1 + W_2 + W_3}{3} \qquad \rho_D = \rho \times \frac{100}{100+w}$$

Fig. 6.14   Compaction test data and calculations

(10)   Repeat stages (4)–(9) for each batch in turn. If necessary, make up another batch and test if another point is required on the compaction curve.

The following stages refer to both procedure (A) and procedure (B)

(11)   *Calculate.*   Calculate the bulk density of each compacted specimen from the equation

$$\rho = \frac{m_2 - m_1}{1000} \ \text{Mg/m}^3$$

where m_1 = mass of mould (and base if included) and m_2 = mass of soil and mould (and base if included). If the volume of the mould is not 1000 cm³ but is V cm³, then

$$\rho = \frac{m_2 - m_1}{V} \ \text{Mg/m}^3$$

Calculate the average moisture content, w%, for each compacted specimen. Calculate the corresponding dry density from the equation

$$\rho_D = \left(\frac{100}{100 + w} \right) \rho \ \text{Mg/m}^3$$

Typical density and moisture content data and calculations are given in Fig. 6.14. Calculate the percentage of stones retained on the 20 mm sieve from the equation

$$\text{stone} \% = \frac{W_2}{W_1} \times 100\%$$

(12)   *Plot graph.*   Plot each dry density, ρ_D, against the corresponding moisture content, w. Draw a smooth curve through the points. The curves for 0, 5 and 10% air voids may be plotted as well.

A typical graph, together with other test data, is shown in Fig. 6.15, which includes three air voids lines.

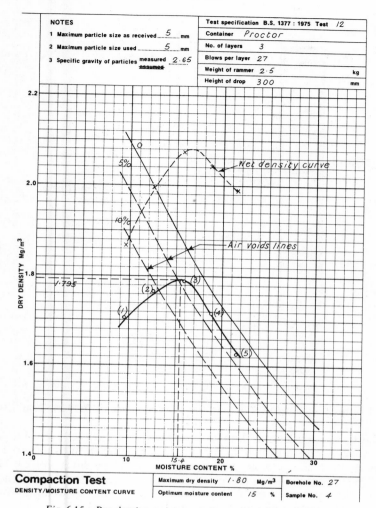

NOTES

1 Maximum particle size as received ....5.... mm
2 Maximum particle size used ....5.... mm
3 Specific gravity of particles measured 2·65 ~~assumed~~

Test specification B.S. 1377 : 1975 Test 12
Container   Proctor
No. of layers   3
Blows per layer   27
Weight of rammer   2·5   kg
Height of drop   300   mm

Compaction Test
DENSITY/MOISTURE CONTENT CURVE

Maximum dry density   1·80   Mg/m³
Optimum moisture content   15   %

Borehole No.   27
Sample No.   4

Fig. 6.15   Dry density–moisture content test results and graph

(13)   *Read off optimum values.*   Ascertain the point of maximum dry density on this curve. Read off the maximum dry density value, and the corresponding moisture content, which is the optimum moisture content for this degree of compaction.

(14)   *Report results.*   The graph plot showing the experimental points is part of the test report, which should include a description of the material.

The maximum dry density for the stated degree of compaction is reported to the nearest 0.01 Mg/m³.

The optimum moisture content is reported as follows:

Below 5%: to the nearest 0.2%;
From 5% to 10%: to the nearest 0.5%;
Exceeding 10%: to the nearest 1%.

The percentage of stones retained on the 20 mm sieve is reported to the nearest 1%.

The procedure is reported as the British Standard 2.5 kg rammer method (Test 12 of BS 1377: 1975). It must be stated whether the test was carried out on a single sample or on separate batches.

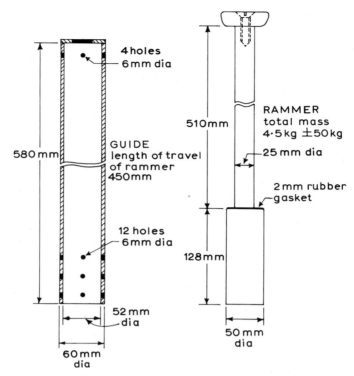

Fig. 6.16   Rammer for BS 'heavy' compaction test

6.5.3   'Heavy' Compaction Test (4.5 kg rammer method) (BS 1377:1975, Test 13)

This test is similar to the American test known as the 'modified AASHO' procedure. It gives the dry density — moisture content relationship for a soil compacted in five layers in the same mould as used in the 'ordinary' compaction test, using 27 blows per layer with a 4.5 kg rammer falling 450 mm. The total compactive energy applied is 4.5 times greater than in the 'ordinary' test. From the density–moisture curve the optimum moisture content, and the maximum dry density, for this heavier degree of compaction can be determined.

As with the 'ordinary' test, this test is suitable for soils no larger than 20 mm, and the detailed procedure depends on whether the particles are susceptible to crushing.

If the soil contains particles larger than 20 mm, refer to Section 6.5.6.

APPARATUS

(1)   Mould — as for the standard test (Section 6.5.2).
(2)   Metal rammer with 50 mm diameter face, weighing 4.5 kg, and a controlled height drop of 450 mm (see Fig. 6.16). Otherwise, it is similar to item (2) of Section 6.5.2.

Items (3)–(9) are as for the 'ordinary' test.

PROCEDURAL STAGES

The stages are similar to those given in Section 6.5.2 for the 'ordinary' test.

TEST PROCEDURE

The procedure is similar as that described for the 'ordinary' test, with the exception of the detailed modifications referred to below.

Stages (1) *and* (2). Sample preparation, as for the 'ordinary' test, depends upon whether the soil particles are susceptible to crushing.

(3)   *Addition of water.*   The quantity of water to be added to the sample initially, or to the first batch, should be approximately as follows:

Sandy and gravelly soils: 3–5% (150–250 ml of water to 5 kg of soil).
Cohesive soils: About 5% for clays of low plasticity, to 15% or more for clays of very high plasticity. If the plastic limit is known, the percentage should be about (PL − 15). Experience is the best guide.

(4)   *Compaction.*   Compaction is carried out in five layers instead of three, the 4.5 kg rammer with a drop of 450 mm being used, with 27 blows for each layer. Take extra care when using this rammer to ensure that it is properly in place before releasing. See stage (4) of Section 6.5.2.

If the correct amount of soil has been used for compacting the first layer, the compacted surface should be at about one-fifth of the height of the mould body — that is, about 90 mm below the top of the mould body, or 140 mm below the top of the extension collar. If significantly different from this, remove the soil and start this stage again.

Compact four more equal layers into the mould as before. The final compacted surface should be about 6 mm above the top of the mould body (see Fig. 6.12). If it is higher than this, remove the soil, break it up and repeat this stage, using slightly less soil in each layer.

Stages (5)–(13).   As for the 'ordinary' test. Moisture content increments are similar to those suggested in stage (9) of the 'ordinary' test.

(14)   *Report results.*   Results are reported as for the 'ordinary' test, except that the procedure is reported as the British Standard 4.5 kg rammer method (Test 13 of BS 1377:1975). State whether the test was carried out on a single sample or on separate batches.

6.5.4   Compaction by Vibration (BS 1377:1975, Test 14; Vibrating hammer method)

This test is applicable to granular soils passing the 37.5 mm sieve. The principle is similar to that of the 'ordinary' and 'heavy' procedure, except that a vibrating hammer is used instead of a drop-weight rammer, and a larger mould (the standard CBR mould) is necessary.

APPARATUS

(1)   Cylindrical metal mould, internal dimensions 152 mm diameter and 127 mm high (CBR mould). The mould can be fitted with an extension collar and baseplate. Details of two types of mould are shown in Figs. 6.17 and 6.18. (Note: The mould shown in Fig. 6.18 must not be confused with the similar ASTM compaction mould which is 116.4 mm high.)

(2)   Electric vibrating hammer, power consumption 600–700 W, operating at a frequency in the range 25–45 Hz. A check test to determine whether the hammer meets the requirements of the British Standard is described in Section 6.5.5. A special supporting frame for the hammer may be used for easier operation, as shown in Fig. 6.19.

(3)   Steel tamper for attaching to the vibrating hammer, with a circular foot 145 mm diameter (see Fig. 6.20).

(4)   37.5 mm BS sieve and receiver.

(5)   Depth gauge or steel rule reading to 0.5 mm.

(6)   Laboratory stop-clock.

Other items required are items (3)–(9) as listed in Section 6.5.2.

PROCEDURAL STAGES

(1)   Prepare apparatus.

(2)   Obtain soil samples.

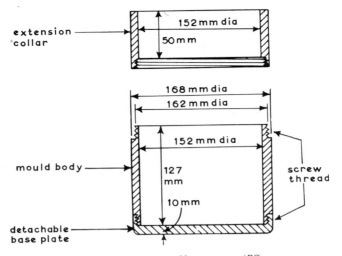

Fig. 6.17   CBR mould, screw type (BS)

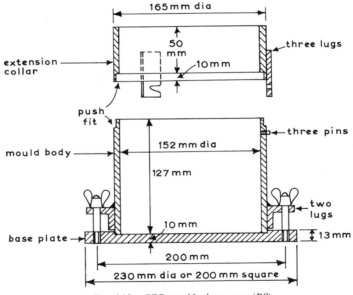

Fig. 6.18   CBR mould, clamp type (BS)

(3)   (A) Prepare test sample, *or* (B) prepare five batches.
(4)   Compact into mould in layers.
(5)   Measure height.
(6)   Weigh.
(7)   Remove soil from mould.
(8)   Measure moisture content.
(9)   Break up and remix.
(10)   Add more water, and repeat stages (4)–(9)
         or
(10B Repeat stages (4)–(8), using the next batch.
         In either case the total number of compaction should be at least five.

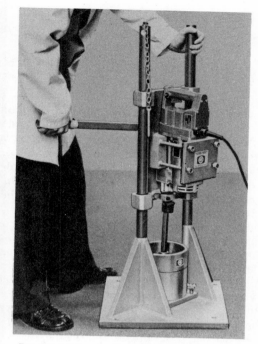

Fig. 6.19   Vibrating hammer in supporting frame

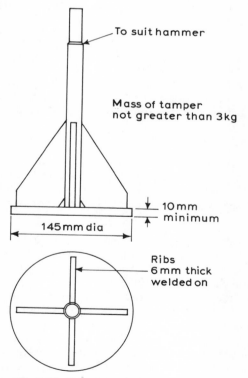

To suit hammer

Mass of tamper
not greater than 3kg

10 mm
minimum

145 mm dia

Ribs
6 mm thick
welded on

Fig. 6.20   Tamper for vibrating hammer

(11) Calculate.
(12) Plot graph.
(13) Read off optimum values.
(14) Report results.

TEST PROCEDURE

(1) *Prepare apparatus.* See that the component parts of the mould are clean and dry. Assemble the mould, baseplate and collar securely, and weigh to the nearest 1 g (m_1). Measure the internal dimensions of the assembly, and calculate the internal volume, as described in Section 6.5.2. The nominal dimensions of the mould give an area of cross-section of 18 146 mm² and a volume of 2304.5 cm³ (say 2305 cm³), but these may change slightly with wear. The inside height of the mould with collar is recorded (h_1 mm).

The metric measurements of the CBR mould are exactly the same as the earlier measurements which were specified in inches.

The comments regarding preparation of the compaction mould given in Section 6.5.2 apply equally to the CBR mould. It is particularly important to ensure that the lugs and clamps holding the mould assembly together are secure and in good condition, in order to withstand the effects of vibration. If the mould has screw-on fittings (Fig. 6.17), the threads must be kept clean and undamaged. Avoid cross-threading when fitting the baseplate and extension collar, and make sure that they are tightened securely as far as they will go without leaving any threads exposed. Screw threads and mating surfaces should be lightly oiled before tightening.

Ensure that the vibrating hammer is working properly, in accordance with the manufacturer's instructions. See that it is properly connected to the mains supply, and that the connecting cable is in sound condition. The supporting frame, if used, must move freely without sticking. The hammer should have been verified as described in Section 6.5.5.

The tamper stem must fit properly into the hammer adaptor, and the foot must fit inside the CBR mould with the necessary clearance (3.5 mm all round).

(2) *Obtain soil sample.* The original bulk sample is air dried and weighed (W_1). If it contains gravel or stones larger than 37.5 mm, these are removed by passing the sample through a 37.5 mm sieve. The material retained is weighed (W_L).

The method of preparation for compaction depends upon whether the soil particles are susceptible to crushing under the action of the vibrating hammer. The two alternative methods are described separately below, but the compaction procedure is the same in either case.

(A) Soil particles not susceptible to crushing

(3) *Prepare test sample.* A sample of 25 kg of air-dried material passing the 37.5 mm sieve is required. Add a suitable amount of water, and mix thoroughly. For a sandy and gravelly soil, a moisture content of 3–5% would be suitable. This requires the addition of about 750 ml of water to 25 kg of soil, but the exact amount can be judged from experience. Thorough mixing of the water is essential.

(4) *Compact into mould.* Place the mould assembly on a solid base, such as a concrete floor or plinth. If the test is performed out of doors because of noise and vibration problems, place the mould on a concrete paved area, not on unpaved ground or on thin tarmac. Any resilience in the base results in inadequate compaction.

Add a quantity of soil to the mould, such that after compaction the mould is one-third filled. A preliminary trial may be necessary to ascertain the correct amount of soil. Compact the layer with the vibrating hammer, fitted with the tamper, for 60 s, applying a firm pressure vertically downwards throughout. The downward force, including that resulting from the mass of the hammer and tamper, should be 300–400 N. This force is sufficient to prevent the hammer bouncing up and down on the soil. The correct force can be determined by standing the hammer, without vibration, on a platform scale and pressing down until a mass of 30–40 kg is indicated. With experience the pressure to be applied can be judged, but an occasional check on the platform scale is advisable. If the hammer-supporting frame is used, the hand pressure required is much less but should be carefully checked.

Repeat the above compaction procedure with a second layer of soil, and then with a third layer. The final thickness of the compacted specimen should be between 127 mm and 133 mm; if it is not, remove the soil and repeat the test.

(5) *Measure height.* After compaction remove any loose material from the surface of the specimen around the edge, so that the surface is reasonably flat. Clean off the top edge of the mould collar. Lay the straight-edge across the top of the collar, and measure down to the surface of the specimen with the steel rule or depth gauge, to an accuracy of 0.5 mm. Take readings at four points spread evenly over the surface, and 15 mm from the side of the mould. Calculate the average depth (h_2 mm). The mean height of the compacted specimen, h, is given by

$$h = h_1 - h_2 \text{ mm}$$

(6) *Weigh.* Weigh the mould with the compacted soil, collar and baseplate to the nearest 5 g, or to 1 g if the balance is accurate enough (m_2).

(7) *Remove soil.* Remove the soil from the mould and place on the tray. A jacking extruder makes this operation easy if fittings to suit the CBR mould are available. But sandy and gravelly (non-cohesive) soil should not be too difficult to break up and remove by hand.

(8) *Measure moisture content.* Take two representative samples in large moisture content containers for measurement of moisture content. This must be done immediately after removal from the mould, before the soil begins to dry out. The moisture content samples must be large enough to give results representative of the maximum particle size of the soil (see Section 2.5.2). The average of the two moisture content determinations is denoted by $w\%$.

(9) *Break up and remix.* Break up the material on the tray and mix in with the remainder of the sample. Add an increment of water so as to raise the moisture content by 1 or 2% (250–500 ml of water for 25 kg of soil). As the optimum moisture content is approached it is preferable to add water in smaller increments.

Mix in the water thoroughly.

(10) *Repeat with added water.* Repeat stages (4)–(9) for each increment of water added. At least five compactions should be made, and the range of moisture contents should be such that the optimum moisture content is within that range. If necessary, carry out one or more additional tests at suitable moisture contents.

Above a certain moisture content the soil may contain an excessive amount of free water, which indicates that the optimum condition has been passed.

(B) Soil particles susceptible to crushing

If the sample contains particles which are likely to be broken down by the action of the vibrator, separate batches should be prepared for compaction at each moisture content. Five or more identical 8 kg samples of air-dried soil passing the 37.5 mm sieve are required. Add a different amount of water to each batch so that the moisture content range is from about 3% to about 10% (240–800 ml of water to 8 kg of soil). The actual moisture contents are best judged from experience. The water must be thoroughly mixed in with the soil.

Compact each batch in turn into the CBR mould, and measure its moisture content, as described in stages (4)–(8).

If additional points are required to define the optimum condition on the compaction curve, make up additional 8 kg batches at appropriate moisture contents and compact each batch as above.

The following stages apply to both procedure (A) and procedure (B)

(11) *Calculate.*   Calculate the bulk density of each compacted speciment from the equation

$$\rho = \frac{m_2 - m_1}{18.15 \times h} \quad \text{Mg/m}^3$$

where m_1 = mass of mould, collar and baseplate; m_2 = mass of mould, collar and baseplate with soil; h = height of compacted soil specimen = $h_1 - h_2$.

The above equation applies only if the average diameter of the mould is 152 mm. If it is not, and is represented by D mm, use the area of cross-section $A (= \pi D^2/4)$ in the equation

$$\rho = \frac{m_2 - m_1}{A \times h} \times 1000 \quad \text{Mg/m}^3$$

Calculate each dry density from the corresponding moisture content, $w\%$, from the equation

$$\rho_D = \frac{100}{100 + w} \times \rho \quad \text{Mg/m}^3$$

Calculate the percentage of coarse material retained on the 37.5 mm sieve.

(12) *Plot graph.*   Plot the values of dry density, ρ_D, against moisture content, w, and draw a smooth curve through the points. The curves corresponding to 0, 5 and 10% air voids may be plotted as well.

(13) *Read off optimum values.*   Read off the maximum dry density, and the corresponding moisture content, from the compaction curve.

(14) *Report results.*   The graph plot showing the experimental points is part of the test report.

The maximum dry density for compaction by vibration is reported to the nearest 0.01 Mg/m³.

The corresponding optimum moisture content is reported as follows

Below 5%: to the nearest 0.2%.
From 5% to 10%: to the nearest 0.5%.
Exceeding 10%: to the nearest 1%.

The percentage of coarse material retained on the 37.5 mm sieve is reported to the nearest 1%.

The procedure is reported as the British Standard vibrating hammer method (Test 14 of BS 1377:1975). It must be stated whether a single sample was used or separate batches.

6.5.5 Verification of Vibrating Hammer

The following procedure may be used to ascertain whether the vibrating hammer used for the test described in Section 6.5.4 complies with the requirements of BS 1377:1975 and is in satisfactory working order. The procedure is outlined in Note 2 of Test 14 in the Standard.

The 'standard sand' used is Leighton Buzzard medium-silica sand, complying with the grading requirements shown in Fig. 6.21. The sand must be dry, and not used previously. A sample of 10 kg is required.

Sieve the sand through a 600 μm sieve and discard the fraction retained. Add water to the sieved sand to bring its moisture content to 2.5% (250 ml of water to 10 kg of dry soil). Mix the

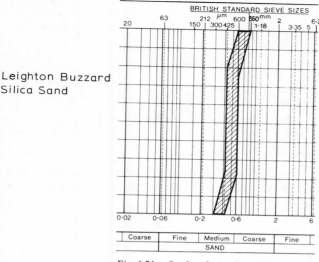

Leighton Buzzard
Silica Sand

specification

100% passing 850 μm sieve
not <75% passing 600 μm sieve
not >25% passing 425 μm sieve
0% passing 300 μm sieve

Fig. 6.21   Grading limits for 'standard sand'

water in thoroughly, and check the actual moisture content, which should not differ from the stated value by more than 0.5%.

Compact the sand into the CBR mould in three layers with the vibrating hammer, using the procedure given in Section 6.5.4, stage (4). Measure the height of the compacted sample, weigh and determine the compacted dry density to the nearest 0.002 Mg/m³ (stages 5–8, 11). Repeat twice on the same sample of sand, making three tests in all.

If the range of values of dry density exceeds 0.01 Mg/m³, repeat the above procedure. The vibrating hammer is satisfactory for the vibrating compaction test if the mean dry density achieved exceeds 1.74 Mg/m³.

This test is valid only for Leighton Buzzard sand of the grading specified. Other types of sand will give different results.

6.5.6   Compaction of Stony Soils

For soils containing gravel-size fragments larger than 20 mm, a calculated correction can be applied to the maximum dry density to estimate the corresponding maximum dry density in the field. The principle is explained in Section 6.3.4, but applies only if the 'stone' content does not exceed about 25%.

For soils containing larger proportions of coarse material, the only satisfactory method of obtaining the compaction characteristics is to carry out a test in a larger container so that a larger maximum particle size can be used. A CBR mould, as used in the vibrating hammer test, is usually used for this purpose, and particles up to 37.5 mm can be included in the compacted sample.

Either the equivalent 'ordinary' or the equivalent 'heavy' standard compaction test may be carried out with the CBR mould. The total quantity of material passing the 37.5 mm sieve required is 25 kg, or five batches each of 8 kg if the particles are susceptible to crushing. The procedures are the same as those described in Sections 6.5.2 and 6.5.3, except that 62 blows are required in each layer instead of 27. This is because of the increased volume of soil compared with the 'Proctor' mould volume (see Table 6.2, page 275).

Application of the first few blows should be done systematically, but the pattern differs from that used for the 'Proctor' mould because of the larger size. The first two blows should be applied at the edge and diametrically opposite each other; the next two half-way between; and

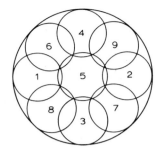

Fig. 6.22   Sequence of blows using hand rammer in CBR mould

the fifth at the centre (see Fig. 6.22). The next four (numbered 6, 7, 8, 9) are placed between those already applied. After that, work systematically around the mould and across the middle so that the whole area is uniformly compacted.

6.5.7   Use of Compaction Machine

A mechanical compaction apparatus eliminates much of the physical effort required for carrying out compaction tests. However, it has been found that the densities achieved by machine are often less than those obtained by hand compaction. This is partly because the blow pattern differs from that recommended in Sections 6.5.2 and 6.5.6, and partly because the base not only rotates but also has to provide horizontal movement when a CBR mould is being used so that its whole area may be covered by the rammer. This results in the mould support being less rigid than a concrete base.

A recently developed machine of the type shown in Fig. 6.23 incorporates the following features:

(1)   The blow pattern follows closely the recommended pattern, by applying widely spaced blows at first, which flatten the soil surface, followed by overlapping blows.

(2)   The area of a CBR mould is covered by shifting the position of the rammer assembly instead of by moving the base.

(3)   The rotating base is supported by an inertia block offering a machined annular surface of large area, which makes for a very rigid support.

6.5.8   Static Compaction

Soil may be compacted into a 'Proctor' or CBR mould under a static pressure, in order to achieve a sample at a known density and moisture content. The usual procedure is to weigh out the exact amount of soil required to just fill the mould at the specified density, and to compress that amount in the mould with a compression machine or other suitable apparatus.

This procedure is used for the preparation of specimens on which other tests (especially CBR tests) are to be carried out. Static compaction is not used for the determination of the moisture–density relationship of soils, and is not further discussed here.

6.6   OTHER COMPACTION TESTS

6.6.1   Types of Test

Some additional procedures for the compaction of soils, other than those given in the British Standard, are described or referred to in this section. These include two small-scale laboratory

Fig. 6.23 Mechanical compaction apparatus

compaction procedures, and two tests which have been developed in recent years to meet certain requirements not adequately covered by the present British Standard tests.

The small-scale laboratory procedures described below are:

(1) Dietert compaction (Sections 6.6.2 and 6.6.3).
(2) Harvard miniature compaction (Section 6.6.4).

Both these procedures require only a small quantity of material, and are limited to the use of the fraction passing a 2 mm sieve. Their main application is as an expedient to obtain an approximation to the compaction characteristics of a soil when the quantity of material available is too small for carrying out one of the standard compaction tests. Results from these tests are not, in general, directly comparable with standard compaction test results, although correlations are possible with particular soils (Wilson, 1950). They are not intended as a substitute for the standard tests described in Section 6.5.

These small-scale procedures can be useful in the laboratory for the preparation of small recompacted specimens for use in other tests. The controlled degree of compaction which they provide gives results which are more reproducible than those obtainable by arbitrary hand tamping methods.

The recently developed procedures which are referred to but not described in detail are as follows:

(1) The 'moisture condition' test (Section 6.6.5), developed by the TRRL for rapid control on earthworks construction projects.
(2) A proposed 'compactability test' (Section 6.6.6) for graded aggregates and road construction materials, which is the subject of a draft new British Standard.

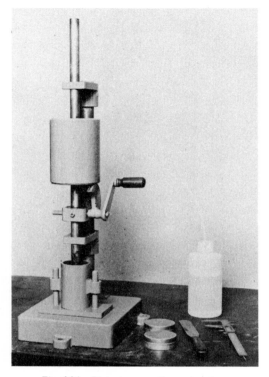

Fig. 6.24   Dietert compaction apparatus

6.6.2   Dietert Compaction

The Dietert compaction test was originally developed in the USA for testing foundry sands, and was adopted in this country for testing soils in the laboratory about 1945. The test provides a useful expedient when only a small quantity of soil is available, or when a large number of tests are required on the fine soil fraction when the presence of stony material is irrelevant; for example, where the effect of the addition of a 'stabiliser' over a wide range of doses is to be measured.

APPARATUS

The Dietert compaction machine, shown in Fig. 6.24, is a hand-operated device, the principle of which is illustrated in Fig. 6.25. A sliding weight A, of about 8 kg, can move freely up and down on the guide-rod B. The weight is actuated by the cam C, which is rotated by the handle H. The cam shaft is attached to the guide rod B, which is free to slide vertically in the guide bushes D. The lower end of the guide rod can be fitted with a detachable foot, which rests on the specimen being compacted. The dimensions of the foot depend upon the size of specimen. Thus, the slide weight always falls through the same distance whatever the thickness of the specimen. The machine is mounted on a massive base, which can be bolted to the bench if necessary.

The standard mould is 2 inches (50.8 mm) internal diameter and $4\frac{3}{4}$ inches (120.65 mm) high, fitted with a base in the form of a plug which protrudes 1 inch (25.4 mm) into the mould. The foot for this mould, which is screwed on to the guide rod of the machine, is about $1\frac{7}{8}$ inches (47.6 mm) diameter.

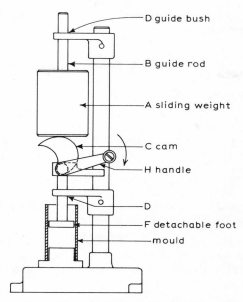

Fig. 6.25   Principle of Dietert apparatus

Other items required are:

Balance, 1 kg capacity reading to 0.1 g.
2 mm BS sieve and receiver.
Mixing tray, trowel, spatulas.
Glass measuring cylinder 250 or 500 ml.
Drying oven and moisture content apparatus.
Vernier depth gauge.

PROCEDURAL STAGES

(1)   Prepare apparatus.
(2)   Prepare soil sample.
(3)   Compact into mould.
(4)   Turn over and compact again.
(5)   Measure height.
(6)   Check weigh.
(7)   Measure moisture content.
(8)   Add increment of water to sample.
(9)   Repeat stages (4)–(9) for a total of at least five compactions.
(10)   Calculate.
(11)   Plot graph.
(12)   Repeat results.

TEST PROCEDURE

(1) *Prepare apparatus.*   Ensure that the mould is clean, and that the base fits securely in place. Screw the compacting foot securely on to the guide rod of the compaction machine. See that the weight falls freely and that the guide rod slides easily in the bushes. A very thin film of oil may be applied to the cam shaft and the sliding parts, but surplus oil must be wiped off.

The machine should be placed on a firm and solid bench and preferably bolted to it.

Weigh the mould, without base, to the nearest 0.02 g (m_1).

Fit the mould securely on the base. Measure the internal height from the top of the mould to the base in three or four positions with the vernier depth gauge to 0.1 mm, and calculate the average height (H_0 mm).

(2) *Prepare soil sample.*   Sieve a quantity of air-dried soil through the 2 mm sieve. About 1.5 kg of sieved soil is required. Measure its moisture content. Add a suitable amount of water from the measuring cylinder to the soil, and mix thoroughly. The volume of water depends on the soil type, as suggested below:

Sandy soils: 4–6% (40–60 ml of water to 1 kg of soil).
Clay soils: About 10% below plastic limit. (For a soil of PL = 20, add 100–120 ml of water to 1 kg soil.)

As for the standard compaction tests, these quantities may be used as a general guide but the amount is best judged by experience. Thorough mixing in of the water is essential, and clay soils should be sealed up and left to mature overnight before testing.

Weigh out 150 g of the wet soil and place it loose into the mould, which it will almost fill.

(3) *Compact into mould.*   Place the mould on the Dietert machine base beneath the rammer, and lower the rammer into the mould so that the foot is supported by the soil. Apply 10 blows of the drop weight by turning the handle on the cam shaft steadily at about one turn per second.

(4) *Turn over and compact again.*   Remove the cylinder from the base, turn it over and replace it. Apply a further 10 blows to the other end of the sample.

(5) *Measure height.*   Using the vernier depth gauge, measure the distance from the top of the mould to the compacted surface of the soil in three or four places, and calculate the average distance (H_1).

(6) *Check weigh.*   Remove the mould with soil from the base, and weigh to 0.1 g (m_2). This is done as a check on the original mass taken, and ($m_2 - m_1$) should be very close to 150 g.

It is preferable not to include the base in the weighings of m_1 and m_2, because its mass is quite considerable compared with the mass of the soil sample.

(7) *Measure moisture content.*   Remove the soil sample from the mould on to the tray, and clean the mould. Use the whole sample for the determination of its moisture content (w%).

(8) *Add water increment.*   Measure an appropriate amount of water in the measuring cylinder, add to the prepared sample and mix thoroughly. The amount of water to be added will range from about 1% (10 ml to 1 kg soil) for sandy soil to 4% (40 ml to 1 kg soil) for clay.

(9) *Repeat compactions.*   Repeat stages (3)–(8) for each increment of water added, so that at least five compactions are made. The range of moisture content should be such as to encompass the peak value of density of the compacted soil. If necessary, carry out additional tests.

(10) *Calculate.*   Calculate the height of soil in the mould, H, from the equation

$$H = H_0 - H_1 \text{ mm}$$

The volume of soil in the 2 inch diameter mould, V cm^3, is given by

$$V = 2.03 \times H \quad \text{cm}^3$$

The mass of soil used, m g, is given by

$$m = m_2 - m_1$$

The wet density, ρ, of the soil can be calculated from the equation

$$\rho = \frac{m}{2.03 \times H} \quad \text{Mg/m}^3$$

If the mass of soil used is exactly 150 g, the density is given by

$$\rho = \frac{74}{H} \quad \text{Mg/m}^3$$

Calculate the moisture content, $w\%$, and the dry density, ρ_D, as in stage (11) of Section 6.5.2.

(11) *Plot graph.* Plot a graph of dry density, ρ_D, against moisture content, w, exactly as for the 'ordinary' compaction test (Section 6.5.2, stage 12). Read off the maximum dry density and the corresponding moisture content.

(12) *Report results.* The graphical plot showing the experimental points is part of the test report. The maximum dry density and optimum moisture content are reported as described in stage (14) of Section 6.5.2.

It must be reported that the Dietert compaction apparatus was used, and that 10 blows were applied to *each* end of the specimen. The size of the mould must also be stated. The results of not necessarily comparable with the results obtained from the British Standard 'ordinary' compacting test.

The proportion of material larger than 2 mm, removed from the original sample before the test, should be reported.

6.6.3 Modifications to Dietert Test

LIMITED QUANTITY OF SOIL

The test described above requires about 1.5 kg of soil, but the amount of soil required, using the 2 inch mould, can be reduced to about 400 g passing the 2 mm sieve by using the following modified procedure.

After removing the compacted soil from the mould (stage 7 of Section 6.6.2), use only a small portion for the determination of moisture content. Break up the rest of the sample on the tray, mix with additional water and recompact. It must be reported that the same soil was reused, and the size of the original sample should be stated.

This procedure is not suitable if the soil particles are susceptible to crushing during compaction.

CLAY SOILS

It has been observed that the maximum dry density achieved in the Dietert test on clay soils of medium to high plasticity is greater than that obtained in the British Standard 'ordinary' compaction test. Some clay soils containing sand have been found to give lower densities. This suggests that clays of medium to high plasticity should receive less than the standard 10 blows at each end, and sandy clay soils somewhat more than 10 blows. The liquid limit of the soil can be used as an indicator for this purpose, and an accepted correlation is given in Fig. 6.26.

The liquid limit of the soil is determined first, and reference to Fig. 6.26 will indicate the number of blows required on each end of the specimen approximate to BS 'ordinary' compactive effort. Otherwise the test is exactly as described in Section 6.6.2. According to Fig. 6.26, the standard 10 blows at each end relates to a clay of liquid limit 33%.

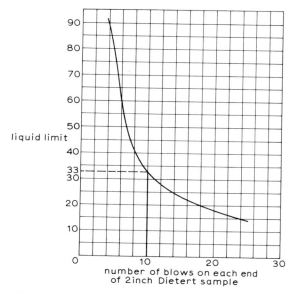

Fig. 6.26   *Liquid limit correction for Dietert compaction test*

PREPARATION OF TEST SPECIMENS

The Dietert apparatus can be adapted for the compaction of small specimens for other laboratory tests, such as triaxial (38 mm diameter) or shear-box (60 × 60 mm) specimens. The appropriate foot should be fitted to the tamping rod, to suit the sample mould being used. For the shear-box test the soil should be compacted directly into the shear-box itself, with the two halves bolted together, in three layers.

Two compaction procedures are possible.

(1)   Determine the degree of Dietert compaction required to equate with one of the standard compaction tests. This can be done only by carrying out trial tests on the soil to be used.
When the specimen is compacted by the Dietert procedure, compaction must be applied equally to both ends or surfaces. Test specimens can then be made up, for instance, to British Standard 'ordinary' compactive effort at various moisture contents.

(2)   Apply sufficient blows of the Dietert apparatus to compact a known mass of soil into a given volume in the mould. Again, compaction must be applied to both ends or surfaces. This procedure is suitable when a soil has to be compacted to a given density, regardless of standard compaction criteria. A process of trial and error will usually be necessary.

The best method for the preparation of recompacted test specimens is to compact the soil in a standard compaction mould from which specimens are cut or trimmed. However, the Dietert procedure is useful when only a limited quantity of soil is available.

6.6.4   Harvard Compaction Apparatus

The Harvard compaction test procedure is given in *ASTM STP* 479 (Wilson, 1970) as a suggested method for determining the compaction characteristics of fine-grained soil when only a small quantity of material is available. The principle of the test is similar to the Dietert test, described in Section 6.6.2, and the full procedure is not discussed here. The action of the apparatus differs from the drop-weight principle of the British Standard and Dietert tests in

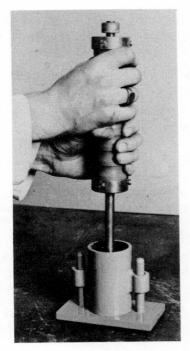

Fig. 6.27   Harvard compaction apparatus

that the soil is subjected to kneading rather than impact. Results from the Harvard test may not be directly comparable with the British Standard tests.

The Harvard compaction apparatus and its use is briefly described below.

APPARATUS

The compaction device consists of a hand-held spring-loaded tamper and special mould, which are shown in Fig. 6.27. The spring is compressed by means of the adjusting nut to a compression of 40 lb (18.2 kgf or 178 N), so that a small increase of force above that value will compress the spring further. Springs of different stiffnesses can be substituted. The metal tamper rod is 0.5 inch (12.7 mm) diameter.

The mould used has an internal diameter of $1\frac{5}{16}$ inches (33.34 mm) and is 2.816 inches (71.5 mm) high. Its volume is $\frac{1}{454}$ ft^3 (62.4 cm^3), and this volume was selected because the mass of soil, in grams, is equal to its density in pounds per cubic foot. An extension collar about $1\frac{1}{2}$ inches (38 mm) high may be added to the mould, both of which can be fitted to a detachable baseplate.

The Harvard compaction procedure, like the Dietert test, can be modified to provide additional or lesser degrees of compaction, but the relationship to the British Standard compaction efforts can only be determined experimentally for a particular soil.

A specially designed jig (the collar remover) enables the compacted soil to be held in place and kept intact while the extension collar is being removed.

A sample ejector quickly and easily removes the compacted specimen from the mould.

SOIL SAMPLE

Soil for use with this apparatus should contain particles no larger than 2 mm. The usual procedure for air drying, sieving, mixing and curing should be followed.

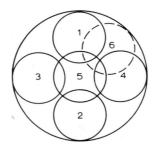

Fig. 6.28   Sequence of tamps using Harvard apparatus

If a complete moisture–density relationship test is to be done, separate batches should be used for each moisture content, and a compacted specimen should *not* be remixed and reused.

COMPACTION PROCEDURE

Compaction is effected by placing the plunger on the soil surface and pressing down with the hand grip until it can be felt that the spring is just starting to compress, and then releasing and moving to the next position. The first four tamps should be applied in opposite quadrants touching the edge of the mould, followed by one at the centre (see Fig. 6.28). The next four should be in a similar pattern but spaced between the first four, then at the centre. This sequence is repeated until the required number of tamps have been applied, at a rate of about one tamp every $1\frac{1}{2}$ s.

With a 40 lb (178 N) spring, compaction in three layers with 25 tamps per layer is roughly equivalent to Proctor effort compaction, but this is given only as a general guide and not an established relationship.

Measurement of density and moisture content, and calculations, are done in the same way as for the compaction tests described above. Results of a moisture–density relationship test should include a note reporting the type of test, size of mould and compression spring used.

Apart from its use for determining the moisture–density relationship, the Harvard tamping device provides a convenient means of preparing small reproducible test specimens for shear strength and other tests on recompacted soil.

6.6.5   Moisture Condition Test

This test is described in the TRRL paper by Parsons (1976). It has been developed for use in construction control of earthworks to assess rapidly whether a material is at a moisture content suitable for placing, in relation to the specified upper limit of moisture content. It could perhaps also be used to provide a 'moisture condition' parameter for correlation with other engineering properties of soil.

The test is based on the principle that the density produced in a given soil depends solely on the moisture content and the compactive effort used. At moisture contents in excess of the optimum, where the zero air-voids line is approached, an increase in compactive effort produces little or no increase in density. If the specified upper limit of moisture content for a given compactive effort is in this zone, a comparison of densities produced by different compactive efforts will indicate immediately whether the moisture content of the material is above or below the specified upper limit.

The main item of apparatus required for this test is a modified version of the machine used for the determination of aggregate impact value (BS 812; Part 3: 1975). Modifications consist of:

(1)   Use of moulds of different heights.

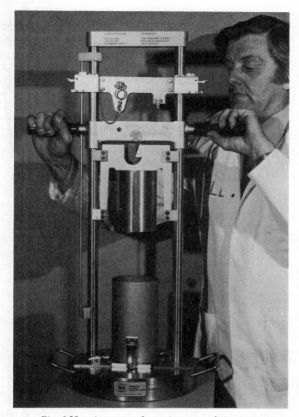

Fig. 6.29   Apparatus for moisture condition test

(2)   Adjustment of the height of drop of the rammer.
(3)   Use of a non-standard rammer.

Details of the apparatus are as follows:

Mass of rammer	6.8 kg
Diameter of rammer	100 mm
Height of drop	adjustable, from 50 to 350 mm
Diameter of mould	102 mm
Height of mould	50 mm or 115 mm (plus 50 mm removable extension)

The test procedure is not described here, because it is not yet covered by British Standards. The modified aggregate impact test apparatus is available commercially, and is shown in Fig. 6.29.

6.6.6   Compactability Test for Graded Aggregates

This test is the subject of draft Standard recommendations (1978), and is a method for assessing the compactability of graded aggregates, particularly those used in road bases and sub-bases. The standard compaction tests used for soils have been found to be unreliable when applied to some of these materials, and this draft Standard aims to provide a standardised approach to compactibility testing.

The principle of the test is similar to the vibrating hammer test (Section 6.5.4; BS 1377:1975, Test 14). However, a more powerful vibrating hammer is used, in a standardised manner,

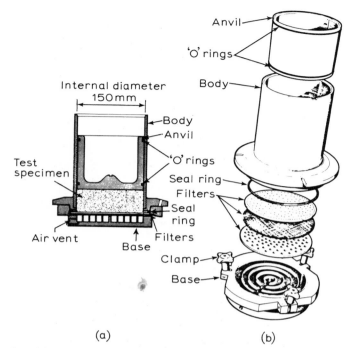

Fig. 6.30   *Mould and anvil for compactability test (Courtesy Transport and Road Research Laboratory, Crowthorne, Berks.)*

mounted in a loading frame, and the sample is compacted in a special heavy-duty mould. The test results are presented in the usual form of a moisture content–dry density relationship, but the dry density can also be expressed in items of a volumetric equivalent.

The following special apparatus is required, in addition to standard soil-testing equipment:

(1)   Compaction mould, comprising body, base, filter assembly and anvil. The latter covers the whole area of the sample, and may be fitted with an optional vacuum release plug. Any excess water is permitted to drain downwards.

(2)   Electric vibrating hammer, power consumption 900 W, frequency about 33 Hz, fitted with a tool to mate with the anvil.

(3)   Loading frame to support the hammer and mould, providing a steady downward force of $360 \pm 10 \, \text{kN}$.

The mould assembly and its component parts are illustrated in Fig. 6.30. The load frame and mould, set up for use in a noise-reducing cabinet, are shown in Fig. 6.31.

The procedure is not described here, but the prototype method will be found in the TRRL papers by Pike (1972) and Pike and Acott (1975).

It is suggested that this apparatus could provide a more satisfactory means for determining the maximum density (minimum porosity) of granular soils, including silty sands, than the procedure described in Sections 3.7.2 and 3.7.3.

BIBLIOGRAPHY

American Association of State Highway Officials (1942). *Standard specifications for highway materials and methods of sampling and testing*. Part II, AASHO. Designation T99-38, 'Standard laboratory method of test for the compaction and density of soil'. AASHO, Washington, DC

American Foundrymen's Association (1944). *Foundry Sand Testing Handbook*, Section 4. American Foundrymen's Association, Chicago

Fig. 6.31 *Equipment for compactability test in noise-reduction cabinet* (*Courtesy Transport and Road Research Laboratory, Crowthorne, Berks.*)

British Standards Institution (1978), Draft Standard recommendations for methods of test on aggregates: Pt. 1, Compactibility test for graded aggregates, Document No. 78/13448

Little, A. L. (1948). 'Laboratory compaction technique'. *Proc. 2nd Int. Conf. on Soil Mech. and Found. Eng., Rotterdam*, Paper II g 1

MacLean, D. J. and Williams, F. H. P. (1948). 'Research on soil compaction at the Road Research Laboratory'. *Proc. 2nd Int. Conf. on Soil Mech. and Found. Eng.*, Rotterdam, Paper 1X 5 b

McLeod, N. W. (1970). 'Suggested method for correcting maximum density and optimum moisture content of compacted soils for oversize particles'. *ASTM STP* 479

Maddison, L. (1944). 'Laboratory tests on the effect of stone content on the compaction of soil mortar'. *Roads and Road Construction*, Vol. 22, Part 254

Markwick, A. H. D. (1944). 'The basic principles of soil compaction and their application'. Inst. Civ. Eng. London. Road Engineering Division, Road Paper No. 16

Odubanjo, T. O. (1968). 'A study of a laboratory compaction test using a Swedish vibratory apparatus'. TRRL Report LR129. Transport and Road Research Laboratory, Crowthorne, Berks.

Parsons, A. W. (1976). 'The rapid measurement of the moisture condition of earthwork material'. TRRL Laboratory Report 750. Transport and Road Research Laboratory, Crowthorne, Berks.

Pike, D. C. (1972). 'Compactability of graded aggregates: I. Standard Laboratory tests'. TRRL Laboratory Report LR 447. Transport and Road Research Laboratory, Crowthorne, Berks.

Pike, D. C. and Acott, S. M. (1975). 'A vibrating hammer test for compactability of aggregates'. TRRL Supplementary Report 140 UC. Transport and Road Research Laboratory, Crowthorne, Berks.

Proctor, R. R. (1933). 'Fundamental principles of soil compaction'. *Engineering News Record*, Vol. 111, No. 9

Taylor, D. W. (1948). *Fundamentals of Soil Mechanics*. Wiley, New York

Transport and Road Research Laboratory (1952). *Soil Mechanics for Road Engineers*, Chapter 9. HMSO, London

Williams, F. H. P. (1949). 'Compaction of soils'. *J. Inst. Civ. Eng.*, Vol. 33, No. 2

Wilson, S. D. (1950). 'Small soil compaction apparatus duplicates field results closely'. *Engineering News-Record*, 2 November 1950

Wilson, S. D. (1970). 'Suggested method of test for moisture-density relations of soils using Harvard compaction apparatus. Special procedures for testing soil and rock for engineering purposes'. *ASTM STP* 479

Chapter 7

Description of soils

7.1 INTRODUCTION

7.1.1 The Nature of Soil

A definition of soil which is appropriate to geotechnical engineering was stated in Section 1.1.5, and is repeated in Section 7.2. Different definitions apply in the fields of geology, pedology, agriculture, and so forth, but in geotechnology the significant features of soils are their particulate nature, and the porous structure which the particles form.

The relatively thin layer of humic topsoil which supports vegetation is usually unimportant unless it extends to exceptionally great depths.

Most natural soils comprise particles known as boulders, cobbles, gravels, sands, silts and clays, as defined in Chapter 4. Extensive deposits of organic matter, notably peats, though usually of fibrous rather than particulate nature, are included within the definition of soils. Natural materials which have been disturbed by man, such as clay or gravel placed in a road embankment, or colliery and quarry waste, are also included. To these may be added man-made materials such as furnace slag; pulverised fuel ash (PFA) from power stations; builders' rubble; domestic refuse.

7.1.2 Origin or Soils

Most natural soils are composed of the breakdown products of rocks which have been attacked by physical, chemical or biological weathering processes. The weathered material may have been transported and deposited elsewhere as sediments, or may remain *in situ* as residual soils.

SEDIMENTS

The principal types of sediments, and their modes of origin, are summarised as follows:

Aeolian deposits are those formed by wind action. They include desert sands and loess silts, and some brick-earths which were probably formed by wind erosion of glacial deposits.

Pyroclastic deposits are granular materials of all sizes blasted from volcanoes and deposited on land or in water. Layers of fine volcanic ash may be found interbedded with loess deposits.

Glacial deposits are materials of all grain sizes, typically unbedded and of various forms and origins deposited by glaciers and their meltwaters in a variety of forms. These include deposits loosely referred to as 'boulder clay' in Britain.

Periglacial deposits are formed in near-glacial conditions by frost heave and solifluxion. An example of these are head deposits (combe rock).

Lacustrine deposits are materials deposited in fresh or salt-water lakes, including glacial lake deposits.

Alluvial deposits are river-laid materials including fine-grained alluvium, levees, river terrace deposits and wadi gravels.

Estuarine deposits are materials deposited in the tidal regions of rivers forming mudflats, and in river delta areas.

Marine deposits are varied materials occurring at all levels in the sea, such as deep-sea muds, shell banks, coral reefs and beach deposits.

RESIDUAL SOILS

These soils are formed *in situ* by deep chemical weathering of rocks, usually in a tropical climate. Some components are removed as a result, usually leaving a clay-based deposit. Important types are laterites and bauxites.

7.1.3 The Need to Identify Soils

Since there is such a wide variety of soil types, it is necessary to be able to describe and classify soils in terms which convey their characteristics clearly and concisely, and which are generally accepted and understood by geotechnical engineers. The engineering properties of a soil are governed to a large extent by its physical properties and behaviour, so a careful visual inspection together with a few simple tests can provide a valuable first appraisal of the soil.

This chapter covers routine visual description of soils as carried out on samples in a soil laboratory. Some soils of an unfamiliar type may require more detailed study, such as mineralogical analyses to identify clay minerals, but this requires more specialised facilities.

7.1.4 Description in the Laboratory

Before a test on a soil sample is started, the sample should be examined and the observations recorded. This initial description, if carefully prepared, can provide valuable information to which field observations and laboratory test results can be related.

The first requirement is to record the identity of the sample in terms of its location, reference number and other details, as outlined in Section 1.4.2. Every description sheet should bear at least the identification number of the sample, the date, and the name and signature of the person describing. A suitable form for recording sample descriptions is shown in Fig. 7.1.

The sample should be described, briefly and clearly, using the recognised terminology given in the following sections. The description should include not only what is seen, but also what is felt by handling the soil, and, where appropriate, what can be smelled. The approach will depend upon which of the four main categories (Section 7.3) the soil can be assigned to. Detailed procedures for each main type are given in Sections 7.4–7.6. They are based on the principles set out in Chapter 8 of the draft revision of CP2001, Code of Practice for Site Investigation. Soil samples should normally be handled and examined by the person who is responsible for describing the soil. An assistant can do the writing of the descriptions and comments dictated by the one who describes.

7.1.5 Equipment for Soil Description

When a batch of samples are to be examined collectively as a separate operation, special facilities are desirable. Description of soils should be done at a well-lit bench with a generous working area. Natural north daylight is best; direct sunlight should be avoided. Artificial light, if needed, should be from fluorescent tubes of 'colour-matching' standard. Ordinary tungsten electric light bulbs cause distortion of colours, especially of the drab brown, green and blue tints which are frequently encountered in soils. The description bench should be within easy reach of a water-tap and sink, the waste from which should incorporate a silt trap.

Sample Description

Location	BRACKNELL				Location No.	2456
Described by	P. D. J.				Borehole No.	6
Date	3. 1. 78				Sheet No.	1 OF 4

Sample No.	Type*	Tube No.	Depth m from	Depth m to	% recovery	VISUAL DESCRIPTION
1	D		G.L.	0.25		Dark brown TOPSOIL with plant roots
2	T		0.25	1.40		Yellow brown fine silty SAND with a little fine to medium gravel
3	D		1.40	1.60		Firm blue-grey CLAY occasionally mottled brown
4	U100	18978	1.60	2.05	100	Top: as 3 Base: firm to stiff blue-grey fissured silty CLAY
5	D		2.05	2.1		As base of 4
6	U100	16373	2.1	2.55	60	Top: as 5 Base: fine to medium grey-green SAND with bands of brown silty sand
7W	W		2.4			Ground water
8	B		2.7	3.4		Brown SAND and GRAVEL with occasional lumps of soft grey clay

*U100, U38 etc.: Undisturbed (number indicates dia. mm)

D: Disturbed   T : Tin   C: Core (state dia. mm)

B: Bulk   W : Water   P : Piston

Remarks

Engineer's signature   P. D. Jones

Fig. 7.1   Sample description form

Equipment and tools which are useful for soil description are listed below:

Tools for unpacking and unsealing samples.
Extruder for jacking out undisturbed samples.
Sample container racks.
Pocket knife.
Cobbler's knife.
Spatulas — large and small.
Tweezers.
Hand lens (× 10 magnification).
Pocket penetrometer, or hand vane.
Watch-glasses.
Small scoop.
Pestle and mortar.
Sieve, 63 μm, and receiver.
Aluminium or galvanised steel trays.
Small brush.
Beakers — glass and plastics.

Glass stirring rod.
Wash bottle containing distilled water.
Hydrochloric acid (N/5) in bottle with dropper.
Steel rule — 150 mm.
Sample description sheets.
Moisture content sheets.
Clip-board.
Plain white paper.
Waxpot, wax, brushes, muslin.
Aluminium foil.
Spare polythene bags, glass jars.
Labels, marker pen (waterproof).
Rubber tubing fitted to water-tap.
Wiping cloths.
Fingernail brush.
Waste-bin.

7.2 DEFINITIONS

SOIL   Any naturally occurring deposit forming part of the earth's crust which consists of an assemblage of discrete particles (usually mineral, sometimes with organic matter), that can be separated by gentle mechanical means, together with variable amounts of water and gas (usually air).

IDENTIFICATION   Establishment of the main characteristic features of soil, either visually or as a result of carrying out index tests.

DESCRIPTION   Verbal presentation of soil characteristics based on visual examination, feel, smell and simple hand tests.

CLASSIFICATION   Division of soils into a number of groups, on the basis of standard index tests, each group being defined by fixed limits of certain characteristics.

INDEX TESTS   Relatively simple tests, including those related to density, specific gravity, particle size and plasticity (as distinct from tests to determine mechanical properties such as strength and compressibility).

SEDIMENTS   Soils made up of materials, derived from pre-existing rocks by weathering, which have been transported by various means and deposited elsewhere.

RESIDUAL SOILS   Soils which have not been transported but are the remains *in situ* from the weathering of rocks.

7.3   IDENTIFICATION OF SOILS

7.3.1   Main Characteristics

The following factors are taken into account when making an engineering description of soils:

(1)   Mass characteristics of the soil formation.
(2)   Material characteristics.
(3)   Geological formation, type and age.
(4)   Classification group.
(5)   Any additional relevant information.

A laboratory description is concerned mainly with item (2), the material characteristics of soils which can be described from undisturbed or disturbed samples. The mass characteristics, item (1), can be described satisfactorily only from exposures on site, although a limited amount of information can be obtained from undisturbed samples. Assignment of soil to a particular geological formation or period, item (3), requires specialist geological knowledge, and any conjecture in this respect should be avoided. Classification groups, item (4), are referred to in Section 2.4.2 for fine soils and in Section 4.4.2 for granular soils.

Under item (5), observations on the general state of the sample, and its packing and preservation, should be recorded at the time of inspection, together with any unusual features or variations in characteristics which may be noticed (see Section 1.4.3).

7.3.2   Soil Groups

Soils can be broadly divided into four main categories:

(1)   **very coarse materials** — boulders and cobbles.
(2)   **coarse soils** — gravels and sands, also called **granular** or **non-cohesive soils.**
(3)   **fine soils** — silts and clays, the latter being **cohesive** soils.
(4)   **organic soils.**

The very coarse materials comprise particles larger than 60 mm across (in practice, particles retained on a 63 mm sieve). Those up to 200 mm across are termed **cobbles**, and particles larger than 200 mm are termed **boulders**.

Before we attempt to classify or describe a soil sample, these very coarse materials should first be removed. The material passing a 63 mm sieve is that which is referred to here as **soil**, and is described according to the principles which follow.

In general, coarse soils contain less than 35% (by dry mass) of particles finer than 0.06 mm — that is, less than 35% of fine soil. The main constituents are **gravel** (particles from 60 mm down to 2 mm) and **sand** (2 mm down to 0.06 mm). Fine soils contain more than 35% of particles finer than 0.06 mm. They consist mainly of **silts** (0.06 mm down to 0.002 mm) and **clays** (smaller than 0.002 mm). Organic soils contain organic matter in significant quantity, and include peat soils, which consist mainly of decomposed plant remains.

Each main soil category is discussed in a separate section below.

7.3.3 Identification Chart

The identification chart given in Table 7.1, which is based on the Table given in CP 2001, provides a key to the visual naming and description of soils. The heavy dividing lines separate soils into the three major categories of coarse, fine and organic soils, as defined in Section 7.3.2. Inorganic soils are further subdivided on the basis of their particle size or their plasticity. These characteristics are described in more detail in Section 7.4 (coarse soils) and 7.5 (fine soils). Organic soils, and other soil types, are covered in Section 7.6.

Table 7.1 may be followed for describing soils on the basis of the following characteristics:

Size of particles.
Plasticity.
Nature of particles.
Compactness, or strength.
Colour.
Structure.
Secondary constituents in mixed basic types.

The terms used in Table 7.1 are those which are generally accepted in the geotechnical engineering sense, and may sometimes differ from geological or colloquial usage.

7.4 DESCRIPTION OF COARSE (GRANULAR) SOILS

7.4.1 Particle Size

The particle size range for **gravel** (from 60 to 2 mm) and for **sand** (from 2 to 0.6 mm)), together with their subdivisions into coarse, medium and fine, are indicated in Table 7.1, and are discussed in Section 4.3.2.

Gravel particles can be easily seen and handled, and their shape and surface texture observed. Sand particles are visible to the naked eye. Sand has a gritty feel, and shows very little or no cohesion when dry.

The main grading characteristics (that is whether the soil is uniformly graded, well graded or poorly (gap) graded), as described in Section 4.4.2, can usually be estimated by eye, and should be noted as part of the visual description.

7.4.2 Nature of Particles

The shape of individual particles should be included in the description of gravel and coarse sand particles. The degree of angularity or roundness is indicated by using the terms *angular*,

IDENTIFICATION AND

	Basic Soil Type	Particle Size mm	Visual Identification	Particle Nature & Plasticity	Composite Soil Types (Mixtures of basic soil type constituents)		
Very Coarse Soils	BOULDERS		Only seen complete in pits or exposures.	Particle Shape :	Scale of Secondary Constituents with Coarse Soils.		
		— 200					
	COBBLES		Often difficult to recover from boreholes.	Angular Subangular	Term		% of clay or silt
		— 60		Subrounded Rounded			
Coarse Soils	GRAVELS	coarse	Easily visible to naked eye: particle shape can be described: grading can be described.	Flat Elongate	slightly clayey } GRAVEL or slightly silty } SAND		under 5
		— 20					
		medium	Well graded - wide range of grain sizes well distributed Poorly graded - not well graded (may be Uniform or Gap graded)		— clayey } GRAVEL or — silty } SAND		5 — 15
		— 6					
		fine		Texture :	very clayey } GRAVEL or very silty } SAND		15 — 35
		— 2					
	SANDS	coarse	Visible to naked eye: very little or no cohesion when dry: grading can be described	Rough Smooth Polished	Sandy GRAVEL Gravelly SAND } Sand or gravel an important second constituent of the coarse fraction.		
		—.6					
		medium	Well graded - wide range of grain sizes, well distributed Poorly graded - not well graded (may be Uniform or Gap graded)		For composite types described as clayey - fines are plastic, cohesive: silty - fines non-plastic or of low plasticity.		
		—.2					
		fine					
		—.06					
Fine Soils	SILTS	coarse	Only coarse silt barely visible to naked eye: exhibits little plasticity and marked dilatancy: slightly granular or silky touch. Disintegrates in water: lumps dry quickly, possess cohesion but can be powdered easily between fingers.	Non-plastic or Low Plasticity	Scale of Secondary Constituents with Fine Soils.		
		—.02					
		medium			Term		% of sand or gravel.
		—.006					
		fine			sandy } CLAY or gravelly } SILT		35 — 65
		—.002					
	CLAYS		Dry lumps can be broken but not powdered between the fingers: they also disintegrate under water but more slowly than silt: smooth touch: exhibits plasticity but no dilatancy: sticks to the fingers and dries slowly: shrinks appreciably on drying usually showing cracks. Intermediate and high plasticity clays show these properties to a moderate and high degree, respectively.	Intermediate Plasticity (Lean Clay)	— CLAY ; SILT		less than 35
				High Plasticity (Fat Clay)	Examples of composite types. (Indicating preferred order for description) Loose, brown, subangular very sandy, fine to coarse GRAVEL with small pockets of soft grey clay.		
Organic	ORGANIC CLAY, SILT, OR SAND.	varies	Contains substantial amounts of organic vegetable matter. Predominantly plant remains usually dark brown or black in colour, often with distinctive smell: low bulk density.		Medium dense, light brown, clayey, fine and medium SAND. Stiff, orange brown, fissured sandy CLAY. Firm, brown, thinly laminated SILT and CLAY. Plastic, brown, amorphous PEAT.		
	PEATS						

DESCRIPTION OF SOILS

Compactness/Strength		Structure			Colour
Term	Field Test	Term	Field Identification	Interval Scales	
Loose	By inspection of voids and particle packing.	Homogen-eous	Deposit consists essentially of one type.	Scale of Bedding Spacing	Red
Dense		Inter-stratified	Alternating layers of varying types or with bands or lenses of other materials: Interval Scale for bedding spacing may be used.	Term / Mean Spacing mm	Pink / Yellow / Brown
				Very thickly bedded / over 2000	Olive
Loose	Can be excavated with a spade: 50 mm wooden peg can be easily driven.	Heterogen-eous	A mixture of types.	Thickly bedded / 2000 - 600	Green / Blue
Dense	Requires pick for excavation: 50 mm wooden peg hard to drive.	Weathered	Particles may be weakened and may show concentric layering.	Medium bedded / 600 - 200	White
				Thinly bedded / 200 - 60	Grey
Slightly Cemented	Visual examination: pick removes soil in lumps which can be abraded.			Very thinly bedded / 60 - 20	Black / etc
				Thickly laminated / 20 - 6	Supplemented as necessary with :
				Thinly laminated / under 6	Light
Soft or Loose	Easily moulded or crushed in the fingers.	Fissured	Break into polyhedral fragments along fissures (Interval Scale for Spacing of Discontinuities may be used).		Dark / Mottled / etc
Firm or Dense	Can be moulded or crushed by strong pressure in the fingers.	Intact	No fissures	Scale of Spacing of Other Discontinuities	and :
Very soft	Exudes between fingers when squeezed in hand	Homogen-eous	Deposit consists essentially of one type.	Term / Mean Spacing mm	Pinkish
Soft	Moulded by light finger pressure.	Inter-Stratified	Alternating layers of varying types. Interval Scale for thickness of layers may be used.	Very widely spaced / over 2000	Reddish / Yellowish
Firm	Can be moulded by strong finger pressure.			Widely spaced / 2000 - 600	Brownish
Stiff	Cannot be moulded by fingers. Can be indented by thumb.	Weathered	Usually has crumb or columnar structure.	Medium spaced / 600 - 200	etc
Very stiff	Can be indented by thumb nail.			Closely spaced / 200 - 60	
Firm	Fibres already compressed together.	Fibrous	Plant remains recognisable and retain some strength.	Very closely spaced / 60 - 20	
Spongy	Very compressible and open structure.			Extremely closely spaced / under 20	
Plastic	Can be moulded in hand, and smears fingers.	Amorphous	Recognisable plant remains absent.		

Table 7.1

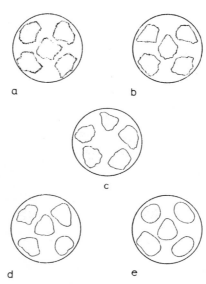

Fig. 7.2   Shape of particles: (a) angular, (b) sub-angular, (c) sub-rounded, (d) rounded, (e) well rounded

sub-angular, sub-rounded, rounded or *well rounded*, as depicted in Fig. 7.2. The general form, if other than approximately equidimensional, may be described as *flat, elongated, flat and elongated* or *irregular*.

An indication of the surface texture should be included, the usual terms being *rough, smooth* or *polished*.

The type of rock which forms the soil particles is not normally identified except by a geologist. However, if the particles are of a type which can be recognised easily, such as fragments of chalk or sandstone, this may be included in the description.

7.4.3   Composite Soils

Many soils consist of particle sizes which span two or more basic soil types. The composition of mixtures of particles of sand and gravel size is indicated by using the following terms in the description. The dominant constituent is printed in bold type.

Term	*Approximate constituents by mass*
Slightly sandy **gravel**	up to 5% sand
Sandy **gravel**	5–20% sand
Very sandy **gravel**	over 20% sand
Gravel/sand	about equal proportions of gravel and sand
Very gravelly **sand**	over 20% gravel
Gravelly **sand**	20–5% gravel
Slightly gravelly **sand**	up to 5% gravel

Coarse soils containing fine material (silt or clay, or both) as a secondary constituent are described by the terms *slightly clayey, clayey* or *very clayey* (or similarly with *silty*), as indicated in Table 7.1. If the soil has no cohesive strength when dry, the fines are non-plastic and contain no clay. Occasionally a deposit may have a secondary constituent which is distinctive because of its composition, in which case this can be mentioned. For example, if a sand contained small shiny plates of mica, it would be described as *micaceous*.

Secondary constituents may include organic matter, which is dealt with in Section 7.6.1.

7.4.4 Compactness

The state of compactness (relative density) of granular soils can be assessed only on the basis of *in situ* tests and experience, so this factor is not included in a laboratory description. However, grains may be bonded together, in which case the soil would be described as *cemented*. If lumps can be abraded easily, the soil is *weakly* or *slightly cemented*.

In some sands a coating of fine particles on the grains causes them to stick together and appear to be cemented. On immersion in water the grains rapidly fall apart, and this type of soil should not be described as cemented.

7.4.5 Colour

The colours listed in the right-hand column of Table 7.1, and the qualifying adjectives listed below them, are sufficient for general description purposes. The 'Munsell Soil Colour Chart' may be used for a more detailed classification of colours.

7.4.6 Structure

Typical terms used to indicate the structure of granular soil are given in Table 7.1. These features cannot usually be observed in disturbed samples.

Bedding and laminations are referred to in Section 7.5.8, in connection with fine soils, but similar characteristics may be present in granular soils.

7.4.7 Summary of Terms

A granular soil is usually described in the following sequence:

Compactness, or whether cemented.
Colour.
Structure.
Secondary constituent.
Particle nature.
Grading.
MAIN SOIL TYPE.
Minor constituents.
Other remarks.

The main soil type is written in capital letters, or underlined.

The terms for description of coarse soils are summarised in Table 7.2. This includes in column 8 some typical descriptions relating to minor constituents which might be present.

Three examples of soil descriptions are given at the foot of Table 7.2.

7.5 DESCRIPTION OF FINE SOILS

7.5.1 Size of Particles

Fine soils consist of silts and clays having partivles smaller than 0.06 mm. Material known as **silt** consists of particles in the size range 0.06 mm to 0.002 mm, and is subdivided into coarse, medium and fine, as indicated in Table 7.1. Particles of coarse silt can just be discerned by the naked eye, but silt particles generally can be seen under a good hand lens (× 10 magnification).

Table 7.2.   TERMS AND DESCRIPTION FOR COARSE SOILS

Relative density (1)	Colour (2)	Lithology (3)	Secondary constituents (4)	Particle nature (5)	Grading (6)	Principal soil type(s) (7)	Minor constituents (8)
	light			angular	fine	**boulders**	containing
Very loose	dark	bedded	clayey	subangular		**cobbles**	many crystals
Loose	mottled	laminated	silty	subrounded	medium	**gravel**	
Medium-dense			sandy	rounded		**sand**	with
Dense	pinkish		gravelly	flat	coarse	**(silt)**	scattered
Very dense	reddish			elongated			gravel
	yellowish	very thickly		platey	well-		occasional
	brownish	thickly		spherical	graded		shells
	olive	medium		tubular			pockets of peat
	greenish	thinly			poorly		bands of clay
	bluish	very thinly	chalky		graded		lenses of clay
	greyish		peaty				some rootlets
			micaceous	rough	gap-		partings of silt,
	pink			smooth	graded		etc.
	red			polished			
	yellow				uniform		
Weakly	brown						
cemented	olive		very				
	green						
	white		slightly				
	grey						
	black						

Examples:

(a)   Loose light grey clayey and peaty fine **sand** with occasional shells
       (1)   (2)   (4)   (6) (7)   (8)

(b)   Medium-dense dark brownish-yellow subangular well-graded **gravel**
       (1)   (2)   (5)   (6)   (7)

(c)   Dense mottled yellow and green medium and coarse sand and subrounded fine gravel
       (1)   (2)   (3)   (7)   (5)   (6) (7)

Particles forming **clay** consist of complex minerals which are mostly flat and plate-like or elongated, and of a size less than 0.002 mm. However, if particles of this size are not true clay minerals, but consist of very fine silt particles (rock flour), the material would behave not as a clay but as a silt.

When describing fine soils, it is first necessary to decide whether the soil is effectively a silt or a clay.

7.5.2   Identification of Silt

Silt has little plasticity, dries quite quickly on the hands and can be dusted off. Silt has a smooth or silky touch, but grittiness can be detected between the teeth. Lumps dry quickly, and when dry have a granular appearance and can be powdered easily. A small lump placed in water disintegrates quickly into its individual particles, which will take several minutes to settle. When loose in a sample tube, silt may tend to slump and the water will tend to drain out of the soil.

Silt exhibits *dilatancy* when subject to the following simple hand test. Moisten a pat of soil so that it is soft but not sticky, and place it in the open palm of one hand. Shake the hand, tapping it against the other hand several times. The appearance of a shiny film of water on the surface of the pat indicates dilatancy. Squeeze the pat by pressing with the fingers, and the surface will dull again as the pat stiffens and finally crumbles. These reactions indicate the presence of predominantly silt-sized material or very fine sand. It is difficult to roll moist silt into threads, which break and crumble easily.

7.5.3 Identification of Clay

The most significant properties of clay are its cohesion and plasticity. If when pressed together in the hands at a suitable moisture content the particles stick together in a relatively firm mass, the soil shows cohesion. If it can be deformed without rupture (i.e. without losing its cohesion), it shows plasticity. Clay dries more slowly than silt and sticks to the fingers; it cannot be brushed off dry. It has a smooth feel, and shows a greasy appearance when cut with a blade. Softer consistencies behave rather like butter, and harder consistencies like cheese. Dry lumps can be broken, sometimes with difficulty, between the fingers, but cannot be powdered. A lump placed in water remains intact for a time, and disintegrates more slowly than silt.

Clay does not exhibit dilatancy. Lumps shrink appreciably on drying, and show cracks which are the more pronounced the higher the plasticity of the clay. At a moisture content within the plastic range, clay can easily be rolled into threads 3 mm diameter (as in the plastic limit test) which for a time can support their own weight. Threads of high-plasticity clay are quite tough; those of low-plasticity clay are softer and more crumbly.

7.5.4 Plasticity

The plasticity of fine-grained soils, and their subdivision in terms of the liquid limit into non-plastic soil and several ranges of plasticity, are discussed in Section 2.4.2. It may not be possible to assign a clay soil to a particular plasticity range by inspection alone, but it is usually possible to assess whether it is of low plasticity (L) or in the upper plasticity range (U), the latter incorporating groups I, H, V and E of Fig. 2.6.

7.5.5 Composite Soils

Fine soils rarely consist of particles entirely of silt size or clay size, but usually contain a mixture of both. The nature of clay minerals is such that a relatively small proportion of clay in a clay–silt mixture is sufficient to impart to the material the properties of a clay, in which case it is described simply as **clay**. For instance, a medium-plasticity (lean) clay may contain only 10–20% of clay-sized particles, and even a high-plasticity (fat) clay may contain not more than 50–70% clay. Examples are given in Section 4.4.3.

The description of soils consisting of mixtures of fine-grained and coarse-grained materials depends upon which constituent predominates. For example, **sand** may be described as slightly clayey, clayey or very clayey, depending on the approximate percentage of clay present. If the clay predominates, the soil would be described as sandy **clay**, or **clay** with a little (or some) sand. The terms used for various proportions of secondary constituents are summarised in Table 7.1, in both the Coarse Soils and Fine Soils sections.

Secondary constituents may include organic matter, which is dealt with in Section 7.6.1.

7.5.6 Strength

The following descriptions provide an indication of the shear strength of clay soils.

Very soft: Exudes between fingers when squeezed in the fist. A finger can be pushed 25 mm or so into it quite easily. Can be cut with a knife from a sample tube very easily. Tends to slump on extrusion.

Soft: Easily moulded in fingers. Finger can be pushed 10 mm or so into it. May tend to slump on extrusion.

Firm: Can be moulded by strong pressure in the fingers. Thumb will make an impression.

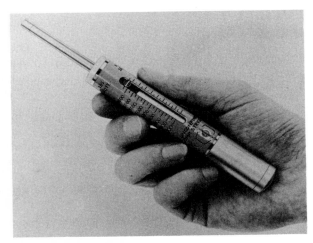

Fig. 7.3   Pocket penetrometer

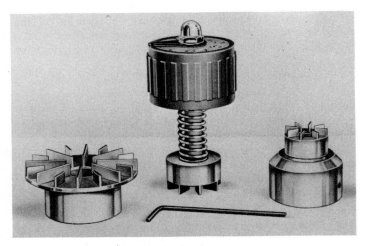

Fig. 7.4   Pocket shear vane

Stiff: Cannot be moulded in fingers and only slight impression possible with thumb. Cutting with a knife requires some effort.

Very stiff: Brittle or very tough. Difficult to cut with a knife.

(Hard): No impression possible with thumb.

Where a sample appears to be borderline, it can be described as soft to firm, or firm to stiff. Some soft or firm clays when worked in the hand will become significantly softer; the true consistency is the original one, provided the material has not already been disturbed by sampling. Very soft clays are prone to this. For a preliminary assessment the consistency can be assessed by using a pocket penetrometer (Fig. 7.3) or a pocket vane shearmeter (Fig. 7.4), and referring to Table 7.4.

Silts are described as:

Soft or loose: Easily moulded in fingers, slumps easily and moisture drains out.

Firm or dense: Can be moulded by strong pressure in the fingers. Does not slump.

7.5.7 Colour

See Section 7.4.5.

In fine soils a mottled pattern probably indicates weathering. A dark-brown, dark-grey or black colour may indicate the presence of organic matter, which would be confirmed by smell. Changes in colour on partial drying can reveal structural (fabric) details which are not obvious at the natural moisture content. Some colours are unstable when exposed to light and air for some time, so colours should be described only from a freshly exposed face.

7.5.8 Structure

A deposit consisting essentially of one soil type is described as *homogeneous*. If laminations or fissures are not present and not reported in the description, it is understood that the soil is *intact*.

The presence of layers within a clay deposit is due to *bedding*. If the bedding spacing is more than 20 mm, the clay is described as *bedded* (thinly, medium or thickly). The scale of bedding spacing is shown in Table 7.1. Laminations in clay may be separated by very thin layers of silt or sand of up to 0.5 mm thickness, known as *partings* or *dustings*. If the laminations are all very regular and grade from sand through silt to clay, each representing a seasonal deposition cycle, the clay is described as *varved*, but this term is normally reserved for glacial lake deposits.

If a clay contains discontinuities not necessarily related to bedding surfaces, it is *fissured*. It will tend to break up into irregular blocks along fissure surfaces, which may be described as polished, striated or dull. The nature of any infilling should be reported. Very close fissures may cause the clay to break down to crumbs when worked in the hand, or a high silt content could have the same effect; this type of clay is *friable*. The scale of fissue spacing is given in Table 7.1.

7.5.9 Summary of Terms

A fine-grained soil is usually described in the following sequence:

    Strength.
    Colour.
    Structure.
    Discontinuities.
    Secondary constituents.
    MAIN SOIL TYPE.
    Minor constituents.
    Other remarks.

The terms used for description of fine-grained soils are summarised in Table 7.3. Three examples of descriptions are given at the foot of the table.

7.6   DESCRIPTION OF OTHER SOIL TYPES

7.6.1   Organic Soils

Organic soils are predominantly plant remains, usually dark brown, dark grey, black or blue-black in colour, often with a distinctive smell and low bulk density. Mineral soils can contain organic matter in finely disseminated form, which produces a similar dark colouring, often oxidising to brown on exposure to air.

Table 7.3.   TERMS AND DESCRIPTIONS FOR FINE SOILS

Strength (1)	Colour (2)	Lithology (3)	Discontinuities (4)	Constituents (5)	Secondary soil type(s) (6)	Minor constituents (7)
Very soft		bedded	fissured			
Soft		laminated		sandy	clay	
Firm	as Table 7.2		very widely	gravelly		as Table 7.2
Stiff		very thickly	widely	silty	(silt)	
Very stiff		thickly	medium	peaty		
Hard		medium	closely	chalky		
		thinly	very closely			
		very thinly	extremely closely			

Examples:

(a)   Very stiff grey thickly bedded silty clay with dustings of silt
     (1)   (2)   (3)   (5) (6)   (7)

(b)   Firm olive-brown thinly laminated and closely fissured clay
     (1)   (2)   (3)   (4)   (6)

(c)   Soft reddish-brown clay thickly interbedded with red fine sand
     (1)   (2)   (6)   (3)   (2) (6) (7)

Table 7.4.   CONSISTENCY OF CLAYS

Consistency of clay	Undrained shear strength (kN/m^2)
Very soft	< 20
Soft	20–40
Soft to firm	40–50
Firm	50–75
Firm to stiff	75–100
Stiff	100–150
Very stiff	> 150

Topsoils contain roots or rootlets and living vegetation, but plant rootlets can penetrate much deeper. Wherever rootlets or other organic remains are found, their attitude and frequency should be reported; for instance, whether as vertical rootlets or as horizontal fibres. These features can have an important effect on drainage characteristics. The diameter of rootlets, and whether they are closed, open or infilled, is also significant.

Organic matter which is partly decomposed may be described as *peat*. This may be *fibrous*, when the structure of individual leaves, roots, twigs and branches is seen: or *amorphous*, where no structure is visible and the sample looks like a dark-brown or black silt. If it is firm, the material has been compressed and it is not easy to indent it with the thumb. If it is *spongy*, identification is easy and the sample may exude brown water. Amorphous peat is often plastic. Peat often occurs in conjunction with clay and so may be *clayey*. *Coal* and *lignite* should be described as such and not simply as organic matter, since they are relatively strong and incompressible materials compared with peat.

Organic soils are difficult to form into a thread, which is very weak. Lumps of organic soil crumble easily.

7.6.2   Man-made Soils

Where these are basically natural soils which have been reworked by man, the above descriptions apply. Where manufactured material is involved, the terms above should only be used if they will not cause confusion with a natural soil. The fact that the material is unnatural can be emphasised in the description: for example, 'cobble-sized angular **slag**', or alternatively (**slag** consisting of cobble-sized angular pieces'.

Pulverised fuel ash from power stations, known as 'fly ash', is fairly easily recognisable. It is usually light grey in colour, consists mainly of silt-size spherical particles, and often contains occasional fragments of unburnt coal. The specific gravity of particles is generally low, owing to entrapped air, and some particles ('floaters') may even be less dense than water.

7.6.3 Composite Soils

Soils which have much secondary material but still retain the basic appearance and behaviour of the principal soil type have been dealt with in the preceding sections.

Some soils, however, particularly *tills* (soils of glacial origin) contain a wide range of particle sizes. The type of glacial till known as *boulder clay* contains particles of all sizes, usually well-graded, ranging from cobbles or gravel down to clay. The principles outlined above still apply but it may be necessary to reappraise the initial description after carrying out field or laboratory tests.

7.6.4 Tropical Soils

The engineering description of soils found in tropical zones follows the same general principles as outlined above, but there are some additional points to be observed, both in description and in the manner in which tests are carried out. Some tropical soils, known as *residual* soils, have been formed by the decomposition of rock *in situ* by chemical decay in humid tropic conditions, and may retain signs of their original structure or fabric. Two of the more important types are laterites and bauxites, but both types are often loosely referred to as *lateritic soils. Laterites* are rich in iron oxide, and are characteristically red-brown in colour. *Bauxites* are rich in aluminium, and are usually dirty-white. The clay particles of these soils tend to aggregate into silt-size flocks, and are difficult to disperse unless a special dispersant is used (see Section 4.5.3).

The *black cotton* soils from another important type which is found extensively in tropical areas. These are usually clays of high plasticity, black or dark brown in colour, and can retain moisture through dry seasons, which is why they are of value for growing crops. The clay fraction contains a large proportion of high-activity minerals of the montmorillonite group, which is responsible for the pronounced shrinkage and swelling capability of these soils. They belong to a category known as expansive clays.

The *sabkha* soils found in the Middle East were formed under arid subtropical conditions. (The Arabic term 'sabkha' refers to the salt-encrusted coastal and inland flats where these soils occur.) Evaporation of groundwater (often sea-water) has left concentrations of salts, mainly chlorides and gypsum, in the pore spaces. The salts contain water of crystallisation which can easily be driven off if dried at too high a temperature, leading to erroneous moisture content and porosity values. (See Section 2.5.2 for special precautions when oven drying.)

Some tropical soils break down further the more they are handled. The results of laboratory tests can vary according to the amount of working, such as by the use of pestle and mortar, sieving or the length of time for which they are worked. Many of these soils are affected by oven drying, or even air drying, and they should never be dried out completely.

It may be necessary, therefore, to carry out a series of comparative tests to assess the effects of sample preparation and method of testing on the end results, and, if necessary, to adopt modified procedures. The procedures laid down in British Standards may not always be suitable, as they stand, for these types of soil.

BIBLIOGRAPHY

BS Code of Practice CP 2001 (1980), 'Site investigation', British Standards Institution, London

Bridges, E. M. (1970). *World Soils*. Cambridge University Press

Dumbleton, M. J. (1968). 'The classification and description of soils for engineering purposes: a suggested revision of the British system'. TRRL Report LR 182. Transport and Road Research Laboratory, Crowthorne, Berks.

Fookes, P. G. and Higginbottom, I. E. (1975). 'The classification and description of near-shore carbonate sediments for engineering purposes'. *Géotechnique*, Vol. 25, No. 2

Manual of Applied Geology for Engineers (1976). Institution of Civil Engineers, London

McFarlane, M. J. (1976). *Laterite and Landscape*. Academic Press, London

Munsell Soil Colour Chart. Reference 6-A. Tintometer Limited, Waterloo Road, Salisbury, Wilts.

Symposium on Airfield Construction on Overseas Soils. Proc. Inst. Civ. Eng. Vol. 8, November 1957:

Clare, K. E. (Paper No. 6243). Part 1, 'The formation, classification, and characteristics of tropical soils'. Part 2, Tropical black clays'.

Tomlinson, M. J. (Paper No. 6239). Part 3, 'Saline calcareous soils'. Part 4, 'Alluvial sands and silts'.

Nixon, I. K. and Skipp, B. O. (Paper No. 6258) Part 5, 'Laterite'. Part 6, 'Tropical red clays'.

Discussion on the above. *Proc. Inst. Civ. Eng.*, Vol. 10, May 1958

Appendix: Units and symbols

A.1 METRIC (SI) UNITS

A.1.1 Units in Use

The units and symbols listed in Table A.1 are used or referred to in this volume. This list is taken from the selection of SI Units for Soil Mechanics and Foundation Engineering which is generally accepted within the industry.

SI is the accepted abbreviation for Système International d'Unités (International System of Units), the modern form of the metric system finally agreed at an international conference in 1960.

A.1.2 Multiplying Prefixes

Multiples and submultiples of SI units are formed by placing prefixes in front of the unit symbol. The most commonly used prefixes are given in Table A.2. Recommended prefixes are those representing 10 raised to a power which is a multiple of ± 3. Use of those marked with an asterisk in Table A.2 should be avoided unless the recommended prefixes are inconvenient.

A.1.3 Definitions and Notes

Length

The metre (m) is defined in terms of a specified number of wavelengths of a particular radiation emitted by the krypton-86 atom.

The former prototype metre (a bar of invar metal) is still kept in the custody of the Bureau International des Poids et Mesures (BIPM) at their laboratories at Sèvres, near Paris.

The millimetre (mm) is used in most laboratory measurements:

$$1 \, mm = 10^{-3} \, m$$

The use of the centimetre (cm) should be avoided.

The micrometer (μm) is often called a micron, but the former is technically correct.

Area

The square millimetre (mm^2) is the unit of area used when areas are calculated from measurements in millimetres.

Volume

The cubic millimetre (mm^3) is the unit of volume derived from volume calculations using linear measurements in millimetres.

Table A.1.

Quantity	Unit	Symbol	Application	Conversions
Length	millimetre	mm	Sample measurements, particle size	$1\,\mu m = 10^{-6}\,m$
	micrometre	μm	Sieve aperture and particle size	$= 10^{-3}\,mm$
Area	square millimetre	mm^2	Area of section	
Volume	cubic metre	m^3	Earthworks	
	cubic centimetre	cm^3	Sample volume	$1\,m^3 = 10^6\,cm^3$
	millilitre	ml	Fluid measure	
	cubic millimetre	mm^3	Sample volume as calculated	
Mass	gram	g	Accurate weighings	
	kilogram	kg	Bulk sample and approximate weights	$1\,kg = 1000\,g$
	megagram	Mg	Alternatively known as tonne	$1\,Mg = 1000\,kg$ $= 10^6\,g$
Density	megagram per cubic metre	Mg/m^3	Sample density and dry density	Density of water $= 1\,Mg/m^3$ $= 1\,g/cm^3$
Temperature	degree Celsius	°C	Laboratory and bath temperatures	Celsius is preferred name for Centigrade**
Time	second	s	Timing of laboratory tests	1 minute $= 60\,s$
Force	newton	N	Load ring calibrations Small-magnitude forces	$1\,kgf = 9.807\,N$ $1\,N = 101.97$ gram f
	kilonewton	kN	Forces of intermediate magnitude	$1\,kN = 1000\,N$ $=$ approx. 0.1 tonne f
Pressure and stress	newton per square metre $=$ pascal	N/m^2 Pa	Very low pressures and stresses	$1\,g/cm^2$ $= 98.07\,N/m^2$ $= 98.07\,Pa$
	kilonewton per square metre $=$ kilopascal	kN/m^2 kPa	Pressure gauges Compressive strength and shear strength of soils	$1\,kgf/cm^2$ $= 98.07\,kN/m^2$ 1 bar $= 100\,kN/m^2$
Pressure (vacuum)	torr*	Torr	Very low pressure under vacuum	$1\,torr = 133.3_2\,Pa$ $= 133.3\,N/m^2$ $= 1\,mmHg$
Dynamic viscosity	millipascal second $=$ millinewton second per square metre	mPas mNs/m^2	Viscosity of water	$1\,mPas = 1\,cP$ (centipoise)

* A non-SI metric unit.
** See Fig. A.1 for temperature conversion chart
Metric sieve aperture sizes used for soil testing are listed in Table 4.5.

Table A.2

Prefix symbol	Name	Multiplying factor
G	giga	$1\,000\,000\,000 = 10^9$
M	mega	$1\,000\,000 = 10^6$
k	kilo	$1\,000 = 10^3$
h	*hecto	$100 = 10^2$
da	*deca	10
d	*deci	$10^{-1} = 0.1$
c	*centi	$10^{-2} = 0.01$
m	milli	$10^{-3} = 0.001$
μ	micro	$10^{-6} = 0.000\,001$
n	nano	$10^{-9} = 0.000\,000\,001$

It usually simplifies calculations if calculated volumes in mm^3 are divided by 1000 at the outset to give cubic centimetres (cm^3):

$$1\,cm = 1000\,mm^3$$

(Although the use of the centimetre is deprecated in SI, the cm^3 is compatible with its recommendations that multiple units should be related by factors of 10^3.)

The litre is recognised as a special name for one cubic decimetre, but should *not* be used to express scientific or high accuracy measurements of volume.

For most practical purposes,

$$1\,litre = 1\,dm^3$$

$$= 1000\,cm^3$$

In precise scientific work

$$1\,litre = 1000.028\,cm^3$$

The millilitre (ml) is one-thousandth of a litre:

$$1\,ml = 1\,cm^3$$

for practical purposes.

Mass

The kilogram (kg) is equal to the mass of the international platinum prototype kept by BIPM at Sèvres. It is the only basic quantity to be a multiple unit:

$$1\,kg = 1000\,g\,(grams)$$

There is no SI unit of 'weight'. When 'weight' is used to mean the force due to gravity acting on a mass, the mass (kg) must be multiplied by g (9.807 m/s^2) to give the force in newtons (N).

Density

The megagram per cubic metre (Mg/m^3) is the density unit adopted for soil mechanics. It is 1000 times larger than the kilogram per cubic metre, the basic SI unit, and is equal to one gram per cubic centimetre:

$$1\,\text{Mg/m}^3 = 1\,\text{g/cm}^3$$

$$= 1000\,\text{kg/m}^3$$

Using Mg/m^3, the density of water is unity, and the specific gravity of soil particles is equal to their density.

Time

The second (s) is defined in terms of the period of radiation of the caesium-133 atom under specified conditions.

Minutes, hours, days and years are used where appropriate. It is useful to note that

$$1\,\text{day} = 1440\,\text{minutes}$$

$$1\,\text{year} = 31.56 \times 10^6\,\text{seconds}$$

Sedimentation test times are usually measured in minutes.

Force

The newton (N) is that force which, applied to a mass of 1 kilogram, gives it an acceleration of 1 metre per second per second:

$$1\,\text{N} = 1\,\text{kg m/s}^2$$

The kilonewton (kN) is the force unit most used in soil mechanics:

$$1\,\text{kN} = 1000\,\text{N}$$

$$= \text{approximately }0.1\,\text{tonne f}$$

Pressure and Stress

The pascal (Pa) is the pressure produced by a force of 1 newton applied, uniformly distributed, over an area of 1 square metre.

The pascal has been introduced as the pressure and stress unit, and is exactly equal to the newton per square metre:

$$1\,\text{Pa} = 1\,\text{N/m}^2$$

In dealing with soils the usual unit of pressure is kilonewton per square metre (kN/m^2), or kilopascal:

$$1\,\text{kN/m}^2 = 1\,\text{kPa} = 1000\,\text{N/m}^2$$

The bar is not an SI unit but is sometimes encountered:

$$1\,\text{bar} = 100\,\text{kN/m}^2 = 100\,\text{kPa}$$

$$= 1000\,\text{mb (millibars)}$$

(about atmospheric pressure).

Standard Gravity

The international standard acceleration due to the Earth's gravity is accepted as

$$g = 9.80665 \, \text{m/s}^2$$

although it varies slightly from place to place. For practical purposes $g = 9.81 \, \text{m/s}^2$, the conventional reference value used as a common basis for measurements made on the Earth.

A.1.4   Conversion Factors, Imperial and SI Units

Table A.3   CONVERSION FACTORS, IMPERIAL AND SI UNITS

Imperial to SI			*SI to imperial*
Length			
0.3048	m	: foot (ft)	3.281
25.4	mm	: inch (in)	0.03937
Area			
0.09290	m²	: square foot	10.76
645.2	mm²	: square inch	0.001550
Volume			
0.02832	m³	: cubic foot	35.31
4.546	litre	: gallon (UK)	0.2200
3.785	litre	: gallon (USA)	0.2642
28.32	litre	: cubic foot	0.03531
16.39	ml	: cubic inch	0.06102
16387	mm³	: cubic inch	
Mass			
1.016	Mg (tonne)	: ton	0.9842
0.4536	kg	: pound (lb)	2.205
453.6	g	: pound	
28.35	g	: ounce (oz)	0.03527
Density			
0.01602	Mg/m³ (g/cm³)	: pound per cubic foot	62.43
Force			
9.964	kN	: ton force	0.1004
4.448	N	: pound force	0.2248
Pressure			
0.04788	kN/m² (kPa)	: lb f/sq ft	20.89
6.895	kN/m²	: lb f/sq in	0.1450
47.88	N/m² (Pa)	: lb f/sq ft	0.02089

Examples: Imperial to SI: to convert feet to m, multiply by 0.3048
SI to imperial: to convert m to feet, multiply by 3.281

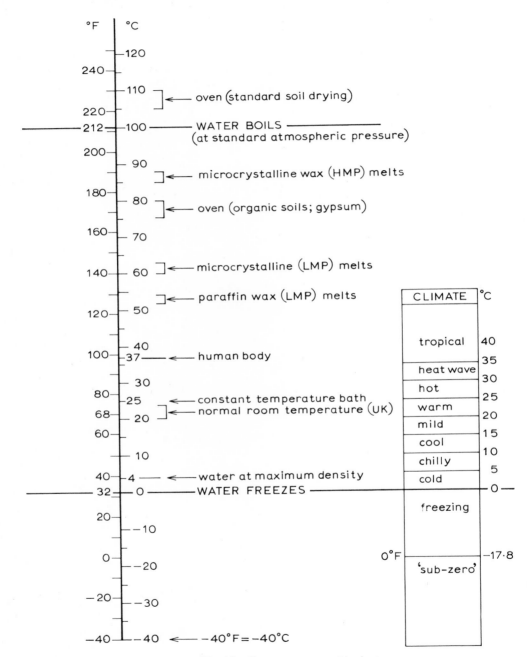

Fig. A.1   Temperature conversion chart

A.2 SYMBOLS

A.2.1 Symbols for Soil and Water Properties

Table A.4

Measured quantity	Symbol	Unit of measurement
Moisture content	w	%
Liquid limit	LL	%
Plastic limit	PL	%
Plasticity index	PI	%
Non-plastic	NP	–
Relative consistency	C_r	–
Liquidity index	LI	–
Shrinkage limit	SL	%
Linear shrinkage	LS	%
Shrinkage ratio	R	–
Unit weight	γ	kN/m³
Bulk (mass) density	ρ	Mg/m³
Dry density	ρ_D	Mg/m³
Saturated density	ρ_s	Mg/m³
Submerged density	ρ'	Mg/m³
Minimum dry density	$\rho_{D\,min}$	Mg/m³
Maximum dry density	$\rho_{D\,max}$	Mg/m³
Density of water	ρ_w	Mg/m³
Optimum moisture content	OMC	%
Specific gravity of soil particles	G_s	–
Specific gravity of liquid	G_L	–
Degree of saturation	S	%
Voids ratio	e	–
Porosity	n	–
Percentage air voids	V_a	%
Particle size	D	μm or mm
Percentage smaller than D	P	%
Effective size	D_{10}	mm
'60% finer than' size	D_{60}	mm
Uniformity coefficient	U	–
Dynamic viscosity of water	η	mPas

A.2.2 The Greek Alphabet

Table A.5

Capital	Small	Name	Capital	Small	Name
A	α	alpha	N	ν	nu
B	β	beta	Ξ	ξ	xi
Γ	γ	gamma	O	o	omicron
Δ	δ	delta	Π	π	pi
E	ε	epsilon	P	ρ	rho
Z	ζ	zeta	Σ	σ	sigma
H	η	eta	T	τ	tau
Θ	θ	theta	Y	υ	upsilon
I	ι	iota	Φ	φ	phi
K	κ	kappa	X	χ	chi
Λ	λ	lambda	Ψ	ψ	psi
M	μ	mu	Ω	ω	omega

A.3   USEFUL DATA

Time
1 day = 1440 minutes
1 week = 10 080 minutes
1 year = 0.526×10^6 minutes = 31.56×10^6 seconds

Density

		g/cm^3
Pure water	15°C	0.999 09
(see Table 4.10)	20°C	0.988 20
	25°C	0.997 04
Sea water	20°C	1.04
Waxes	20°C	
Paraffin	(m.p. 52°–54°C)	0.912
Microcrystalline	(m.p. 60°–63°C)	0.915
Mercury	20°C	13.546

Viscosity
Dynamic viscosity of water at 20°C   1.0019 mPas
  (see Table 4.10)
   1 mPas = 1mNs/m^2 = 1 cP (centipoise)

Nominal Container Sizes

	Diameter (mm)	Height (mm)	Volume (cm^3)	Approximate mass of soil contained
Compaction mould	105	115.5	1000	1.8–2.2 kg
CBR mould	152	127	2305	4.0–5.0 kg
U-100 tube (per 100 mm)	100	100	785.4	1.4–1.7 kg
U-100 tube (full)	100	450	3534	6.3–7.7 kg
Sample tube	38	76	86.2	150–190 g

General
Circumference/diameter of circle $\pi = 3.142$
Base of natural logarithms $e = 2.718$
Standard acceleration due to gravity $g = 9.81$ m/s^2

BIBLIOGRAPHY

Anderton, P. and Bigg, P. H. (1972). *Changing to the Metric System.* National Physical Laboratory, HMSO, London
British Geotechnical Society Sub-committee on the Use of SI units in Geotechnics (1973). 'Report of the sub-committee'. News Item, *Géotechnique*, Vol. 23, No. 4, pp. 607–610
BS 3763:1970, 'The International system of units (SI)'. British Standards Institution, London
Metrication Board (1976). *Going Metric — The International Metric System.* Leaflet UM1, 'An outline for technology and engineering'. Metrication Board, London
Metrication Board (1977). *How to Write Metric — A Style Guide for Teaching and Using SI Units.* HMSO, London
Page, C. H. and Vigoureux, P. (1977). *The International System of Units* (approved translation of *Le Systeme International d'Unités,* Paris, 1977). National Physical Laboratory, HMSO, London
Walley, F. (1968). 'Metrication' (Technical Note). *Proc. Inst. Civ. Eng.,* Vol. 40, May 1968. Discussion includes contribution by Head, K. H., Vol. 41, December 1968.

Index